Theory of Elasticity and Plasticity

Valentin Molotnikov • Antonina Molotnikova

Theory of Elasticity and Plasticity

A Textbook of Solid Body Mechanics

Translation: Subachev Yu. V.

 Springer

Valentin Molotnikov
Don State Technical University
Rostov-on-don, Russia

Antonina Molotnikova
Institute of Management and Entrepreneur
Rostov-on-don, Russia

ISBN 978-3-030-66624-8   ISBN 978-3-030-66622-4   (eBook)
https://doi.org/10.1007/978-3-030-66622-4

This Springer imprint is published by the registered company Springer Nature Switzerland AG
The registered company address is: Gewerbestrasse 11, 6330 Cham, Switzerland

The authors are grateful to D.Sc. in Physics and Mathematics, Professor L. M. Zubov and D.Sc. in Engineering, Professor V. P. Zabrodin for their selfless work in reviewing the manuscript, valuable advice and comments that were taken into account when preparing the final edition of the book. We express special gratitude to our son Zaur who inspired us for creating this work and took up most communications with the Springer publishing house. We are also grateful to Candidate of Technical Sciences Yu. V. Subachev who translated this book.

Dedication

This book is dedicated to our children, Zaur and Alexandra. Thanks to them, we sought to enrich the world with a particle of new knowledge.

Valentin and Antonina Molotnikovs

Preface to the English-Language Edition

This book was written based on the course of lectures given by the authors for many years to students of industrial and military universities of the Soviet Union and post-Soviet Russia. It also reflects the experience of the authors' work at academic institutes of the USSR.

The book was published in Russian by "Lan," St. Petersburg, in 2017. Since then, the authors and the publisher have received several opinions of students, post-graduates, and colleagues indicating the relevance of translating the book into foreign languages and publishing it abroad.

The first step in implementing these wishes was a translation into the Vietnamese language on the initiative of Mr. Le Quang Hoa and colleagues from the Hanoi Technical University. Knowing that about 100 million people communicate in Vietnamese in the modern world, the authors supported this initiative.

Unlike the first edition, the proposed English version of the book includes revised sections highlighting the history and development of the mechanics of deformable solid bodies. Materials whose utility is regarded as low have been removed. The book design has also been modified taking into account the Springer's rules.

When writing this book, the authors intended to help students in mastering the methods of elasticity and plasticity theories, by demonstrating their use with practically important examples and preparing readers for the conscious use of multiple computer programs for analysis of engineering structures. Research methods used in the book require the knowledge of mathematics and the strength of materials in the scope of the university training program.

Rostov-on-Don, Russia   Antonina Molotnikov

Rostov-on-Don, Russia   Valentin Molotnikov

Preface

As defined by M. Ya. Leonov, "...the mechanics of a real solid body is not only engineering, technology, and mathematics, but also a concentrate of rationality, a powerful engine of social progress, and an example of classical natural science."

A fundamental and the most complete section of the mechanics of deformations is classical elasticity theory occupying about a quarter of the book. The subject matter is described with an increasing complexity of topics. The existing boundaries between the educational and scientific literature are erased. The book includes the authors' solutions to a number of new problems important for engineering applications.

A quarter of the book is dedicated to principal variants of plasticity theory. It is preceded by a short overview of plasticity theory from the first experiments of Coulomb to endochronic variants of the theory and unconventional (synergetic) models of our days. Primary attention in this part is paid to the theory of small elastic–plastic deformations by Hencky–Nadai–Ilyushin, since this theory allows solving the most important problems of engineering. It also discusses the flaws in strain theory and the limits of its applicability.

The second part of the book is represented by the authors' studies in the development of the sliding conception in plasticity theory as set by M. Ya. Leonov. The Batdorf–Budiansky sliding mechanism is taken as the basis. Solutions to some problems eliminate primary objections against the Batdorf–Budiansky model and rehabilitate the sliding concept in plasticity theory.

The conclusion is made that due to the complexity of this matter, it must be admitted that creating a unified, universal, and rather complete theory of plasticity has not been successful by using either classical methods of mechanics of continuous media or semi-physical synergetic methods. Therefore, one must expect that in the nearest decades, high attention will be paid to simplified partial models. A model of complex deformation of materials with a residual change of volume is given as an example of such simplified setting.

Rostov-on-Don, Russia   Valentin Molotnikov

Rostov-on-Don, Russia   Antonina Molotnikov

Abstract

At the beginning, the book was conceived as a textbook for the classical course in solid mechanics. However, at the exit, it turned out that most of the book was written in the form of a monograph, contains many references and broadly represents the authors' developments. In addition, the book describes the main achievements of the modern development of the mechanics of elastic and inelastic materials. The book is intended not only for getting interested persons familiar with the primary results of mechanics development, from the first Coulomb experiments to the late twentieth or the early twenty-first century, but also for stimulation of practical use of new achievements of the deformation theory in geomechanics, mining, and construction industry. The book is presented for senior students majoring in structural analysis and experts in the field of solid mechanics. The book materials can be used as a topic for theses, and the solutions given are the methodological basis for students and post-graduates of the "Construction," "Applied Mechanics" courses, etc.

Contents

Part I   Basis of Elasticity Theory

1   Summary of Elasticity Theory: Basic Concepts 3
    1.1   From the History of Elasticity Theory 3
    1.2   Elasticity of Solid Bodies 5
    1.3   Homogeneous Strain 5
    1.4   Internal Forces: Method of Sections 6
    1.5   Homogeneous Body 7
    1.6   Stress Vector 7
    1.7   Elongation of Steel Specimens 8
    1.8   Permanent Deformations 10
    1.9   Elastic Limit 10
    1.10   Elastic Shear Deformation 10
    1.11   Law of Twoness of Tangential Stresses 12
    1.12   Homogeneous Stressed State 13
    1.13   Generalized Hooke's Law 14
    1.14   Another Form of Hooke's Law 15
    1.15   Plane Stress-Strain State 16
    1.16   Homogeneous Model of a Solid Body 17
    1.17   Axisymmetric Plane Strain 17
    1.18   Lame Task 18
    1.19   Phenomenon of Stress Concentration 21
    1.20   Saint-Venant Principle 21
    References 22

2   The First Basic Problem of Elasticity Theory 23
    2.1   Equilibrium Equations 23
    2.2   Expression of Strains Through Movements 24
    2.3   Definition of Movements 25
    2.4   Saint-Venant Identities 26
    2.5   Compatibility Conditions 26
    2.6   Boundary Conditions 27

2.7   The First Basic Problem of Elasticity Theory 28
References.. 28

3   The Second Primary Problem of Elasticity Theory..................... 29
3.1   Definition of Stresses Through Deformations 29
3.2   Equations of Elastic Body Strain................................. 30
3.3   Application of Harmonic Functions................................ 31
3.4   Trefftz Integral 33
3.5   Grodsky–Neyber–Papkovich Integral 34
References.. 36

4   Three-Dimensional Harmonic Function 37
4.1   Simplest Examples of Harmonic Functions 37
4.2   Green Function.. 39
4.3   Green's Spatial Functions 39
4.4   Boundary Problems for Half-Space 42
4.5   Other Properties of Harmonic Functions 43
References.. 44

5   Elastic Half-Space 45
5.1   Volumetric Expansion on Surface................................. 45
5.2   Stress on Surface... 47
5.3   Strain of Elastic Half-Space 48
        5.3.1   Integral Operator of Formulas (5.18)–(5.20) 49
5.4   Examples 50
References.. 53

6   Herz's Task 55
6.1   Deformation of Adjoining Bodies................................. 55
6.2   Primary Assumptions 56
6.3   Axisymmetric Hertz Problem 57
6.4   Compression of Orthogonal Cylinders............................. 59
        6.4.1   Simplest Case 59
        6.4.2   Primary Case 61
6.5   Compression of Barrel-Shaped Bodies 63
        6.5.1   Rotation Bodies with Parallel Axes 63
        6.5.2   Case of Intersecting Axes 64
6.6   Elongated Contact Area 66
6.7   Compression of Parallel Cylinders............................... 67
References.. 68

7   Stressed State in a Body Point....................................... 69
7.1   Principal Stresses 69
7.2   Maximum Stresses... 70
7.3   Intensity of Stresses... 73
7.4   Some Properties of Tangential Stresses 74
References.. 75

8   Linear Elastic Systems 77
      8.1   General Comments... 77
      8.2   Linear System 78
      8.3   Potential Energy of a Helical Spring 79
      8.4   Principle of Mutuality of Works 80
      8.5   Castigliano's Theorem 82
      8.6   Specific Potential Energy of Elastic Deformation 83
      References.. 85

9   Plane Problem of Elasticity Theory 87
      9.1   Functions of Stresses 87
               9.1.1   Example 1: Concentrated Force in the Wedge Apex ... 88
               9.1.2   Example 2: Wedge Bending by Uniform Pressure 89
      9.2   Complex Representation of a Bi-Harmonic Function 90
      9.3   Kolosov Displacement Integral................................... 91
      9.4   Action of Concentrated Force 92
      9.5   Solution of the First Principal Problem for a Circle.............. 93
      9.6   Annex to the Brazilian Test....................................... 95
      References.. 100

10   Mathematical Structural Imperfections 101
      10.1   Mathematical and Physical Theories of Structural Imperfections 101
      10.2   Edge Dislocation in an Infinite Body 103
      10.3   Mathematical Wedge-Shaped Dislocation 105
      10.4   Mathematical Biclination.. 106
      10.5   Flat Dislocation of Somigliana.................................... 106
      10.6   Somigliana Dislocation in Half-Plane 107
               10.6.1   Functions Φ, Ψ for the Plane with Dislocation 108
               10.6.2   Functions Φ, Ψ for a Half-Plane with Dislocation 109
               10.6.3   Calculation of Galin Functions.......................... 111
               10.6.4   Completion of Problem Solution........................ 112
               10.6.5   Addition to Geomechanics 113
      10.7   Pair of Fislocations in a Plane 116
      10.8   Edge Dislocation in a Half-Plane 117
      10.9   Half-Plane with a System of Dislocations 118
      References.. 119

11   The Beginning of the Theory of Stability of Equilibrium 121
      11.1   Stability and Instability 121
      11.2   Work and Classification of Forces 121
      11.3   Stability with Conservative and Dissipative Forces 123
      11.4   Lyapunov–Chetaev Theorem...................................... 124
      11.5   Instability in the First Approximation 125
      11.6   Critical Load 125
      11.7   The Theorem on Stability by the First Approximation............ 126
      11.8   The Raus–Hurwitz criterion 126

11.9   Main Types of Stability Loss 127
11.10   Methods for Determining Critical Load 128
11.11   The Perturbed Motion of the Compressed Rod.................. 129
11.12   Stability Under Non-conservative Load (Example) 131
         11.12.1   Equations of Perturbed Motion......................... 131
         11.12.2   Area of Valid Stability 134
         11.12.3   Investigation of the Value $\Delta\mu$, (Formula (11.31)) 135
         11.12.4   Investigation of the Effect of Friction.................. 137
         11.12.5   The influence of the spacing of the End Masses 140
References... 142

Part II   Principal Variants of Mathematical Plasticity Theory

12   Origin and Development of Plasticity Theory 145
     12.1   Primary Definitions 145
     12.2   The Subject and Tasks of the Theory of Plasticity 146
     12.3   Early Development Stages of Plasticity Theory 147
     12.4   Development of Plasticity Theory in the Twentieth Century 148
     12.5   Soviet Period of Plasticity Theory Development 150
     12.6   Russian Mechanics in the Post-Soviet Period 152
              12.6.1   General Situation and Dangerous Trends.............. 152
              12.6.2   Plasticity Theory in Russia in the Post-Soviet Period.. 155
     12.7   Abstract.. 158
     References... 158

13   Initial Concepts of Plasticity Theory 165
     13.1   Second-Rank Tensor in Euclidean Space 165
     13.2   Tensors in Plasticity Theory 166
     13.3   Decomposition of Stress and Strain Tensors 169
     13.4   Other Invariants in Plasticity Theory........................... 171
     13.5   On the Criterion of Similarity of Stress and Strain Deviators 174
     13.6   Stress Diagrams and Their Idealization 177
     References... 179

14   On the Plasticity Conditions of an Isotropic Body 181
     14.1   General Considerations 181
     14.2   General Notes 181
     14.3   Tresca Plasticity Condition 183
     14.4   Huber–Mises Plasticity Condition 185
     14.5   Experimental Study of Elastic–Plastic Materials................. 187
     14.6   Volumetric Elasticity of Materials 190
     14.7   Invariant Form of Hooke's Law 191
     References... 193

15   Plasticity Theory of Henky–Nadai–Ilyushin............................ 195
     15.1   Laws of Active Elastic–Plastic Deformation 195
     15.2   Defining the Universal Hardening Function 197

15.3 Some Properties of the Hardening Function 198
15.4 Another Form of Strain Ratios 200
15.5 Unloading Laws 202
15.6 Work of Stresses, Potential Energy, and Potentials 203
  15.6.1 Stress Potential 205
  15.6.2 Potential of Strains 206
15.7 Theorem of the Minimal Work of Inner Forces 207
15.8 Lagrange Equilibrium Variation Equation 208
15.9 Setting Boundary Problems of Plasticity Theory 211
15.10 Theorem of Simple Loading 213
15.11 Theorem of Unloading 215
References 217

16 Solution of the Simplest Problems for the Strain Theory of Plasticity 219
16.1 Pure Bending of a Straight Beam 219
16.2 Torsion of a Round-Section Beam 223
16.3 Elastic–Plastic Inflation of a Spherical Vessel 225
16.4 Symmetric Strain of a Cylindrical Tube 230
16.5 Torsion of a Beam of Ideally Plastic Material 236
  16.5.1 Elastic Torsion: Prandtl Analogy 236
  16.5.2 Elastic–Plastic Beam Torsion 238
16.6 Rod of a Variable Section: Method of Elastic Solutions 240
  16.6.1 Preparation of Initial Ratios 240
  16.6.2 Specification of Problem Setting 242
  16.6.3 Algorithm of the Elastic Solutions Method 243
References 244

17 Additions and Generalizations to the Strain Theory of Plasticity 245
17.1 Generalizations of Goldenblatt and Prager 245
17.2 Tensor–Linear Ratios in Plasticity Theories 246
17.3 Vector Representation of Tensors 248
17.4 Transformations of Rotation and Reflection 250
17.5 Ilyushin's Isotropy Postulate 252
17.6 Delay Law 254
17.7 Loading Surface 256
17.8 Drucker Postulate 257
17.9 On the Applicability Limits of the Strain Theory of Plasticity ... 260
References 264

18 Theories of Plastic Yield 267
18.1 General Ratios 267
18.2 Prandtl–Reuss Yield 269
18.3 Saint-Venant–Mises Yield Theory 269
18.4 Plastic Yield in Isotropic Hardening 270
18.5 Handelman–Lin–Prager Plasticity Theory 271
18.6 Yield for Plane Loading Surfaces 273

18.7   Yield for Some Loading Surfaces 275
18.8   Kadashevich–Novozhilov Plasticity Theory 278
18.9   Singular Loading Surfaces 278
References 281

19   Other Variants of Plasticity Theories 283
19.1   Batdorf–Budiansky Slip Theory 283
19.2   Two-Dimensional Klyushnikov Model 287
19.3   Endochronic Plasticity Theory 291
19.4   On the Methods of Physical Mesomechanics and Synergetics ... 294
References 298

Part III   Development of the Slip Concept in Plasticity Theory

20   Problem Setting 303
20.1   Initial Concepts and Definitions 303
20.2   Shift Resistance 304
20.3   Slip Synthesis 306
20.4   Definition of Principal Strains 309
References 312

21   Strain Specifics of Plastic Bodies 313
21.1   Elongation Diagram of a Plastic Material Specimen 313
21.2   Delay of Yield 316
21.3   Yield Stress and Loading Rate 317
References 319

22   Axioms of the Inelastic Body Model 321
22.1   Deformational Softening 321
22.2   Initial Shear Resistance 321
22.3   Function of Elastic Softening 323
References 325

23   The Fluidity at the Finite Speed of Loading 327
23.1   Yield Strength at the Final Loading Speed 327
23.2   Defining the Aging Function 328
        23.2.1   Example 329
23.3   Components of Deformational Softening 330
23.4   Almost Simple Strain 331
References 332

24   Specimen Elongation with Yield Drop 333
24.1   Original Assumption 333
24.2   Occurrence of Non-elastic Strain 333
24.3   Origins of Boundary Layer Theory 334
24.4   Simplified Model of Non-elastic Strain Growth 335
24.5   Definition of the Plastic Zone Growth Rate 337

24.6   Steady-State Yield 338
24.7   Building an Elongation Diagram.................................. 338
References.. 339

25   Building a Shear Resistance Operator 341
25.1   General Form of the Shear Resistance Operator 341
25.2   Boundary Condition 342
25.3   Special Cases... 343
References.. 344

26   Full Bauschinger Effect 345
26.1   Secondary Yield Stress 345
26.2   Proportional Primary Loading....................................... 346
26.3   Proportional Loading of an Opposite Sign 346
26.4   Function Ψ in Almost Simple Strain.............................. 349
References.. 349

27   Non-elastic Uniaxial Elongation–Compression........................ 351
27.1   Calculating Slip Intensity... 351
27.2   Calculation of the Integral (27.5) 352
27.3   Solving the Integral Equation 354
27.4   Study of the Tensor Intensity of Slips............................. 355
27.5   Determinant Equations in Uniaxial Elongation 357
27.6   Plastic Strain in Loading and Compression 360
        27.6.1   Increment of Non-elastic Strain in Loading 360
        27.6.2   Strain in Compression 360
27.7   Strain Creep and Stress Relaxation 362
27.8   Examples of Building Diagrams in an Uniaxial Stressed State... 364
References.. 367

28   Module of Additional Orthogonal Load 369
28.1   Problem Statement.. 369
28.2   Determining the Intensity of Additional Slips..................... 369
28.3   Calculation of the Strain Increments and Additional
        Loading Modulus 371
28.4   Analysis of Results and Conclusions 373
References.. 374

29   Plane-Plastic Strain... 377
29.1   Theorem of Strain in Pure Shear 377
29.2   General Dependencies in Pure Shear 379
29.3   Monotonous Plane-Plastic Strain 381
        29.3.1   Preparation of Initial Dependencies.................... 381
        29.3.2   Determinant Ratios 382
        29.3.3   Continuity Condition 384
        29.3.4   Monotony Conditions.................................... 387
References.. 389

Part IV   Non-elastic Strain of Geomaterials

30   Complex Strain of Soils 393
   30.1   Real State of the Mechanics of Non-elastic Strains 393
   30.2   Simple Strain Model of Hardening Dense Soils................... 394
   30.3   Defining the Form of the Function G 396
            30.3.1   Building the G Function for a Material with
                     High Hardening 397
            30.3.2   Universal G Function for Hardening Soils 397
   References... 398

31   Simple Loadings of Geomaterials....................................... 401
   31.1   Uniaxial Compression 401
   31.2   Creep in Uniaxial Compression 401
   31.3   Uniaxial Elongation... 402
   31.4   Pure Shift.. 402
   31.5   Determination of Model Parameters 403
   31.6   Comparison of Experimental and Calculation Results............ 403
   References... 405

32   On Boundary Value Problems of Inelastic Body Mechanics........... 407
   32.1   General Formulation of the Problem of Inelastic Solid
            Mechanics 407
   32.2   More About the Method of Elastic Solutions...................... 409
   32.3   An Example of Using the Birger Method........................... 412
            32.3.1   The Initial Stage of the Process with Linear Hardening 412
            32.3.2   Case of Semi-Infinite Plastic Zone..................... 413
            32.3.3   Auxiliary Task 414
            32.3.4   Final Length of the Plastic Zone 416
            32.3.5   The Dependence of the Tensile Force and
                     Pressure p on the Length of the Plastic Zone 417
   32.4   Perfectly Plastic Body Case 420
   32.5   Using the Kröner Theory of Residual Stresses 421
   32.6   Kröner Method for Plane Deformation 423
   32.7   More About Incompatible Deformations 424
            32.7.1   Distributed Wedge Dislocations......................... 424
            32.7.2   Strain Incompatibility Tensor 426
   32.8   The Application of Kröner's Method to the Brazilian Test 429
            32.8.1   Zero Approximation 429
            32.8.2   Green's Tensor Function for a Circle 431
            32.8.3   Definition of Deformation in a First Approximation... 434
   References... 436

Index 439

Notation Conventions

Loads and Stresses

A_{ij}	Tensor of second rank
A_{ijkl}, C_{ijkl}	Fourth-rank tensor
$\alpha_{1,2}$	Boundaries set of sliding planes
$\vec{\beta}(\beta_1, \beta_2, \beta_3)$	Unit vector of octahedral shear stress
δA, δW	Variation of external and internal forces operation
D_σ, D_ε	Stress and strain deviators
$E(e)$, $K(e)$	Elliptic integrals
$(\vec{e}_1, \vec{e}_2, \vec{e}_3)$	Orta of coordinate axes
I_1, $I(_2, I_3)$	Tensor invariants
I_σ, II_σ, III_σ	Main invariants of stress tensor
J_1, J_2, J_3	Different forms of stress tensor invariants
$\Phi(x)$	Integral of probabilities
$\Phi(\varepsilon_{\grave{e}})$	Universal hardening function
$\Pi(x, y)$	Prandtl's function
P_e, P_i	Set of external (internal) forces
$\Phi(z)$, $\Psi(z)$	Holomorphic Muskhelishvili functions
p	Intensity of superficial loading
q	Load per unit of length
S_{ij}	Green's tensor function
S_σ	Ball stress tensor
Σ	Total stress at body point
σ_k	Normal component of stress vector in the direction of k-axis
σ_0	Average normal tension
σ_1, σ_2, σ_3	$(\sigma_1 \geqslant \sigma_2 \geqslant \sigma_3)$—main stress
σ_k	Normal component of stress vector in the direction of k-axis
σ_r, σ_t	Normal stresses in polar coordinates

$\sigma_è$, $\varepsilon_è$	Stress intensity and strain intensity
$\dot{\sigma}_{ij}$, $\dot{\varepsilon}_{ij}$	Stress and strain speed components
T_σ, T_ε	Stress tensor and strain tensor
τ_{kl}	Tangent component of stress vector in axes k, l
τ_{max}	Maximum shear stress
τ_s	Yield point in shear
τ	Shear stress
$\omega_1(z)$, $\omega_2(z)$	Functions Galina

Deformations and Movements

Δ	Laplacian
Δl	Absolute linear strain in the l direction
δ_{ij}	Kronecker symbol
$d\Omega$	Elementary space angle
$\varepsilon_{\Pi p}$	Inelastic deformation of prefluidity
Γ	Green function (source function)
γ	Angle (deformation) of shift
γ_{kl}	Shear deformation (change of initially right angle) between axes k and l
H	Neumann's function
l	Distance between points of deformed body
$\mathbf{S}$, $\acute{\mathbf{Y}}$	Vector representation of tensors
$S_{\nu\lambda}$	Operator from gliding intensity
T_{ε^y}, T_{ε^e}	Elastic and plastic strain tensor
$\Theta_{1,2}(t)$	Slip region boundaries in a flat model
$\varphi(z)$, $\psi(z)$	Muskhelishvili functions for movements
u, v, w	Components of linear movement in three orthogonal directions
ε	Linear relative deformation
ε_0	Hydrostatic (average) deformation
ε_l	Linear relative deformation in the direction l
ω_l	Angular movement around axis l

Physical and Mechanical Characteristics of Materials

A, W	Work of external (A) and internal (W) forces
E	Module of elasticity of Jung
E_s, E_t	Secant and tangent module
F	Area of the bar normal section

φ_{nl}	Intensity of slides
G	Shear modulus; hardening function for soils
G_i	Module of orthogonal additional load
K	Module of volume elasticity
$\lambda,\ \mu$	Constant Lame
ν	Poisson's ratio
ρ	Material density
r_{jj}	Normal slip tensor intensity
R	Area of slidings
σ_Π	The limit of proportionality
σ_B	Uniaxial tensile strength
σ_Γ	Durability of a boundary layer
Θ	Volume expansion
U	Potential energy of a system; Erie's function
$U_{\nu\lambda}$	Reduced strength from aging of material
u_Φ	Specific potential energy of shape change
V	Volume of the body or its element
$\Phi_{n\lambda}$	Vector intensity of slidings

Part I
Basis of Elasticity Theory

Chapter 1
Summary of Elasticity Theory: Basic Concepts

1.1 From the History of Elasticity Theory

Elasticity theory as one of the most important parts of mathematical physics was formed in the first half of the nineteenth century; however, many of its sections had been developed much earlier. The works of the scientists of the seventeenth and eighteenth centuries, Galileo, Mariotte, Hooke, Bernoulli, Euler, Coulomb, etc., significantly developed the bending theory for thin elastic rods.

Early in the nineteenth century, Lagrange and Sophie Jermain solved the task of bending and oscillations of thin elastic plates. Strain features of thin elastic bodies allowed simplifying the setting and finding the solution on their straining under the action of external forces.

The early nineteenth century also saw a rampant development of mathematical analysis following the occurrence of multiple important tasks in physics and engineering. The existing situation became a basis for a new specific field of physics called *mathematical physics*. New tasks assigned to this young discipline, first of all, included the need to thoroughly study properties of elastic materials and to build the mathematical theory based on them that would permit determining internal forces occurring in an elastic body under the action of external forces, as well as body strain, e.g. changes in its dimensions and/or shape. These studies were extremely necessary to satisfy the demands of rapidly developing engineering due to construction of railroads, bridges, steam machines, metalworking machines, etc.

Some of these issues were solved in 1825 when the French engineer A. Navier issued the "Course of Lectures on Strength of Materials" based on existing experimental data and including approximated theories mentioned above. A similar course by N. F. Yastrzhembsky appeared in 1837 in Russia.

Other important engineering tasks (contact tasks, strain of massive bodies, etc.) were still to be solved. There was a relevant need to create theoretic methods for strength analysis of structural parts and machines of arbitrary shape.

© The Author(s), under exclusive license to Springer Nature Switzerland AG 2021
V. Molotnikov, A. Molotnikova, *Theory of Elasticity and Plasticity*,
https://doi.org/10.1007/978-3-030-66622-4_1

The first researcher who acquired general equations of equilibrium and oscillations of elastic bodies was Navier. He used the Newton concept of the
discrete structure of substances and believed that the interaction force between
two molecules, the distance between which changed due to the body strain, was
proportional to the product of distance increment and some function of the initial
distance. However, the Navier line of reasoning was not commonly accepted. The
legality of using the integration operation in the area (discrete) occupied by the body
was argued.

As we know, mathematical analysis of that time was built on the concept of
continuous geometric space (continuum) where infinitely short sections could be
considered and where differentiation and integration processes could be conducted
based on that. Newton's molecular theory of body structure represented them as
discrete media consisting of individual particles interrelated by mutual attraction
and repulsion forces. Therefore, the applicability of the mathematical analysis
apparatus to such media, which was substantially associated with the concept of
continuous functions, seemed illegal and unjustified.

However, a large number of particles is contained even in an extremely small
volume mentally separated from a body. This circumstance made researchers think
of using the law of large numbers and the method that would be called statistical
thereafter. Using the statistical method helped to build a bridge between the
continuous space of mathematical analysis and the solid body as a discrete medium
and resulted in the rehabilitation of using the powerful apparatus of mathematics for
creating a new field of physics. This was also supported by light wave propagation
theory reported by Fresnel to the French Academy of Science in 1821.

The importance of applying elasticity theory in physics and engineering and the
high complexity of assigned tasks in terms of mathematical analysis drew attention
to this new field of science on part of the largest mathematicians and mechanics of
that time, O. Cauchy and S. Poisson.

Cauchy substantiated the legality of Navier's approach as follows [6, p. 23]. For
the volume containing a multitude of molecules and having small dimensions as
compared to the radius of the sphere where noticeable molecular action is expressed,
the number of molecules can be deemed proportional to the volume. If we leave
aside the molecules adjacent to the considered molecule, the action of all molecules
contained in one of the small volumes described above is equivalent to the force,
whose line of action crosses the center of gravity of the volume, and the value is
proportional to this volume and to some function from the distance of the center of
gravity of the volume from the considered molecule.

By the autumn of 1822, Cauchy introduced the most basic concepts in modern
elasticity theory. In particular, the concept of stress in a body point as a ratio between
forces and the area of various flat elements built in this point is of fundamental value.

Poisson's first memoir dedicated to the matter under consideration was read
by the French Academy in April 1828. When considering the issue in general
equations of elasticity theory, Poisson (just as Cauchy) started with the derivation of
equilibrium equations expressed in stress components. The formulas expressing the
stress components through strains contain sums taken for all "molecules" in the area

of action of this "molecule." Poisson does not find it possible to substitute all sums with integrals and thinks that such substitution is possible only when summing upon the solid angle around this "molecule," but not when summing upon the distance counted from the "molecule." Equations of equilibrium and motion of an isotropic elastic body derived in this manner coincide with the Navier equations.

Apart from the above-mentioned founders of elasticity theory (Cauchy, Navier, and Poisson), we may also name such prominent scientists as M. V. Ostrogradsky, G. Lame, B. Clapeyron, A. Saint-Venant, B. Green, D. Maxwell, V. Thompson (Lord Kelvin), D. Rayleigh, D. Michell, Mathieu, F. S. Yasinsky, S. P. Timoshenko, G. V. Kolosov, N. I. Muskhelishvili, and many others. Those readers who want to get familiar with the history of occurrence and development of elasticity theory may refer to a comprehensive feature placed in the introduction to the book by A. Love "Mathematical Theory of Elasticity" [6], as well as books by S. P. Timoshenko "History of Science on Strength of Materials," [8] A. T. Grigoryan [1–3], etc.

1.2 Elasticity of Solid Bodies

In the case of interaction of forces, distances between particles of a solid body change. This change defines *strain* of a body.[1] The body's property to take initial dimensions and shape after forces are removed is referred to as *elasticity*. The physical nature of elasticity lies in the following.

It is known [4] that the location of atoms in a solid body is characterized by a specific order (short-range order for amorphous bodies and long-range order for crystalline bodies). The property of a solid body to keep the atom ordering in a loaded state and the distance between atoms in a non-loaded state defines elasticity in physical terms. Disturbances in the initial ordering of atoms are referred to as [4] *structural imperfections.* They are found in all real solid bodies, except maybe artificial single crystals. When exposed to forces, structural imperfections are slowly re-distributed. If strains caused by the above re-distribution of defects in the material structure disappear some time after removing the load, they are referred to as *reversible.* The disappearance of such strains is called *elastic after-effect.*

1.3 Homogeneous Strain

Assume that a multitude of pairs of material points of a solid body satisfies the following conditions.

---

[1] Temperature strains and fluctuations are not considered.

(A) Straight lines connecting each pair of points are parallel to an arbitrarily defined direction.
(B) The distance between these points is low as compared to the body dimensions.

A *strain* is called homogeneous when the distance (l) between these points is increased by the value (Δl), proportional to l. The relation of

$$\varepsilon = \frac{\Delta l}{l}$$

is referred to as *relative elongation* in the specified direction. Negative relative elongation is defined as *shortening*.

1.4 Internal Forces: Method of Sections

Interaction forces of solid body particles form a system of internal forces. Some of them act upon the Newton law of gravitation and represent interest only in the case of large astronomic bodies. Such forces between individual parts of engineering structures are negligibly low and we will not take them into account. Other internal forces of a solid body for its two elementary particles have a significant value only if the considered particles are spaced by no more than the distance between molecules. Therefore, by mentally separating one part of the body from the other, we will represent the interaction of these parts by forces distributed over the section surface.

For any dissected parts of the body, the system of internal forces replaces the action of the discarded part on the considered part. Thus, the method of imagined sections, firstly, allows detecting internal forces and, secondly, transfers internal forces to the category of external loads applied to each of the dissected parts of the body. Using the interrelation principle of action and counter-action, these loads applied to each of the sides of the made section are equal in value and opposite in direction.

Let us explain the above using Fig. 1.1. Assume an arbitrary solid body loaded by internal forces. Some of these forces (or even all of them) can be reactive. Let us designate the aggregate of external forces with (P_e).

Let us mentally dissect the body by any surface (for example, by the plane α, Fig. 1.1a) into two parts—right (R) and left (L). Let us detach these parts (Fig. 1.1b).

Let us replace the action of each part with the other one with internal forces applied to each of the section sides (see Fig. 1.1b). Internal forces are continuously distributed over the section surface in a complicated manner. Let us designate the aggregate of these forces by (P_i).

For any distribution over the section surface, internal forces must be such that they satisfy the equilibrium conditions for the left and right parts of the body individually. Symbolically, this can be written as follows:

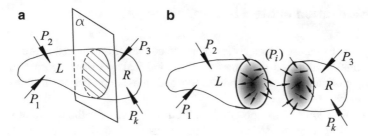

Fig. 1.1 Internal forces

$$(P_e)_R + (P_i) = 0;$$
$$(P_e)_L + (P_i) = 0, \tag{1.1}$$

where symbols $(P_e)_R$, $(P_e)_L$ mean the aggregate of internal forces applied to the right and left parts, respectively.

In this manner, the system of internal forces (P_i) can be successfully defined from the equilibrium condition of the right and left dissected part of the body.

1.5 Homogeneous Body

A body is called *homogeneous* if the physical attributes (properties) of all its particles are identical. These properties can be seen in most solid bodies when the dimensions of these particles exceed some limits. For ideal crystals [4], these limits are comparable with inter-atom distances. For polycrystalline materials such as steel, the dimensions of these particles reach many micrometers, or even centimeters for concrete.

1.6 Stress Vector

Internal forces in the case of homogeneous strain define the homogeneous stressed state in a homogeneous body. Assume that there is homogeneous stressed state in a body. Let us mentally dissect the body by planes parallel to some arbitrarily defined plane. It is obvious that internal forces in sufficiently large parts of these sections will have resultants parallel to each other.

By dividing the resultant of internal forces by the area of a respective part of the section, we obtain a vector called the *stress vector*. In a homogeneous stressed state, the stress vector remains constant for any parallel areas.

1.7   Elongation of Steel Specimens

Primary mechanical characteristics of most construction materials are defined in experiments for uniaxial compression of cylindrical or prismatic specimens. The appearance of such a specimen before (dashed lines) and after (solid lines) the test is given in Fig. 1.2.

Figure 1.3 shows an exemplary dependency diagram for the resultant value (P) of forces elongating the specimen of structural steel, and elongation (Δl) of the specimen. This diagram is automatically plotted using the diagrammatic device during elongation experiments almost in all existing pull test machines.

In such tests, we can see a homogeneous strain of the specimen everywhere at a sufficient distance from its ends. Figure 1.3 shows, in a dashed line, that part of the diagram corresponds to the homogeneous strain of the specimen due to the formation of local narrowing (neck) shown in Fig. 1.2.

Relative longitudinal strain in the direction of the elongating force is the value

$$\varepsilon_1 = \frac{\Delta l}{l},$$

whereas l is the working length of the specimen (Fig. 1.2). *Transverse relative deformation* (ε_2) is the value of relative elongation in the direction perpendicular to the specimen axis. *Normal stress* is the force attributable to the unit of area F of normal section of a non-strained rod, e. g.

$$\sigma_1 = \frac{P}{F}. \tag{1.2}$$

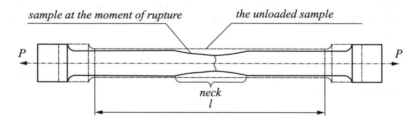

Fig. 1.2 Specimen for elongation test

Fig. 1.3 Dependency of specimen elongation on elongating force

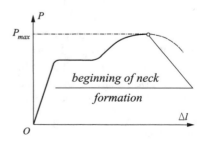

Fig. 1.4 Elongation diagram
in coordinates
"stress—strain"

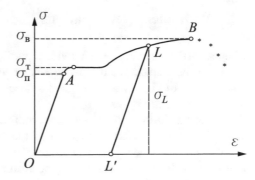

Graphic representation of the dependency between σ_1 and ε_1 (Fig. 1.4) is referred
to as the *elongation diagram.*

Before neck formation, the elongation diagram does not depend on the cross-
section shape of the specimen and its dimensions usually used in tests.

Elongation experiments give the following results in the first approximation.

(A) Before some stress σ_Π, (Fig. 1.4) called the *proportionality limit,* strain is
proportional to stress. This dependency is defined as follows:

$$\varepsilon_1 = \frac{\sigma_1}{E}, \quad \varepsilon_2 = -\nu\varepsilon_1 = -\nu\frac{\sigma_1}{E}. \tag{1.3}$$

The coefficient E having the dimension of stress is called the Jung elasticity
modulus, and the non-dimensional parameter ν is referred to as the Poisson
coefficient.

The dependencies (1.3) mathematically express the Hooke's law during
elongation-compression.

(B) At some stress, strain increases without significant change in stress. This stress
is called *the yield strength* of a material σ_T (Fig. 1.3). They say that at stress σ_T,
the material flows, and the horizontal part of the elongation diagram is referred
to as the *yield plateau.*

(C) After loading the specimen beyond the yield strength, for example, to the point
L (Fig. 1.4), the dependency between ε_1 and σ_1 in the case of further unloading
in the first approximation can be represented by a straight line LL'. Before
neck formation, this straight line is almost parallel to AO.

(D) The highest stress $\sigma_?$ corresponding to the point B in the diagram is referred to
as the *ultimate tensile strength.* At this point, the specimen is destroyed in the
area of the formed neck (Fig. 1.2).

1.8   Permanent Deformations

Changes in distances between individual particles of a body before and after loading ceases to act characterize the permanent or *plastic deformation.* For example, the section OL' in Fig. 1.4 represents, at a specific scale, relative permanent (plastic) deformation caused by stress σ_L.

Due to elastic after-effect, permanent deformations almost always decrease with time. This process is accelerated when the body is heated. Deformations disappearing due to elastic after-effect are referred to *unstable permanent deformations.* Permanent deformations that cannot be eliminated by the above method are called *stable permanent deformations.*

1.9   Elastic Limit

Maximum stress when stable permanent deformations are still seen is called [5] *the natural elastic limit.* At stresses not exceeding the natural elastic limit, unstable permanent deformations occur, which have specific limit values for each material.

For practical determination of the elastic limit, the permanent deformation is conditionally set based on the technical capabilities of measuring it. The value determined in this manner is called *the conditional elastic limit. In engineering, the elastic limit is frequently identified with the proportionality limit.*

1.10   Elastic Shear Deformation

Assume that the homogeneous stressed state in a body is created by elongation in a single direction by stress $\sigma_1 = p$ and compression in a perpendicular direction by stress $\sigma_2 = -p$ (Fig. 1.5a).

By projecting all forces to the normal line to the section equally inclined to the directions of elongation and compression, we can easily make sure that there will be no normal stresses in these sections. This indicates that the stress vector on these areas is located in the section plane.

The forces acting on the element $ABCD$ cut from the body by the above-mentioned equally inclined sections will look as shown in Fig. 1.5b, (strain state). Internal forces act on the surface of this element, which lie in the planes it is confined by. Such forces are called *tangential,* and their values belonging to the unit of surface area are referred to as *tangential stresses.*

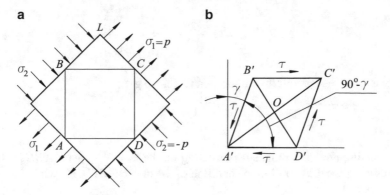

Fig. 1.5 Elastic shear deformation

Assume that the elastic properties of the material in the case of elongation to any direction are identical.[2] By designating relative elongation through ε_1 in the direction of elongation forces and through ε_2 in the direction of compression and using formulas (1.3) in a respective way, we obtain

$$\varepsilon_1 = \frac{1+\nu}{E}p, \quad \varepsilon_2 = -\frac{1+\nu}{E}p. \tag{1.4}$$

Assume that before deformation, the rectangle $ABCD$ was a square. If we know the relative deformation ε_1 in the direction of its diagonal $A'C'$, we will find the length of this diagonal after deformation

$$A'C' = AC(1 + \varepsilon_1). \tag{1.5}$$

Taking into account formulas (1.4), we can write (Fig. 1.5b) that

$$A'B' = AC\left(1 + \frac{1+\nu}{E}p\right), \quad B'D' = BD\left(1 - \frac{1+\nu}{E}p\right);$$

then

$$\mathrm{tg}\angle OA'B' = \frac{OB'}{OA'} = \frac{1 - \dfrac{1+\nu}{E}p}{1 + \dfrac{1+\nu}{E}p}. \tag{1.6}$$

Using γ we designate the angle that the straight angle BAD (Fig. 1.5a) has initially changed, and then we have

---

[2]This material is called *isotropic*. This book describes only initially isotropic materials.

$$\text{tg}\angle OA'B' = \text{tg}\frac{90° - \gamma}{2} = \frac{1 - \text{tg}\dfrac{\gamma}{2}}{1 + \text{tg}\dfrac{\gamma}{2}}.$$

Comparing the last result with formula (1.6), we get

$$\text{tg}\frac{\gamma}{2} = \frac{1+v}{E}p. \tag{1.7}$$

Let us find now tangential stresses acting on the faces of square $ABCD$. To do it, let us compare the equilibrium condition of the element BCL shown in Fig. 1.5a. We notice that

$$\angle BCL = \angle CBL = 45°, \quad BL = LC = \frac{BC}{\sqrt{2}} \tag{1.8}$$

and zero the sum of projections of all forces to the direction BC, we get $\tau = p$, and formula (1.7), provided low deformation

$$\text{tg}\frac{\gamma}{2} \approx \frac{\gamma}{2},$$

can be written as

$$\gamma = \frac{\tau}{G} \tag{1.9}$$

that designates

$$G = \frac{E}{2(1+v)}. \tag{1.10}$$

The formula (1.9) expresses *the Hooke's law in shear.* It defines increment (γ) of the straight angle as a result of action of tangential stresses τ as shown in Fig. 1.5b. This increment is called *shear deformation,* and the coefficient G included in formula (1.9) is called *the shear modulus.*

1.11 Law of Twoness of Tangential Stresses

Assume that a body shaped as a rectangular parallelepiped (Fig. 1.6) is exposed to the action of only tangential forces in the conditions of a homogeneous stressed state [7].

Provided that the projections of forces onto coordinate axes are equal to zero xy, it goes that $\tau'_x = \tau_x$, $\tau'_y = \tau_y$, e. g., tangential stresses on parallel faces are equal

Fig. 1.6 To the law of
twoness of tangential stresses

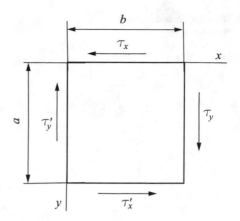

between each other and directed oppositely. This trivial result can be considered as
a condition of homogeneity of a stressed state.

Now let us satisfy the third condition of equilibrium—zero equality of the sum of
moments. If we deem the size of the parallelepiped shown in Fig. 1.6 in the direction
perpendicular to the plane xy to be equal to zero, we find that tangential forces acting
on two horizontal areas are equal to $\tau_x \cdot b \cdot 1$. The moment of the pair of horizontal
forces will be $M_1 = \tau_x \cdot b \cdot a$. In the same manner, let us find the moment M_2 of
forces parallel to the axis y. For the directions of tangential forces given in Fig. 1.6,
the moment M_2 will have the sign opposite to the sign of M_1, e. g. $M_2 = -\tau_y \cdot a \cdot b$.

By equating the sum of moments of all forces applied to the element to zero,
we get $\tau_x = \tau_y$. The acquired equality expresses the so-called *law of twoness of
tangential stresses.* For an arbitrary stressed state, it can be generalized as follows:
*tangential stresses on perpendicular areas, perpendicular to the intersection line of
these areas, equal between each other.*

These stresses are always directed as shown in Fig. 1.6 or opposite to the given
one, simultaneously for all four areas.

1.12 Homogeneous Stressed State

When studying the elongation of a material specimen previously, we already spoke
(p. 7) of the homogeneous stressed state. In a general case, let us mentally separate
from a body a rectangular parallelepiped whose faces are parallel to coordinate
planes (Fig. 1.7). The stress vector on any of its faces can be decomposed into
components in three directions coinciding with the direction of the coordinate axes
x, y, z.

One of the components will be directed along the normal line to the specified
face. This component of the stress vector is called normal, and its vector is called
normal stress. Let us designate normal stresses by the symbol σ_k, $(k \sim x, y, z)$,

Fig. 1.7 Homogeneous
stressed state

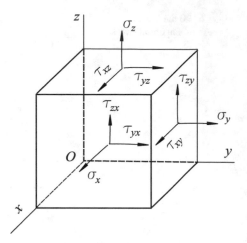

whereas the index k coincides with the designation of the coordinate axis parallel to
the respective normal components of the stress vector. In the case of a homogeneous
stressed state, normal stresses on opposite faces are equal in magnitude.

Two other components of the stress vector will be located in the face plane. They
are called *tangential stresses*. Due to the law of twoness, tangential stresses are fully
defined by three values: τ_{xy}, τ_{yz}, τ_{zx}. The first of the indexes in the designation
of tangential stresses indicates the axis in parallel to which this component of the
stress vector acts, while the second index shows the normal line to the area under
consideration.

Figure 1.7 shows the components of the stress vector on three (visible) areas only.
On each of three other (non-visible) areas, the directions of stress vector components
will be opposite to those that have place on the opposite (visible) area.

Let us describe the rules of signs for stresses. Let us assume that the direction
of the external normal line to the surface of the element under consideration of a
body coincides with the direction of the coordinate axis. In this case, stress vector
components are deemed positive when their direction coincides with the coordinate
direction. Alternatively, when the external normal line to the surface is directed
opposite to the coordinate axis direction, stress vector components are directed
opposite to the directions of coordinate axes.

1.13 Generalized Hooke's Law

Homogeneous deformation is characterized by three relative elongations (ε_x, ε_y, ε_z)
in the direction of coordinate axes and three shear deformations of an element
parallel to the coordinate axes (γ_{xy}, γ_{yz}, γ_{zx}). In an elastic body, these deformations
are associated with stresses *by the generalized Hooke's law:*

$$\varepsilon_x = \frac{1}{E}[\sigma_x - \nu(\sigma_y + \sigma_z)],$$

$$\varepsilon_y = \frac{1}{E}[\sigma_y - \nu(\sigma_z + \sigma_x)], \qquad (1.11)$$

$$\varepsilon_z = \frac{1}{E}[\sigma_z - \nu(\sigma_x + \sigma_y)];$$

$$\gamma_{xy} = \frac{\tau_{xy}}{G}, \quad \gamma_{yz} = \frac{\tau_{yz}}{G}, \quad \gamma_{zx} = \frac{\tau_{zx}}{G}. \qquad (1.12)$$

Formulas (1.11) and (1.12) are applicable in the case of specific limitations applied to stress values $\sigma_x, \ldots, \tau_{zx}$. In engineering designs, these limitations are often met by satisfying the requirements dictated by the conditions of strength and/or rigidity.

1.14   Another Form of Hooke's Law

Let us consider the change of the body element volume due to deformation. If rib lengths of an infinitely small parallelepiped before deformation are designated through dx, dy, dz, its volume will be $dV = dxdydz$. After deformation, rib lengths will change and become equal to $(1 + \varepsilon_x)dx$, $(1 + \varepsilon_y)dy$, $(1 + \varepsilon_z)dz$. By designating the element volume after deformation through $dV_\prime$, we can calculate the element volume increment due to deformation:

$$\Delta(dV) = (1 + \varepsilon_x)(1 + \varepsilon_y)(1 + \varepsilon_z)dxdydz - dxdydz.$$

In the case of small deformations, we neglect here the products of linear deformations as infinitely small and of higher order, and we obtain

$$\Delta(dV) = (\varepsilon_x + \varepsilon_y + \varepsilon_z)dxdydz.$$

The relation of the volume increment to the initial element volume is referred to as *the volumetric expansion*. By designating it through Θ, we can write

$$\Theta = \frac{\Delta(dV)}{dV} = \varepsilon_x + \varepsilon_y + \varepsilon_z. \qquad (1.13)$$

Let us call the hydrostatic part of stress (σ_0) as the average of normal stresses for three arbitrary orthogonal areas routed through the body point:

$$\sigma_0 = \frac{1}{3}(\sigma_x + \sigma_y + \sigma_z). \qquad (1.14)$$

In a similar way, the average of linear relative deformations (ε_0) in the directions of normal lines to these areas will be called *the hydrostatic strain:*

$$\varepsilon_0 = \frac{1}{3}(\varepsilon_x + \varepsilon_y + \varepsilon_z) = \frac{\Theta}{3}. \tag{1.15}$$

By directly checking, we can define that the dependencies (1.11) and (1.12), taking into account the designations (1.14) and (1.15), are equivalent to the following:

$$\varepsilon_0 = \frac{\sigma_0}{K}, \quad D_\varepsilon = \frac{1}{2G}D_\sigma, \tag{1.16}$$

where

$$K = \frac{E}{1 - 2v}, \tag{1.17}$$

and D_σ, D_ε are *deviators* of stresses and strains, respectively, which are defined using tables:

$$D_\sigma = \begin{pmatrix} \sigma_x - \sigma_0 & \tau_{xy} & \tau_{xz} \\ \tau_{yx} & \sigma_y - \sigma_0 & \tau_{yz} \\ \tau_{zx} & \tau_{zy} & \sigma_z - \sigma_0 \end{pmatrix},$$

$$\tag{1.18}$$

$$D_\varepsilon = \begin{pmatrix} \varepsilon_x - \varepsilon_0 & \gamma_{xy} & \gamma_{xz} \\ \gamma_{yx} & \varepsilon_y - \varepsilon_0 & \gamma_{yz} \\ \gamma_{zx} & \gamma_{zy} & \varepsilon_z - \varepsilon_0 \end{pmatrix}.$$

1.15   Plane Stress-Strain State

Assume that relative elongation in the direction of one of the axes equals zero (for example, $\varepsilon_z = 0$). In this case, movements of all body points in the case of its deformations occur in parallel to the plane $z = 0$. Such deformation is referred to as *plane strain.*

In the assumption $\varepsilon_z = 0$, the last formula (1.11) gives

$$\sigma_z = v(\sigma_x + \sigma_y), \tag{1.19}$$

and the two first formulas can be written as

$$\varepsilon_x = \frac{1}{E^*}(\sigma_x - v^*\sigma_y); \quad \varepsilon_y = \frac{1}{E^*}(\sigma_y - v^*\sigma_x), \tag{1.20}$$

where

$$E^* = \frac{E}{1 - \nu^2}, \quad \nu^* = \frac{\nu}{1 - \nu}. \tag{1.21}$$

Noting that

$$\frac{E^*}{1 + \nu^*} = \frac{E}{1 + \nu}, \tag{1.22}$$

we come to the following conclusion:

the dependency between relative elongations and normal stresses in the case of plane strain ($\varepsilon_z = 0$) can be obtained from the respective dependencies in the case of a plane stressed state ($\sigma_z = 0$) by substituting E and ν through E^* and ν^* using formulas (1.21).

In the case of such substitution, the dependency between tangential stresses and shears towards the ratios of (1.22) and (1.10) will not change.

1.16 Homogeneous Model of a Solid Body

During mathematical studies of a deformable solid body, we consider infinitely small elements of the body that are attributed with mechanical properties found in experiments with relatively large specimens. Thus, a real body is substituted by its ideally homogeneous continuous model.

Movements of points of such a model are represented by functions of coordinates of these points differentiated for the required number of times. In this case, the stress-strain condition in sufficiently small volumes of a body can be deemed homogeneous.

In some cases, let us assume to associate each atom of a solid body with a geometric point of an ideal model. For specific loading, relative movements of any two respective points of a real body and its continuous model must almost coincide only if the distance between points is sufficiently long as compared to distances between atoms.

This section suggests that strains and stresses of a continuous model of a solid body are connected by Hooke's law unless there are special exceptions. *A real solid body is substituted by the specified ideally homogeneous model.*

1.17 Axisymmetric Plane Strain

Assume that during deformation, all points of a body move in the directions perpendicular to a straight line being the symmetry axis, whereas the value of these movements (u) is only a function of the distance (r) of the point from this axis.

Movements u are deemed positive if the distance of the point from the symmetry axis is increased. Otherwise $u \leqslant 0$.

If the movement equals u in the point located at the distance of r from the symmetry axis, the movement in the point $r + dr$ can be represented as

$$u(r + dr) = u(r) + \frac{du}{dr} dr. \tag{1.23}$$

If two points located on the same perpendicular to the symmetry axis were distanced before deformation from each other to the distance of $l = dr$, after deformation this distance will be $dr + \frac{du}{dr} dr$. Consequently, the increment of the distance between points will be $\Delta l = \frac{du}{dr} dr$, and relative elongation $(\Delta l / l)$ in the radial direction will be

$$\varepsilon_1 = \frac{du}{dr}. \tag{1.24}$$

Due to deformations, distances between body points in the direction perpendicular to the symmetry axis will change along with the lengths of coordinate circumferences. The length a circumference with radius r with the center on the symmetry axis will get an increment $\Delta l = 2\pi (r + u) - 2\pi r$. Relative elongation $\Delta l / l$, $l = 2\pi r$, $\Delta l = 2\pi u$) of this circumference will be equal to

$$\varepsilon_2 = \frac{u}{r}. \tag{1.25}$$

1.18   Lame Task

Let us consider deformation of a round hollow cylinder with walls of permanent thickness that is subject to the action of evenly distributed internal (p_a) and external (p_b) pressures (Fig. 1.8a). In the following, we will consider a round ring cut from a cylinder by two planes perpendicular to its axis and located from each other at the distance equal to one.

The symmetry condition follows that only normal stresses will act on the sides of the element mnn_1m_1 shown in Fig. 1.8b and limited by two coaxial surfaces and the axial plane. Let us designate annular normal stresses on the sides mm_1 and nn_1 as σ_t, and normal radial stresses on the sides mn and m_1n_1 as σ_r.

When the radius r is increased by dr, the radial stress will change for $\frac{d\sigma_r}{dr} dr$. Therefore, the normal radial stress on the side m_1n_1, (Fig. 1.8b) will be

$$\sigma_r + \frac{d\sigma_r}{dr} dr. \tag{1.26}$$

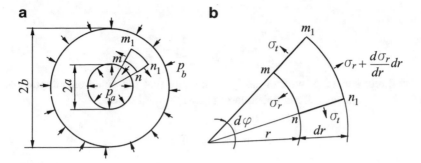

Fig. 1.8 To derivation of equilibrium equations in the Lame tasks

By projecting all forces acting on the element mm_1n_1n to the direction of the bisecting line of a small angle $d\varphi$, we will obtain the following equilibrium condition of the separated element:

$$\sigma_r r d\varphi + \sigma_t dr d\varphi - \left(\sigma_r + \frac{d\sigma_r}{dr} dr\right)(r + dr)d\varphi = 0. \tag{1.27}$$

By neglecting the infinitely small values of higher orders in the equality (1.27), we will obtain

$$r\frac{d\sigma_r}{dr} = \sigma_t - \sigma_r. \tag{1.28}$$

Let us express stresses through strains: For the considered case, when using expressions (1.25) and (1.24), formulas (1.20) take the following form

$$\frac{u}{r} = \frac{1}{E^*}(\sigma_t - v^*\sigma_r), \quad \frac{du}{dr} = \frac{1}{E^*}(\sigma_r - v^*\sigma_t).$$

By subtracting the respective parts of these equations, we obtain

$$\frac{u}{r} - \frac{du}{dr} = \frac{1 + v^*}{E^*}(\sigma_t - \sigma_r). \tag{1.29}$$

In a similar way, we find

$$\sigma_r = \frac{E^*}{1 - v^{*2}}\left(\frac{du}{dr} + v^*\frac{u}{r}\right). \tag{1.30}$$

By substituing (1.30) in the Eq. (1.28) and using formula (1.29), we obtain

$$r\frac{E^*}{1 - v^{*2}}\left(\frac{d^2u}{dr^2} + \frac{v^*}{r}\frac{du}{dr} - \frac{v^*}{r^2}u\right) = \frac{E^*}{1 + v^*}\left(\frac{u}{r} - \frac{du}{dr}\right)$$

or

$$\frac{d^2u}{dr^2} + \frac{1}{r}\frac{du}{dr} - \frac{u}{r^2} = 0. \tag{1.31}$$

By direct checking, we can easily make sure that the common integral of the linear homogeneous differential equation (1.31) is the function

$$u = Ar + \frac{B}{r}, \quad (A, B - const). \tag{1.32}$$

The constant values A and B are defined from the conditions

$$\sigma_r\Big|_{r=a} = -p_a, \quad \sigma_r\Big|_{r=b} = -p_b,$$

whereas p_a and p_b are set (see. p. 18) values. Using formula (1.30), we obtain

$$A = \frac{1-\nu^*}{E^*}\frac{a^2 p_a - b^2 p_b}{b^2 - a^2}; \quad B = \frac{1+\nu^*}{E^*}\frac{p_a - p_b}{b^2 - a^2}a^2b^2. \tag{1.33}$$

By substituting the function (1.32) and (1.33) and its derivative to formula (1.30), we obtain

$$\sigma_r = \frac{a^2 p_a - b^2 p_b}{b^2 - a^2} - \frac{p_a - p_b}{b^2 - a^2}\frac{a^2b^2}{r^2}. \tag{1.34}$$

With the known radial stress σ_r, we find the annular stress from formula (1.29)

$$\sigma_t = \frac{a^2 p_a - b^2 p_b}{b^2 - a^2} + \frac{p_a - p_b}{b^2 - a^2}\frac{a^2b^2}{r^2}. \tag{1.35}$$

It was suggested above (p. 18) that the cylinder was in the conditions of plane strain. e. g. there was no strain towards the cylinder axis. By combining the coordinate axis Oz with the cylinder axis, we have $\varepsilon_z = 0$ and find the axial stress using formula (1.19)

$$\sigma_z = 2\nu\frac{a^2 p_a - b^2 p_b}{b^2 - a^2}. \tag{1.36}$$

From the last result, we see that the axial stress in the considered case does not depend on the coordinate r. Hence, axial stresses can be eliminated by imposing a respective stress or compression of the cylinder in the axial direction. In this case, the cylinder will be in the conditions of a plane stressed state.

1.19   Phenomenon of Stress Concentration

Assume that in a cylinder considered in the previous paragraph $p_a = 0$, $b \to \infty$. In this case, by dividing numerators and denominators of formula (1.35) by b^2 and then making a passage to the limit with $b \to \infty$, we will find

$$\sigma_t(r) = -p_b - p_b\frac{a^2}{r^2}, \quad (p_a = 0,\ b = \infty). \tag{1.37}$$

The first term in the right part of the last formula represents annular stress that would be present in the case of all-around equal compression of a continuous disc. The second term expresses the change in annular stresses that is caused by the appearance of a round hole with a radius of a, which is small as compared to the disc dimensions. We have $\sigma_t(a) = -2p_b$ on the outline of the hole, e. g. the stress on the hole outline is twice as much as in those that could have appeared in the continuous disc. The same result for $\sigma_t(a)$ will be obtained for final b if we bring the hole radius (a) to zero, e. g.

$$\lim_{a \to 0} \sigma_t(a) = -2p_b, \quad (p_a = 0), \tag{1.38}$$

whereas for $a = 0$, formula (1.35) gives $\sigma_t = -p_b$.

This paradox is explained by the fact that there is no specific bond in sufficiently small volumes between internal forces in a real body and in an ideally homogeneous model. For real internal forces, the influence of the hole at $a \to 0$ disappears despite formula (1.38), since there is no difference in the micro-structure of a continuous disc and the same disc with a sufficiently small hole.

A rise in stresses near relatively small holes and cut-outs on the body surface is called *the stress concentration.*

1.20   Saint-Venant Principle

Assume $p_b = 0$, $b \to \infty$ in formulas (1.34) and (1.35). We obtain

$$\sigma_r = -\frac{a^2}{r^2}\, p_a, \quad \sigma_t = \frac{a^2}{r^2}\, p_a.$$

Stress from internal pressure in the case of $r \gg a$ becomes relatively low. By summarizing this observation in any case of a plane stress-strain state, we come to the following conclusion:

> if a system of mutually equilibrating forces is applied in a specific area, stresses from such loading have a substantial value only in such vicinity of the loaded part of the body whose dimensions are commensurable with the dimensions of the loaded area.

The same concept can be formulated as the following principle proposed by Saint-Venant:

statically equivalent forces distributed in various ways in a relatively small area of the body cause stresses insignificantly different when distancing from the loaded area for the distance of the order of the maximum size of this area.

References

1. A. Grigor'yan, *Ocherki istorii mekhaniki v Rossii* [Essays on the history of mechanics in Russia] (Izd-vo AN SSSR Publ, Moscow, 1961)
2. A. Grigor'yan, *Mekhanika v Rossii* [Mechanics in Russia] (Nauka Publ., Moscow, 1978)
3. A. Grigor'yan, B. Fradlin, *Istoriya mekhaniki tverdogo tela* [History of solid state mechanics] (Nauka, Moscow, 1982)
4. L. Landau, E. Lifshits, *Teoreticheskaya fizika : ucheb. posobie. V 10-ti tomakh. T. 7. Teoriya uprugosti. 5-e izd.* [Theoretical physics: textbook. stipend. In 10 volumes. Vol. 7. Theory of elasticity. 5th edn.] (Fizmatlit Publ., Moscow, 2003)
5. M. Leonov, *Osnovy mekhaniki uprugogo tela* [Fundamentals of elastic body mechanics] (Izd-vo AS Kirg. SSR, Frunze, 1963)
6. A. Lyav, *Matematicheskaya teoriya uprugosti* [Mathematical theory of elasticity] (ONTI NKTP SSSR Publ., Moscow, Leningrad, 1935)
7. V. Molotnikov, *Kurs soprotivleniya materialov* [The course of strength of materials] (SPb., Moscow/Lan' Publ., Krasnodar, 2006)
8. S. Timoshenko, *Istoriya nauki o soprotivlenii materialov s kratkimi svedeniyami iz istorii teorii uprugosti i teorii sooruzhenii. 2-e izd.* [History of resistance science materials with brief information from the history of the theory of elasticity and the theory of structures. The 2nd prod.] (GITTL Publ., Moscow, 1957)

Chapter 2
The First Basic Problem of Elasticity Theory

2.1 Equilibrium Equations

Assume that an equilibrated, arbitrarily distributed system of forces acts on the solid body surface. Strains and stresses caused by this system of forces are continuous unrestrictedly differentiated functions of coordinates of all internal points of the body. On the surface confining the body, strains and stresses can be diverse. When approaching boundary surfaces from inside the body, strains and stresses must satisfy some *boundary conditions* that will be considered separately.

Refer to Fig. 2.1. It shows an infinitely small parallelepiped with ribs dx, dy, dz separated from the vicinity of an arbitrary point inside a body. Internal forces act on the face of the separated element. The figure shows only components of these forces in the direction of the coordinate axis Ox. By zeroing their geometric sum, we obtain

$$\left(\frac{\partial \sigma_x}{\partial x} + \frac{\partial \tau_{xy}}{\partial y} + \frac{\partial \tau_{xz}}{\partial z} \right) dx \, dy \, dz = 0.$$

By reducing by $dx \, dy \, dz$ and making a circular permutation of indexes of coordinate axes, we obtain

$$\frac{\partial \sigma_x}{\partial x} + \frac{\partial \tau_{xy}}{\partial y} + \frac{\partial \tau_{xz}}{\partial z} = 0,$$

$$\frac{\partial \sigma_y}{\partial y} + \frac{\partial \tau_{yz}}{\partial x} + \frac{\partial \tau_{yz}}{\partial z} = 0, \qquad (2.1)$$

$$\frac{\partial \sigma_z}{\partial z} + \frac{\partial \tau_{zx}}{\partial x} + \frac{\partial \tau_{zy}}{\partial y} = 0.$$

© The Author(s), under exclusive license to Springer Nature Switzerland AG 2021
V. Molotnikov, A. Molotnikova, *Theory of Elasticity and Plasticity*,
https://doi.org/10.1007/978-3-030-66622-4_2

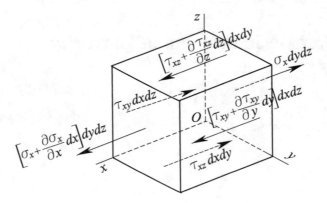

Fig. 2.1 Equilibrium of internal forces

2.2   Expression of Strains Through Movements

Assume that an arbitrary point A of a body before deformation had coordinates (x, y, z). As a result of deformation, the point had movement whose components in the direction of the coordinate axes Ox, Oy, Oz are designated as u, v, w. In this manner, the point $A(x, y, z)$ after deformation will have the coordinates $(x + u, y + v, z + w)$. The point $B(x + dx, y + dy, z + dz)$ infinitely close to it will also go to the new position defined by the coordinates

$$x + dx + u + \frac{\partial u}{\partial x}dx, \quad y + v + \frac{\partial v}{\partial x}dx, \quad z + w + \frac{\partial w}{\partial x}dx.$$

By calculating the distance l between the points A and B after deformation, we obtain

$$l = \sqrt{\left(1 + \frac{\partial u}{\partial x}\right)^2 + \left(\frac{\partial v}{\partial x}\right)^2 + \left(\frac{\partial w}{\partial x}\right)^2}\, dx.$$

In the case of low deformations, mixed certain derivatives v_x, w_x are low as compared to one, therefore $l = dx + u_x dx$. Then the absolute elongation of the element AB will be $u_x dx$, and its relative elongation $\varepsilon_x = u_x$. In a similar way, we find relative elongation in the direction of the axes Oy and Oz, e. g.

$$\varepsilon_x = \frac{\partial u}{\partial x}, \quad \varepsilon_y = \frac{\partial v}{\partial y}, \quad \varepsilon_z = \frac{\partial w}{\partial z} \tag{2.2}$$

provided that all six mixed derivatives

$$\left(\frac{\partial u}{\partial y}, \frac{\partial u}{\partial z}, \frac{\partial v}{\partial x}, \frac{\partial v}{\partial z}, \frac{\partial w}{\partial x}, \frac{\partial w}{\partial y}\right)$$

are small as compared to one. If the conditions are met, the angles between elements parallel to the coordinate axes change as a result of deformation by the values

$$\gamma_{xy} = \frac{\partial u}{\partial y} + \frac{\partial v}{\partial x}, \quad \gamma_{yz} = \frac{\partial v}{\partial z} + \frac{\partial w}{\partial y}, \quad \gamma_{zx} = \frac{\partial w}{\partial x} + \frac{\partial u}{\partial z}. \tag{2.3}$$

2.3 Definition of Movements

Let us associate a body with the system of rectangular coordinates $Oxyz$ whose origin is located in the point O inside the body. Assume that the body endures elongation in three directions parallel to the coordinate directions. Let us use (u, v, w) to designate the movement components of a point $A(x, y, z)$ sufficiently distanced from the point O. Using formulas (2.2), they can be represented as

$$u \approx \varepsilon_x x, \quad v \approx \varepsilon_y y, \quad w \approx \varepsilon_z z. \tag{2.4}$$

These formulas define movement due to deformation without rotation (no rotations of areas that have no tangential stresses).

As shown in Fig. 1.5, the action of tangential stress, for example, τ_{xy}, can be represented as elongation and compression in the plane xOy at the angle $\pm 45°$ with the coordinate axes. In this case, formula (1.9) defines the rotation angles of sections that were parallel to the respective coordinate axes before deformation. If we assume that there are rotations of square diagonals (Fig. 1.5), the linear element parallel to the axis Ox will turn to the angle γ_{xy}, and the linear element orthogonal to it will turn in the opposite direction by the same value, e. g. by the angle $-\gamma_{xy}$. By summing movements due to these (small) shears, we can write

$$u \approx y\gamma_{xy} + z\gamma_{xz}, \quad v \approx x\gamma_{yx} + z\gamma_{yz}, \quad w \approx x\gamma_{zx} + y\gamma_{zy}. \tag{2.5}$$

Apart from movements caused by linear (2.4) deformations and shears (2.5), the body can turn around coordinate axes by the angles $\omega_x, \omega_y, \omega_z$ as a rigid body. Let us recall [3] that positive rotation occurs from the axis Ox to the axis Oy, from Oy to Oz and from Oz to Ox. Movements due to these rotations will be

$$u \approx z\omega_y - y\omega_z, \quad v \approx x\omega_z - z\omega_x, \quad w \approx y\omega_x - x\omega_y. \tag{2.6}$$

By summing the respective components of movement (2.4)–(2.6), we obtain the movement components of sufficiently distanced points in the case of homogeneous deformation:

$$u \approx \varepsilon_x x + y\gamma_{xy} + z\gamma_{xz} + z\omega_y - y\omega_z,$$

$$v \approx \varepsilon_y y + x\gamma_{yx} + z\gamma_{yz} + x\omega_z - z\omega_x, \qquad (2.7)$$

$$w \approx \varepsilon_z z + x\gamma_{zx} + y\gamma_{zy} + y\omega_x - x\omega_y.$$

2.4   Saint-Venant Identities

By directly checking, we can make sure that the following identity is true for a combination of certain derivatives:

$$\frac{\partial}{\partial x}\left(\frac{\partial v}{\partial z} + \frac{\partial w}{\partial y}\right) + \frac{\partial}{\partial y}\left(\frac{\partial w}{\partial x} + \frac{\partial u}{\partial z}\right) - \frac{\partial}{\partial z}\left(\frac{\partial u}{\partial y} + \frac{\partial v}{\partial x}\right) \equiv 2\frac{\partial^2 w}{\partial x \partial y}.$$

The right part of the identity is obtained after reduction of similar terms in its left parts. By differentiating using z, we will find that

$$2\frac{\partial^3 w}{\partial x \partial y \partial z} \equiv \frac{\partial}{\partial z}\left[\frac{\partial}{\partial x}\left(\frac{\partial v}{\partial z} + \frac{\partial w}{\partial y}\right) + \frac{\partial}{\partial y}\left(\frac{\partial w}{\partial x} + \frac{\partial u}{\partial z}\right) - \frac{\partial}{\partial z}\left(\frac{\partial u}{\partial y} + \frac{\partial v}{\partial x}\right)\right],$$

or, taking into account formulas (2.2) and (2.3),

$$2\frac{\partial^2 \varepsilon_z}{\partial x \partial y} \equiv \frac{\partial}{\partial z}\left(\frac{\partial \gamma_{yz}}{\partial x} + \frac{\partial \gamma_{xz}}{\partial y} - \frac{\partial \gamma_{xy}}{\partial z}\right). \qquad (2.8)$$

By differentiating formulas (2.2) and (2.3), we can easily obtain that

$$\frac{\partial^2 \gamma_{xy}}{\partial x \partial y} \equiv \frac{\partial^2 \varepsilon_x}{\partial y^2} + \frac{\partial^2 \varepsilon_y}{\partial x^2}. \qquad (2.9)$$

From formulas (2.8) and (2.9) by circular permutation of indexes, we can obtain four more similar ratios. These six ratios are called [2] *Saint-Venant identities*.

2.5   Compatibility Conditions

Substituting formulas (1.11) and (1.12) into identities (2.8) and (2.9) gives

$$\frac{2}{E}\frac{\partial^2}{\partial x \partial y}\left[\sigma_z - \nu(\sigma_y + \sigma_x)\right] = \frac{1}{G}\frac{\partial}{\partial z}\left(\frac{\partial \tau_{yz}}{\partial x} + \frac{\partial \tau_{xz}}{\partial y} - \frac{\partial \tau_{xy}}{\partial z}\right), \qquad (2.10)$$

$$\frac{1}{G}\frac{\partial^2 \tau_{xy}}{\partial x \partial y} = \frac{1}{E}\left[\frac{\partial^2 \sigma_x}{\partial y^2} + \frac{\partial^2 \sigma_y}{\partial x^2} - v\left(\frac{\partial^2 \sigma_x}{\partial x^2}\right.\right.$$

$$\left.\left. + \frac{\partial^2 \sigma_y}{\partial y^2} + \frac{\partial^2 \sigma_z}{\partial x^2} + \frac{\partial^2 \sigma_z}{\partial y^2}\right)\right].$$ (2.11)

By circular permutation of indexes of coordinate axes, we obtain four more formulas of a similar kind. This system of six Eqs. (2.10) and (2.11) is called *compatibility conditions*.

2.6 Boundary Conditions

Let us separate an infinitely small triangle on the surface of a body and assume that the element endures the action of only normal stresses σ (Fig. 2.2). Angles between the vector σ and axes Ox, Oy, Oz are designated, respectively, by $\widehat{nx}$, $\widehat{ny}$, $\widehat{nz}$.

Assume the separated element to be the basis of a three-face pyramid whose faces are located in coordinate planes. The areas of these faces will be

$$S_{xy} = \frac{1}{2}dxdy, \quad S_{xz} = \frac{1}{2}dxdz, \quad S_{yz} = \frac{1}{2}dydz.$$ (2.12)

Here, the values S_{xy}, S_{xz}, S_{yz} are the areas of projections of the pyramid basis onto coordinate planes. They can be expressed using the area S of the body surface element:

$$S_{xy} = S\cos\widehat{nz}, \quad S_{yz} = S\cos\widehat{nx}, \quad S_{zx} = S\cos\widehat{ny}.$$ (2.13)

By zeroing the sum of projections onto the axis Ox of all forces acting on the separated pyramid and using formulas (2.13), we find

$$S(\sigma_x \cos\widehat{nx} + \tau_{xy}\cos\widehat{ny} + \tau_{xz}\cos\widehat{nz} - \sigma\cos\widehat{nx}) = 0.$$

Fig. 2.2 Equilibrium of a boundary element

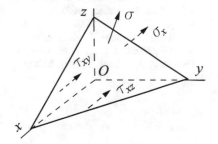

By circular permutation of indexes of coordinate axes, we can obtain two more similar equilibrium equations. After increment by S, we will have three following boundary conditions:

$$(\sigma_x - \sigma)\cos\widehat{nx} + \tau_{xy}\cos\widehat{ny} + \tau_{xz}\cos\widehat{nz} = 0,$$

$$\tau_{yx}\cos\widehat{nx} + (\sigma_y - \sigma)\cos\widehat{ny} + \tau_{yz}\cos\widehat{nz} = 0, \qquad (2.14)$$

$$\tau_{zx}\cos\widehat{nx} + \tau_{zy}\cos\widehat{ny} + (\sigma_z - \sigma)\cos\widehat{nz} = 0.$$

When deriving the last formulas, we assume that there are no tangential stresses on the body surface. If this limitation is abandoned, we can easily obtain three boundary conditions with their rights parts representing the known functions on the body surface.

2.7   The First Basic Problem of Elasticity Theory

Now we have all necessary ratios for formulate the boundary problem of elasticity theory in stresses. The first primary problem of elasticity theory is as follows.

> There is an elastic body and a self-equilibrated system of forces at its boundary. It is required to define the stressed state in an arbitrary point inside the body.

Mathematically, this means that six functions must be defined σ_x, σ_y, σ_z, τ_{xy}, τ_{yz}, τ_{zx} that satisfy three equilibrium equations (2.1), six compatibility equations (2.10) and (2.11), and three boundary conditions of type (2.14).

The solution to this problem is unambiguous for simple-connected bodies. In the case of a multiply-connected body, it is necessary to use an additional condition of the potential energy minimum of elastic strains [1].

References

1. M. Leonov, *Osnovy mekhaniki uprugogo tela* [Fundamentals of elastic body mechanics] (Izd-vo AS Kirg. SSR, Frunze, 1963)
2. A. Lyav, *Matematicheskaya teoriya uprugosti* [Mathematical theory of elasticity] (ONTI NKTP SSSR Publ., Moscow, Leningrad, 1935)
3. V. Molotnikov, *Osnovy teoreticheskoi mekhaniki* [Fundamentals of theoretical mechanics] (Feniks Publ., Rostov on Don, 2004)

Chapter 3
The Second Primary Problem of Elasticity Theory

3.1 Definition of Stresses Through Deformations

Let us re-write the first of formulas (1.11) as

$$\varepsilon_x = \frac{\partial u}{\partial x} = \frac{1}{E}\left[\sigma_x(1+v) - v(\sigma_x + \sigma_y + \sigma_z)\right].$$

Then using formulas (1.15) and (1.16), we find

$$\frac{\partial u}{\partial x} = \frac{1}{E}\left[\sigma_x(1+v) - \frac{vE}{1-2v}\Theta\right]. \tag{3.1}$$

Formula (3.1) can be written as follows

$$\sigma_x = 2G\frac{\partial u}{\partial x} + \lambda\Theta, \tag{3.2}$$

where G is expressed through the Jung modulus E and the Poisson coefficient v using formula (1.10), while λ designates

$$\lambda = \frac{vE}{(1+v)(1-2v)}. \tag{3.3}$$

The constant values λ and G are referred to as [1] *Lame coefficients.*
  For other components of stress, we can obtain:

$$\sigma_y = 2G\frac{\partial v}{\partial y} + \lambda\Theta, \quad \sigma_z = 2G\frac{\partial w}{\partial z} + \lambda\Theta,$$

$$\tau_{xy} = G\left(\frac{\partial u}{\partial y} + \frac{\partial v}{\partial x}\right), \quad \tau_{yz} = G\left(\frac{\partial v}{\partial z} + \frac{\partial w}{\partial y}\right), \tag{3.4}$$

$$\tau_{zx} = G\left(\frac{\partial w}{\partial x} + \frac{\partial u}{\partial z}\right).$$

3.2 Equations of Elastic Body Strain

Let us substitute formulas (3.2) and (3.4) into equilibrium equations (2.1). We will obtain a system of three differentiated equations of relative movement u, v, w of elastic body points. The first of these equations will look as follows:

$$2G\frac{\partial^2 u}{\partial x^2} + \lambda\frac{\partial \Theta}{\partial x} + G\frac{\partial}{\partial y}\left(\frac{\partial u}{\partial y} + \frac{\partial v}{\partial x}\right) + G\frac{\partial}{\partial z}\left(\frac{\partial u}{\partial z} + \frac{\partial w}{\partial z}\right) = 0$$

or

$$G\left(\frac{\partial^2 u}{\partial x^2} + \frac{\partial^2 u}{\partial y^2} + \frac{\partial^2 u}{\partial z^2}\right) + G\frac{\partial}{\partial x}\left(\frac{\partial u}{\partial x} + \frac{\partial v}{\partial y} + \frac{\partial w}{\partial z}\right) + \lambda\frac{\partial \Theta}{\partial x} = 0.$$

By applying formula (1.13), we will obtain

$$G\left(\frac{\partial^2 u}{\partial x^2} + \frac{\partial^2 u}{\partial y^2} + \frac{\partial^2 u}{\partial z^2}\right) + (\lambda + G)\frac{\partial \Theta}{\partial x} = 0.$$

Similar equations can be obtained from two last equations of equilibrium (2.1). By using the common designation for Laplacian [3]

$$\Delta = \frac{\partial^2}{\partial x^2} + \frac{\partial^2}{\partial y^2} + \frac{\partial^2}{\partial z^2}, \tag{3.5}$$

let us represent these equations as follows:

$$\begin{aligned} \Delta u + \frac{1}{1 - 2v}\frac{\partial \Theta}{\partial x} &= 0, \\ \Delta v + \frac{1}{1 - 2v}\frac{\partial \Theta}{\partial y} &= 0, \\ \Delta w + \frac{1}{1 - 2v}\frac{\partial \Theta}{\partial z} &= 0. \end{aligned} \tag{3.6}$$

In this manner, the second primary problem of elasticity theory is as follows.

An elastic body is given (area D). Three components u, v, w of the movement vector are required that satisfy equations (3.6) inside D. In each non-specific point of the body surface, movements must satisfy three boundary conditions.

Boundary conditions can be formulated in three variants of the problem: movements are set; combinations of stresses are set, which are written through normal and tangential derivatives from movements; combinations of stresses and movements are set, which are written through normal and tangential derivatives from movements and through movements themselves.

3.3   Application of Harmonic Functions

Any function $\varphi(x, y, z)$ having in some area D continuous second derivatives and satisfying in this area Laplace's equation [2]

$$\Delta\varphi = \frac{\partial^2\varphi}{\partial x^2} + \frac{\partial^2\varphi}{\partial y^2} + \frac{\partial^2\varphi}{\partial z^2} = 0 \tag{3.7}$$

is referred to as *the function harmonic in D*. Speaking of harmonic functions in the future, we will consider D as a non-enclosed area occupied by a body, excluding boundary body surfaces.

Let us show that volumetric expansion Θ is the function harmonic in D. For this purpose, let us differentiate the first Eq. (3.6) upon x, the second equation upon y, and the third equation upon z and let us sum up differentiated equations. We obtain

$$\frac{\partial}{\partial x}\Delta u + \frac{\partial}{\partial y}\Delta v + \frac{\partial}{\partial z}\Delta w + \frac{1}{1-2v}\Delta\Theta = 0.$$

By changing the order of differentiation, we have

$$\Delta\left(\frac{\partial u}{\partial x} + \frac{\partial v}{\partial y} + \frac{\partial w}{\partial z}\right) + \frac{1}{1-2v}\Delta\Theta = 0$$

or

$$\frac{2(1-v)}{1-2v} \cdot \Delta\Theta = 0.$$

The multiplier in front of $\Delta\Theta$ is not converted into zero since for real solid bodies $0 < v < 0,5$. The last equation follows that $\Delta\Theta = 0$, which means that Θ is the function harmonic in D.

Any function Γ harmonic in D is unrestrictedly differentiated. Therefore, we can write

$$\Delta \frac{\partial \Gamma}{\partial x} = \frac{\partial}{\partial x} \Delta \Gamma,$$

e. g. the derivative in the function harmonic in D is also the function harmonic in D.

By directly checking, we can make sure that for the function harmonic in D, there is equation

$$\Delta[(z+c)\Gamma] = 2\frac{\partial \Gamma}{\partial z}, \quad (c = const). \tag{3.8}$$

Suggesting in (3.8) $\Gamma \equiv \Theta$, we have

$$\frac{\partial \Theta}{\partial z} = \frac{1}{2}\Delta[(z+c)\Theta].$$

By substituting this expression into the last Eq. (3.6), we will find that

$$\Delta\left(w + \frac{z+c}{2(1-2v)} \cdot \Theta\right) = 0.$$

The last equation means that the expression in brackets is the function harmonic in D, e. g.

$$w = f_z - \frac{z\Theta}{2(1-2v)}, \tag{3.9}$$

where f_z is the function harmonic in D.

By substituting the coordinate z with x, and then with y, we obtain two other similar ratios:

$$u = f_x - \frac{x\Theta}{2(1-2v)},$$
$$v = f_y - \frac{y\Theta}{2(1-2v)}. \tag{3.10}$$

Here, the functions harmonic in D f_x, f_y, f_z ? Θ are connected, according to formulas (1.13) and (2.2), by the dependency

$$\frac{5-4v}{2(1-2v)}\Theta + \frac{1}{2(1-2v)}\left(x\frac{\partial \Theta}{\partial x} + y\frac{\partial \Theta}{\partial y} + z\frac{\partial \Theta}{\partial z}\right)$$
$$= \frac{\partial f_x}{\partial x} + \frac{\partial f_y}{\partial y} + \frac{\partial f_z}{\partial z}. \tag{3.11}$$

3.4   Trefftz Integral

Let us consider some function harmonic in D Π_0 satisfying the condition

$$\frac{\partial \Pi_0}{\partial z} = \Theta. \tag{3.12}$$

The condition (3.12) in D is also satisfied by the sum

$$\Pi = \Pi_0 + C(x, y), \tag{3.13}$$

where $C(x, y)$ is the arbitrary function harmonic in D of two variables x and y.
We notice that

$$\Delta(x\Theta) \equiv \Delta\left(x\frac{\partial \Pi}{\partial z}\right) = 2\frac{\partial^2 \Pi}{\partial x \partial z} = \Delta\left(z\frac{\partial \Pi}{\partial x}\right), \quad (\Delta \Pi = 0), \tag{3.14}$$

we can write as

$$x\Theta = z\frac{\partial \Pi}{\partial x} + f_1,$$

where f_1 is the function harmonic in D.
Taking into account this dependency, the first of formulas (3.10) can be written as

$$u = f_x + f_1 - \frac{z}{2(1 - 2\nu)}\frac{\partial \Pi_0}{\partial x}. \tag{3.15}$$

By expressing Π_0 from formula (3.13) and substituting this expression in the dependency (3.15), we obtain

$$u = \varphi_x - \frac{z}{2(1 - 2\nu)}\frac{\partial \Pi_0}{\partial x} \tag{3.16}$$

that designates

$$\varphi_x = f_x + f_1 - \frac{z}{2(1 - 2\nu)} \cdot \frac{\partial C(x, y)}{\partial x}.$$

Since f_x, f_1 and $\dfrac{z\partial C(x, y)}{\partial x}$ are functions harmonic in D, the function φ_x is also a function harmonic in D. Consequently, the total integral of the equation system (3.6) can be represented as

$$u = \varphi_x - \frac{z}{2(1 - 2v)} \frac{\partial \Pi}{\partial x},$$

$$v = \varphi_y - \frac{z}{2(1 - 2v)} \frac{\partial \Pi}{\partial y}, \qquad (3.17)$$

$$w = \varphi_z - \frac{z}{2(1 - 2v)} \frac{\partial \Pi}{\partial z},$$

based on formula (3.12), the functions harmonic in D φ_x, φ_y, φ_z, and Π are connected by the dependency

$$\frac{\partial \Pi}{\partial z} = \Theta = \frac{\partial u}{\partial x} + \frac{\partial v}{\partial y} + \frac{\partial w}{\partial z} = \frac{\partial \varphi_x}{\partial x} + \frac{\partial \varphi_y}{\partial y} + \frac{\partial \varphi_z}{\partial z} - \frac{1}{2(1 - 2v)} \frac{\partial \Pi}{\partial z},$$

or

$$\frac{\partial \Pi}{\partial z} = \frac{2(1 - 2v)}{3 - 4v} \left(\frac{\partial \varphi_x}{\partial x} + \frac{\partial \varphi_y}{\partial y} + \frac{\partial \varphi_z}{\partial z} \right). \qquad (3.18)$$

The total integral of the equation system (3.6) in the form of formulas (3.17) provided that (3.18) is referred to as the Trefftz integral

We note that formulas (3.17) can be obtained directly from Eqs. (3.6). Indeed, by using expression (3.14), we have

$$\frac{\partial \Theta}{\partial x} = \frac{\partial^2 \Pi}{\partial x \partial z} = \frac{1}{2} \Delta \left(z \frac{\partial \Pi}{\partial x} \right).$$

In this case, the first Eq. (3.6) will look as:

$$\Delta \left(u + \frac{z}{2(1 - 2v)} \frac{\partial \Pi}{\partial x} \right) = 0,$$

hence we obtain the first of formulas (3.17). The second formula is obtained in the case of respective substitution of the axes Ox and Oy, and the third formula from (3.17) is identical to the previously obtained formula (3.9).

3.5   Grodsky–Neyber–Papkovich Integral

Let F be the function, which is an arbitrary solution of the equation

$$\Delta F = \Theta. \qquad (3.19)$$

Strain equations (3.6) are represented as:

$$\Delta\left(u + \frac{1}{1-2v}\frac{\partial F}{\partial x}\right) = 0,$$

$$\Delta\left(v + \frac{1}{1-2v}\frac{\partial F}{\partial y}\right) = 0, \tag{3.20}$$

$$\Delta\left(z + \frac{1}{1-2v}\frac{\partial F}{\partial z}\right) = 0.$$

It follows from Eqs. (3.20) that

$$u = -\frac{1}{1-2v}\frac{\partial F}{\partial x} + \Phi_x,$$

$$v = -\frac{1}{1-2v}\frac{\partial F}{\partial y} + \Phi_y, \tag{3.21}$$

$$w = -\frac{1}{1-2v}\frac{\partial F}{\partial z} + \Phi_z,$$

where Φ_x, Φ_y, and Φ_z are functions harmonic in D. Due to formulas (1.13) and (2.2), they are related with volumetric expansion Θ by the ratio:

$$\Theta = \frac{\partial\Phi_x}{\partial x} + \frac{\partial\Phi_y}{\partial y} + \frac{\partial\Phi_z}{\partial z} - \frac{1}{1-2v}\Delta F.$$

By using designations (3.19), we obtain:

$$\Delta F = \frac{1-2v}{2(1-v)}\left(\frac{\partial\Phi_x}{\partial x} + \frac{\partial\Phi_y}{\partial y} + \frac{\partial\Phi_z}{\partial z}\right). \tag{3.22}$$

By directly checking, we can make sure that a certain solution of Eq. (3.22) is the function

$$F_1 = \frac{1-2v}{4(1-v)}\left(x\Phi_x + y\Phi_y + z\Phi_z\right). \tag{3.23}$$

By adding the function harmonic in D Φ_0 to the certain solution (3.23), we obtain the general solution of Eq. (3.22):

$$F = \frac{1-2v}{4(1-v)}\left(x\Phi_x + y\Phi_y + z\Phi_z + \Phi_0\right). \tag{3.24}$$

Representations of movements by formulas (3.21) where the functions Φ_x, Φ_y, Φ_z are harmonic and F is found using the formula (3.24) is proposed by Grodsky and Neyber and detailed study of these formulas was given by Papkovich.

References

1. A. Lyav, *Matematicheskaya teoriya uprugosti* [Mathematical theory of elasticity] (ONTI NKTP SSSR Publ., Moscow, Leningrad, 1935)
2. A. Tikhonov, A. Samarskii, *Uravneniya matematicheskoi fiziki: uchebnoe posobie. 6-e izd., ispr. i dop.* [Mathematical physics equations: tutorial. 6th ed., correct and additional] (Izd-vo MGU Publ., Moscow, 1999)
3. I.E.A. Vinogradov, *Matematicheskaya ehntsiklopediya, t.3, red. I. M. Vinogradov [i dr.] [Mathematical Encyclopedia, vol. 3, editor I.M. Vinogradov (and others)]* (Sovetskaya ehntsiklopediya Publ., Moscow, 1982)

Chapter 4
Three-Dimensional Harmonic Function

4.1 Simplest Examples of Harmonic Functions

In the simple case, a harmonic function depends only on one variable x. Wquation (3.7) goes to a regular homogeneous differential equation of the second order

$$\frac{d^2 f}{dx^2} = 0.$$

Hence it follows that

$$f = ax + b, \quad (a, b - const),$$

e. g. the harmonic function depending on one coordinate in the Cartesian system is linear.

In the cylindrical system of coordinates (ρ, φ, z), we have

$$\rho = \sqrt{x^2 + y^2}, \quad \varphi = arctg\frac{y}{x},$$

and Laplace's equation (3.7) looks as follows:

$$\frac{\partial}{\partial \rho}\left(\rho\frac{\partial f}{\partial \rho}\right) + \frac{1}{\rho}\frac{\partial^2 f}{\partial \varphi^2} + \rho\frac{\partial^2 f}{\partial z^2} = 0. \tag{4.1}$$

By analyzing the last formula, we may notice that the simplest harmonic function depending only on φ or only on z will be the linear function. If the harmonic function (f_1) in cylindrical coordinates depends only on the variable ρ, we will obtain from Eq. (4.1)

© The Author(s), under exclusive license to Springer Nature Switzerland AG 2021
V. Molotnikov, A. Molotnikova, *Theory of Elasticity and Plasticity*,
https://doi.org/10.1007/978-3-030-66622-4_4

$$\frac{d}{d\rho}\left(\rho\frac{df_1}{d\rho}\right) = 0, \quad \rho\frac{df_1}{d\rho} = a, \quad f_1 = a\ln\rho + b, \quad (a, b - const). \tag{4.2}$$

In formula (4.2), ρ means the distance of the considered point from some axis Oz. In a general case, the harmonic function (f_2) depending on the distance (r) to some point (ξ, η, ζ)

$$r = \sqrt{(x - \xi)^2 + (y - \eta)^2 + (z - \zeta)^2}$$

satisfies Laplace's equation (3.7) that looks as follows:

$$\frac{d}{dr}\left(r^2\frac{df_2}{dr}\right) = 0, \quad \Rightarrow \quad r^2\frac{df_2}{dr} = const,$$

e. g., in a general case

$$f_2(r) = \frac{a}{r} + b. \tag{4.3}$$

Elementary solutions (4.2) and (4.3) can be written as

$$f_1 = a\ln\sqrt{(x - \xi)^2 + (y - \eta)^2} + const, \tag{4.4}$$

$$f_2 = \frac{a}{\ln\sqrt{(x - \xi)^2 + (y - \eta)^2 + (z - \zeta)^2}} + const, \tag{4.5}$$

where ξ, η, ζ are some parameters.

We should note that the function f_2 satisfies Laplace's equation (3.7) and is limited only if the point (x, y, z) does not coincide with the point (ξ, η, ζ) where f_2 goes into infinity. Consequently, the function f_2 is harmonic everywhere except for the point (ξ, η, ζ).

We should also note that in the cylindrical system of coordinates, there is a solution for Laplace's equation (4.1), which is linear relative to the angle φ

$$f_3 = a\varphi + b.$$

If the angle φ gets an increment 2π, the function f_3 increases by the value $2\pi a$, e. g. it is a multivalued function of a point space. However, this function is single-valued in the space with semi-infinite section including the straight line $r = 0$. A semi-plane $\varphi = 0$ or any other plane obtained by bending this semi-plane can be used as such a surface, provided there is no axis bending $r = 0$.

4.2   Green Function

Let us consider a function harmonic in D F. Let us use S to designate the boundary D and assume that any internal point (x, y, z) in the area D arbitrarily tends to the surface S, the function F tends to

$$F|_S = -\frac{1}{r}, \quad \left(r = \sqrt{(x - \xi)^2 + (y - \eta)^2 + (z - \zeta)^2}\right), \tag{4.6}$$

where ξ, η, ζ are some parameters.

The Green function is the [2] function defined by the formula

$$\Gamma = \frac{1}{r} + F. \tag{4.7}$$

The definition (4.7) means that the Green function is, firstly, harmonic everywhere in D except the point $x = \xi$, $y = \eta$, $z = \zeta$ and, secondly, goes to zero on the surface S. Physically, it represents an electrostatic potential of a point single charge placed to the point (ξ, η, ζ) inside the grounded conductive surface S.

The case when the surface S coincides with the entire plane $z = 0$ is of specific interest. In this case, the Green function for the positive half-space $(z > 0)$ can be composed by placing an additional negative single electrical charge in the point $(\xi, \eta, -\zeta)$. Then the total potential of two single charges located in the points $(\xi, \eta, -\zeta)$? (ξ, η, ζ) will be

$$\Gamma = \frac{1}{\sqrt{(x - \xi)^2 + (y - \eta)^2 + (z - \zeta)^2}}$$
$$- \frac{1}{\sqrt{(x - \xi)^2 + (y - \eta)^2 + (z + \zeta)^2}}. \tag{4.8}$$

The function (4.8) satisfies all requirements placed on the Green function for a half-space.

4.3   Green's Spatial Functions

Let us take the formula [1] Green

$$\iiint\limits_{(D)} (\psi \Delta\varphi - \varphi \Delta\psi)\,dx\,dy\,dz = \iint\limits_{(S)} \left(\varphi \frac{\partial\psi}{\partial n} - \psi \frac{\partial\varphi}{\partial n}\right) dS, \tag{4.9}$$

where D is some three-dimensional area confined by the surface S, n is the internal normal line S, and φ and ψ are scalar functions having derivatives up to the second order, inclusively.

Let φ be the function harmonic in (D), and the function ψ is represented as

$$\psi = \frac{1}{r} + \Psi, \quad (r = \sqrt{(x-\xi)^2 + (y-\eta)^2 + (z-\zeta)^2}, \tag{4.10}$$

where Ψ is the function harmonic in (D). In particular, the Green function can be taken as Ψ. The function ψ set by formula (4.10) will be harmonic in the area $(\overline{D})$ that is obtained by subtracting from the area (D) internals of the sphere (σ) of an infinitely small radius ρ with a center in the point (ξ, η, ζ).

For the selected φ and ψ, the integral in the right part of formula (4.9) will be equal to zero, e. g.

$$\iint\limits_{(S)} \left[\varphi \frac{\partial}{\partial n} \left(\frac{1}{r} + \Psi \right) - \left(\frac{1}{r} + \Psi \right) \frac{\partial \varphi}{\partial n} \right] d\overline{S} = 0, \tag{4.11}$$

whereas the surface $(\overline{S})$ consists of two parts: (S) and (σ).

We should note that

$$\frac{\partial}{\partial n} \frac{1}{r} \bigg|_{\sigma} = -\frac{1}{\rho^2},$$

we have

$$\iint\limits_{(\sigma)} \varphi \frac{\partial \frac{1}{r}}{\partial n} d\sigma = -\frac{1}{\rho^2} \varphi(\xi, \eta, \zeta) \iint\limits_{(\sigma)} d\sigma = -4\pi \varphi(\xi, \eta, \zeta).$$

By tending ρ to zero in formula (4.11), we obtain

$$\varphi(\xi, \eta, \zeta) = \frac{1}{4\pi} \iint\limits_{(S)} \left[\varphi \frac{\partial}{\partial n} \left(\frac{1}{r} + \Psi \right) - \left(\frac{1}{r} + \Psi \right) \frac{\partial \varphi}{\partial n} \right] dS. \tag{4.12}$$

Substituting the function F harmonic in D defined by formula (4.6) in the last formula instead of Ψ, we will obtain

$$\varphi(\xi, \eta, \zeta) = \frac{1}{4\pi} \iint\limits_{(S)} \varphi \frac{\partial \Gamma}{\partial n} dS, \tag{4.13}$$

where Γ is the Green function

With some limitations applied to the behavior of the function $\varphi(\xi, \eta, \zeta)$ in the infinities, formula (4.13) can be used for unlimited surfaces S, for example, when S coincides with the plane $z = 0$. The area D in this case can be considered as a limited part of the plane $z = 0$ and the sphere of an unrestrictedly large radius with the center in the point (ξ, η, ζ). In the case of a sufficiently fast decrease in functions φ and ψ, the integral over the spherical part of the surface S in formula (4.9) with an unrestricted increase in the radius tends to zero, and only integral over the plane $z = 0$ will remain in the right part of formula (4.9).

Let us consider the case when the area D includes also an infinitely distanced point. Let us suggest that some harmonic function Ψ on the surface of S satisfies the condition of

$$\frac{\partial \Psi}{\partial n} = -\frac{\partial \left(\frac{1}{r}\right)}{\partial n}\bigg|_{S}. \tag{4.14}$$

With limited φ and Ψ in infinities, in the considered case we can use formula (4.12) that results in

$$\varphi(\xi, \eta, \zeta) = -\frac{1}{4\pi} \iint\limits_{(S)} H \frac{\partial \varphi}{\partial n} dS, \tag{4.15}$$

where

$$H = \frac{1}{r} + \Psi, \quad (r = \sqrt{(x - \xi)^2 + (y - \eta)^2 + (z - \zeta)^2}). \tag{4.16}$$

Due to the property (4.14), the function H harmonic in D satisfies the condition of

$$\frac{\partial H}{\partial n}\bigg|_{S} = 0. \tag{4.17}$$

The function H harmonic in the area D having a pole (single charge) in the point $\varphi(\xi, \eta, \zeta)$ whose normal derivative on the boundary surface S of the area D turns to zero is referred to as the [2] Neumann function. It can be defined as a potential of movement speeds of ideal liquid enveloping the body of a defined shape provided there is a source in the point $\varphi(\xi, \eta, \zeta)$ and a drain in infinitely distanced points.

For the case of half-space $z > 0$, the Neumann function can be built as a potential of single masses located in the points $\varphi(\xi, \eta, \zeta)$? $\varphi(\xi, \eta, -\zeta)$, e. g.

$$H = \frac{1}{r} + \frac{1}{r^*}, \tag{4.18}$$

where

$$r = \sqrt{(x-\xi)^2 + (y-\eta)^2 + (z-\zeta)^2},$$
$$r* = \sqrt{(x-\xi)^2 + (y-\eta)^2 + (z+\zeta)^2}.$$

On the surface $z = 0$, we have

$$H|_S = \frac{2}{\sqrt{(x-\xi)^2 + (y-\eta)^2 + \zeta^2}}. \tag{4.19}$$

4.4   Boundary Problems for Half-Space

Substituting the function (4.19) into formula (4.15) gives

$$\varphi(\xi, \eta, \zeta) = -\frac{1}{2\pi} \iint\limits_{(\infty)} \frac{Z(x, y, +0)dxdy}{\sqrt{(x-\xi)^2 + (y-\eta)^2 + \zeta^2}}, \tag{4.20}$$

where

$$Z(x, y, +0) = \lim_{z \to 0} \frac{\partial \varphi}{\partial z}, \quad (z > 0), \tag{4.21}$$

and the symbol (∞) means that the integration area coincides with the infinite plane $z = 0$.

Formula (4.20) defines the function harmonic in half-space $z > 0$ disappearing in the infinity and taking the specified values of the normal derivative on the surface $z = +0$. The problem of finding such a function is referred to as the [2] Neumann problem. In this manner, formula (4.20) gives the solution for the Neumann problem for a positive half-space $z > 0$.

Another boundary problem of the theory of harmonic functions is the definition of the function φ harmonic in the half-space and satisfying the conditions

$$\varphi(x, y, z)\big|_{z=+0} = \varphi_0(x, y), \quad \lim_{z \to \infty} \varphi(x, y, z) = 0, \tag{4.22}$$

where φ_0 is the defined continuous function. Finding this function is called [2] the Dirichlet problem.

For the considered case of a half-space, the solution of the Dirichlet problem can be obtained by using the function (4.8) and formula (4.13). We obtain

$$\varphi(\xi, \eta, \zeta) = \frac{1}{2\pi} \iint\limits_{(\infty)} \frac{\zeta \varphi_0(x, y)dxdy}{\sqrt{|(x-\xi)^2 + (y-\eta)^2 + \zeta^2|^3}}. \tag{4.23}$$

The solution (4.23) can also be represented as

$$\varphi(\xi, \eta, \zeta) = -\frac{1}{2\pi} \lim_{s \to \infty} \frac{\partial}{\partial \zeta} \iint\limits_{(S)} \frac{\varphi_0(x, y)dxdy}{\sqrt{(x - \xi)^2 + (y - \eta)^2 + \zeta^2}}. \tag{4.24}$$

4.5   Other Properties of Harmonic Functions

Let φ be some harmonic function defined in the area D limited by an arbitrary surface S. Assume that in formula (4.9) $\psi \equiv 1$; we obtain

$$\iint\limits_{(S)} \frac{\partial \varphi}{\partial n} dS = 0. \tag{4.25}$$

Calculating in formula (4.12) $\Psi \equiv 0$, we have

$$\varphi(\xi, \eta, \zeta) = \frac{1}{4\pi} \iint\limits_{(S)} \left(\varphi \frac{\partial}{\partial n} \frac{1}{r} - \frac{1}{r} \frac{\partial \varphi}{\partial n} \right) dS.$$

In particular, if S is a sphere with a radius R and with a center in the point (ξ, η, ζ), we have

$$\frac{\partial}{\partial n} \frac{1}{r} = -\frac{\partial}{\partial r} \left(\frac{1}{r} \right) \Big|_{r=R} = \frac{1}{R^2}, \quad (dn = -dr),$$

and the previous formula for the function $\varphi(\xi, \eta, \zeta)$ gives

$$\varphi(\xi, \eta, \zeta) = \frac{1}{4\pi^2 R^2} \iint\limits_{(S)} \varphi dS - \frac{1}{4\pi R} \iint\limits_{(S)} \frac{\partial \varphi}{\partial n} dS.$$

Taking into account formula (4.25), we obtain from the last result

$$\varphi(\xi, \eta, \zeta) = \frac{1}{4\pi R^2} \iint\limits_{(S)} \varphi dS. \tag{4.26}$$

The expression in the right part of formula (4.26) is the average value of the harmonic function $\varphi(\xi, \eta, \zeta)$ on the sphere surface. In this manner, formula (4.26) expresses that *the value of the harmonic function in the sphere center equals its average value on the surface of this sphere.*

Let us consider the integral

$$Q = \iiint\limits_{(V)} \varphi dV, \tag{4.27}$$

where (V) is the area inside the sphere with a radius of R, and φ is the function harmonic in (V). We will calculate the integral (4.27) by integrating inside a sphere of some radius r, $(r < R)$ and then along the normal line to such sphere. Taking into account formula (4.26), we obtain

$$Q = \frac{4}{3}\pi R^3 \varphi(\xi, \eta, \zeta),$$

or

$$\varphi(\xi, \eta, \zeta) = \frac{1}{V} \iiint\limits_{(V)} \varphi dV, \tag{4.28}$$

where V is the volume of such sphere.

In this manner, *the value of the harmonic function in the center of the sphere* (ξ, η, ζ) *equals the average value of this function upon the sphere volume.*

References

1. B. Budak, S. Fomin, *Kratnye integraly i ryady* [Multiple integrals and series] (Nauka Publ., Moscow, 1965)
2. A. Tikhonov, A. Samarskii, *Uravneniya matematicheskoi fiziki: uchebnoe posobie. 6-e izd., ispr. i dop.* [Mathematical physics equations: tutorial. 6th ed., correct and additional] (Izd-vo MGU Publ., Moscow, 1999)

Chapter 5
Elastic Half-Space

5.1   Volumetric Expansion on Surface

Let the loading application area be small in dimensions as compared to the body dimensions, and we are interested in a stressed state in the vicinity of applied forces. In this case, we can represent that the body is limited by only the plane tangential to the body surface in the middle point of the load application area. This semi-indefinite body whose material conforms to Hooke's law is referred to as the [1] *elastic half-space*.

Let us assume that external loads represent normal pressure $p(x, y)$. Let us use formulas (3.17) to find movements. By changing designations, let us write these formulas as follows

$$u = \varphi + z\frac{\partial P}{\partial x},$$

$$v = \psi + z\frac{\partial P}{\partial y}, \tag{5.1}$$

$$w = f + \frac{\partial P}{\partial z},$$

where the harmonic functions φ, ψ, f and P are connected by the following condition due to formula (3.17)

$$\frac{\partial P}{\partial z} = -\frac{1}{3 - 4v}\left(\frac{\partial \varphi}{\partial x} + \frac{\partial \psi}{\partial y} + \frac{\partial f}{\partial z}\right). \tag{5.2}$$

Four harmonic functions φ, ψ, f and P are found from the condition that they disappear at the infinity boundary together with their derivatives, and there are known stresses at the half-space boundary $z = 0$

V. Molotnikov, A. Molotnikova, *Theory of Elasticity and Plasticity*,
https://doi.org/10.1007/978-3-030-66622-4_5

45

$$\tau_{xz} = 0, \quad \tau_{yz} = 0, \quad \sigma_z = -p(x, y). \qquad (5.3)$$

Four conditions (5.2) and (5.3) allow finding the sought functions φ, ψ, f, and P. The first condition (5.3) gives

$$\left.\frac{\partial \varphi}{\partial z}\right|_{z=0} = -\frac{\partial (f + P)}{\partial x}. \qquad (5.4)$$

Harmonic functions are in the last formula in the right and left parts, since derivatives from harmonic functions are also harmonic functions. We should also note that if this equation takes place at $z = 0$, then it is also true for any z, since the harmonic function is totally defined by its value on the surface. By differentiating formula (5.4) upon x, we obtain

$$\frac{\partial^2 \varphi}{\partial x \partial z} = -\frac{\partial^2}{\partial x^2}(f + P).$$

A similar condition $\tau_{yz}(x, u, 0) = 0$ gives

$$\frac{\partial^2 \psi}{\partial y \partial z} = -\frac{\partial^2}{\partial y^2}(f + P).$$

By summing the left and right parts of the two last formulas taking into account the harmonic nature of the function $(f + P)$

$$\left(\frac{\partial^2}{\partial x^2} + \frac{\partial^2}{\partial y^2} + \frac{\partial^2}{\partial z^2}\right)(f + P) = 0$$

we find

$$\frac{\partial}{\partial z}\left(\frac{\partial \varphi}{\partial x} + \frac{\partial \psi}{\partial y}\right) = \frac{\partial^2}{\partial z^2}(f + P). \qquad (5.5)$$

By integrating Eq. (5.5) and taking into account the condition on the infinity, we have

$$\frac{\partial \varphi}{\partial x} + \frac{\partial \psi}{\partial y} = \frac{\partial}{\partial z}(f + P).$$

Substituting this expression into the right part of formula (5.2) gives

$$\frac{\partial P}{\partial z} = -\frac{1}{3 - 4\nu}\frac{\partial}{\partial z}(2f + P)$$

or

$$P = -\frac{f}{2 - 2v}. \tag{5.6}$$

The third condition (5.3) when using the second formula from (3.4) looks as follows:

$$2G \left[\frac{\partial f}{\partial z} + \frac{\partial P}{\partial z} \right]_{z=0} + \lambda \Theta|_{z=0} = -p(x, y).$$

By substituting here the expression f from formula (4.3), we obtain

$$2G(1 - 2v) \frac{\partial P}{\partial z} \bigg|_{z=0} + \lambda \Theta(x, y, 0) = -p(x, y). \tag{5.7}$$

Let us note that the comparison of formulas (5.1) and (3.17) shows

$$\Pi = -2(1 - 2v)P.$$

Having this in mind, we now use formula (3.12) to determine volumetric expansion on the surface $\Theta(x, y, 0)$. We have

$$\Theta = -2(1 - 2v) \frac{\partial P}{\partial z}. \tag{5.8}$$

In this manner, the condition (5.7) taking into account formula (5.8) results in the following formula for volumetric expansion on the surface of the elastic half-space

$$\Theta(x, y, 0) = -\frac{1 - 2v}{G} p(x, y). \tag{5.9}$$

5.2 Stress on Surface

The representation of volumetric expansion as (3.17) allows formula (5.9) to look as follows

$$\Theta(x, y, 0) = \frac{1 - 2v}{E} \left[\sigma_x(x, y, 0) + \sigma_y(x, y, 0) - p(x, y) \right] = -\frac{1 - 2v}{G} p(x, y),$$

e.g.

$$\sigma_x(x, y, 0) + \sigma_y(x, y, 0) = -(1 + 2v) p(x, y). \tag{5.10}$$

In individual cases, formula (5.10) allows partially determining stresses on the surface upon the defined pressure $p(x, y)$. For example, in the case of axisymmetric loading, we have $\sigma_x(0, 0, 0) = \sigma_y(0, 0, 0)$. Then we will find from formula (5.10) as follows

$$\sigma_x(0, 0, 0) = \sigma_y(0, 0, 0) = -\frac{1 + 2\nu}{2} p(x, y). \tag{5.11}$$

Furthermore, the symmetry condition in the considered case of axisymmetric loading results in

$$\tau_{xy}(0, 0, 0) = \tau_{yz}(0, 0, 0) = \tau_{zx}(0, 0, 0) = 0.$$

In this manner, the stressed state in the center of axisymmetric loading is determined fully.

Another example is plane strain. In this case

$$\sigma_y = \nu(\sigma_x + \sigma_z), \tag{5.12}$$

formula (5.10) shows that

$$\sigma_x|_{z=0} = -p. \tag{5.13}$$

5.3   Strain of Elastic Half-Space

We have already said (p. 45) that to determine movements, it is required to find the functions φ, ψ, f and P. To determine them, we have the following conditions on the surface.

Formulas (5.8) and (5.9) result in

$$\frac{\partial P}{\partial z}\bigg|_{z=0} = \frac{1}{2G} p(x, y).$$

Using formula (5.6, we will present this condition also as

$$\frac{\partial f}{\partial z}\bigg|_{z=0} = -2\frac{1 - \nu^2}{E} p(x, y). \tag{5.14}$$

In this manner, the determination of the functions f and P narrowed down to the solution of the Neumann problem. Using this solution (4.20), we have

$$f(x, y, z) = \frac{1 - \nu^2}{\pi E} \iint\limits_{(\infty)} \frac{p(\xi, \eta)d\xi d\eta}{\sqrt{(x - \xi)^2 + (y - \eta)^2 + z^2}}. \tag{5.15}$$

To perform similar steps relative to the functions φ and ψ from formulas (5.4) and (5.6), we find

$$\frac{\partial \varphi}{\partial z} = -\frac{1 - 2\nu}{2 - 2\nu} \frac{\partial f}{\partial x} \text{ ??? } z = 0. \tag{5.16}$$

Using formula (5.15) and the designation

$$\rho = \sqrt{x - \xi)^2 + (y - \eta)^2}, \tag{5.17}$$

we find

$$\frac{\partial \varphi}{\partial z}\bigg|_{z=0} = \frac{1 - 2\nu}{\pi G} \frac{\partial}{\partial x} \iint\limits_{(\infty)} \frac{p(\xi, \eta)d\xi d\eta}{\rho}. \tag{5.18}$$

Let us find in a totally similar way as follows

$$\frac{\partial \psi}{\partial z}\bigg|_{z=0} = \frac{1 - 2\nu}{\pi G} \frac{\partial}{\partial y} \iint\limits_{(\infty)} \frac{p(\xi, \eta)d\xi d\eta}{\rho}. \tag{5.19}$$

Noting that $w(x, y, 0) = f(x, y, 0)$, let us find the half-space surface subsidence from formula (5.15):

$$w(x, y, 0) = \frac{1 - \nu^2}{\pi E} \iint\limits_{(\infty)} \frac{p(\xi, \eta)d\xi d\eta}{\sqrt{(x - \xi)^2 + (y - \eta)^2}}. \tag{5.20}$$

The last formula has a number of important applications in both strain mechanics and engineering.

5.3.1   Integral Operator of Formulas (5.18)–(5.20)

Formulas (5.18)–(5.20) include an integral

$$\Re = \iint\limits_{(\infty)} \frac{p(\xi, \eta)d\xi d\eta}{\rho}, \tag{5.21}$$

where ρ is defined by formula (5.17). The calculation of $\Re$ is reasonable to be performed in polar coordinates. To do it, apart from the polar radius ρ, let us introduce the angle α between some fixed beam and the beam ρ connecting the points (x, y) and (ξ, η). Then formula (5.21) can be written as follows

$$\Re = \iint\limits_{(\infty)} \frac{p[\rho, \alpha] dS}{\rho},$$

where

$$p[\rho, \alpha] = p[\xi(\rho, \alpha), \eta(\rho, \alpha)].$$

Since the area element dS in polar coordinates is expressed through the polar coordinates ρ, α under the formula $dS = \rho d\rho d\alpha$, then

$$\Re = \iint\limits_{(\infty)} p[\rho, \alpha] d\rho d\alpha. \tag{5.22}$$

Let us note that the integral

$$I = \int\limits_{(\infty)} p d\rho,$$

taken upon an unlimited straight line, represents an area (ω) of cross-section of some body limited by the coordinate plane $z = 0$ and surface $z = p[\rho, \alpha]$. With the parameters x, y being constant, the area (ω) is a function of only the angle α. Consequently, formula (5.22) can be written as follows

$$\Re = \int\limits_0^\pi \omega(\alpha) d\alpha, \tag{5.23}$$

where

$$\omega(\alpha) = \int\limits_0^\infty |p[\rho, \alpha] + p[\rho, \alpha + \pi]| \, d\rho. \tag{5.24}$$

5.4   Examples

Case 1.   Let the pressure p on the half-space boundary be set by the formula

$$p(x, y) = \begin{vmatrix} \sqrt{a^2 - x^2 - y^2} & \text{at } x^2 + y^2 \leqslant a^2, \\ 0 & \text{at } x^2 + y^2 > a^2. \end{vmatrix} \tag{5.25}$$

From the structure of formula (5.25), we can see that in the considered case, the function p in the rectangular coordinates (x, y, p) depicts a semi-ball located above the plane xOy, whereas the big circle of the semi-ball lies in the plane xOy. Any cross-section of this semi-ball perpendicular to the plane of the big circle is a semi-circle.

Let us find the radius R of this semi-circle (Fig. 5.1). To do it, let us draw a beam ON through the reference point O to the point $N(x, y)$ under the angle α to the line ON and designate the trace of crossing the above-mentioned semi-circle and plane xOy as l. The section $EP/2$ of the straight line l is the sought radius R. Let us calculate it from triangles OED and OND:

$$R^2 = a^2 - (x^2 + y^2) \sin^2 \alpha. \tag{5.26}$$

This formula makes sense if the following condition is met

$$|\alpha| < \arcsin \frac{a}{\sqrt{x^2 + y^2}} \text{ for } x^2 + y^2 > a^2. \tag{5.27}$$

The area of the semi-circle with a radius R will be

$$\omega(\alpha) = \frac{\pi}{2} \left[a^2 - (x^2 + y^2) \sin^2 \alpha \right]. \tag{5.28}$$

Beyond (5.27), the external load corresponds to $(p = 0)$ and, consequently, $\omega(\alpha) = 0$. By using the obtained results and formula (5.23), we obtain

Fig. 5.1 To building of the function $\omega(\alpha)$

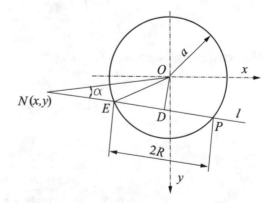

$$\Re = \begin{vmatrix} \dfrac{\pi^2}{4}(2a^2 - x^2 - y^2) & \text{for } x^2 + y^2 \leqslant a^2, \\[2ex] \dfrac{\pi}{2}\left[(2a^2 - x^2 - y^2)\arcsin\dfrac{a}{\sqrt{x^2 + y^2}} + \varrho\right] & \text{for } x^2 + y^2 > a^2 \end{vmatrix}, \quad (5.29)$$

where

$$\varrho = a\sqrt{x^2 + y^2 + a^2}.$$

Case 2.   By generalizing case 1, assume that the pressure curve $p(x, y)$ is limited by the plane xOy and half of the ellipsoid located under this plane, e. g.

$$p(x, y) = \begin{vmatrix} \sqrt{1 - \dfrac{x^2}{a^2} - \dfrac{y^2}{b^2}} & \text{for } \dfrac{x^2}{a^2} + \dfrac{y^2}{b^2} \leqslant 1, \\[2ex] 0 & \text{for } \dfrac{x^2}{a^2} + \dfrac{y^2}{b^2} > 1. \end{vmatrix} \quad (5.30)$$

Let us determine the value of the operator $\Re$ at

$$\frac{x^2}{a^2} + \frac{y^2}{b^2} \leqslant 1. \quad (5.31)$$

Let us introduce the polar coordinates (ρ, α) upon formulas

$$\xi = x + \rho\cos\alpha, \quad \eta = y + \rho\sin\alpha. \quad (5.32)$$

When replacing (5.32), the function p inside the ellipse (5.31) is written as

$$p(\xi, \eta) = p[\rho, \alpha] = \sqrt{1 - \frac{(x + \rho\cos\alpha)^2}{a^2} - \frac{(y + \rho\sin\alpha)^2}{b^2}}.$$

This equation can be converted as follows:

$$p[\rho, \alpha] = A\sqrt{1 - \left(\frac{\rho + \varepsilon}{B}\right)^2}, \quad (5.33)$$

where

$$A = \sqrt{1 - \frac{(x\sin\alpha - y\cos\alpha)^2}{a^2\sin^2\alpha + b^2\cos^2\alpha}};$$

$$B = A\frac{ab}{\sqrt{a^2\sin^2\alpha + b^2\cos^2\alpha}};$$

$$\varepsilon = \frac{b^2 x \cos\alpha + a^2 y \sin\alpha}{a^2 \sin^2\alpha + b^2 \cos^2\alpha}. \tag{5.34}$$

If we consider x, y, and α in formula (5.33) as parameters, formula (5.33) in the plane (ρ, p) represents an equation of the ellipse half with semi-axes (5.34). The area of this semi-ellipse will be

$$\omega(\alpha) = \frac{\pi AB}{2} = \frac{\pi ab}{2}\left[1 - \frac{(x\sin\alpha - y\cos\alpha)^2}{\vartheta^2}\right]\frac{1}{\vartheta}, \tag{5.35}$$

where designation is introduced

$$\vartheta = \sqrt{a^2 \sin^2\alpha + b^2 \cos^2\alpha}. \tag{5.36}$$

By assuming that $a > b$ and by designating

$$\frac{a^2 - b^2}{a^2} = e^2, \quad (a > b), \tag{5.37}$$

let us make the operator $\Re$ to look as follows

$$\Re = \pi b\left[K(e) - \frac{x^2}{a^2}\frac{K(e) - E(e)}{e^2} - \frac{y^2}{a^2}\frac{E(e) - (1 - e)^2 K(e)}{e^2(1 - e^2)}\right], \tag{5.38}$$

where

$$E(e) = \int_0^{\pi/2}\sqrt{1 - e^2 \sin^2\varphi}\,d\varphi,$$

$$K(e) = \int_0^{\pi/2}\frac{d\varphi}{\sqrt{1 - e^2 \sin^2\varphi}} \tag{5.39}$$

are elliptical integrals of the first and second type, respectively. Their values can be easily calculated using a computer or can be taken from literature (for example, see [2, 3]).

References

1. M. Leonov, *Osnovy mekhaniki uprugogo tela* [Fundamentals of elastic body mechanics] (Izd-vo AS Kirg. SSR, Frunze, 1963)

2. N. Samoilova-Yakhontova, *Tablitsy ehllipticheskikh integralov* [Tables of elliptic integrals] (ONTI Publ., Moscow, Leningrad, 1935)
3. E. Yanke, F. Ehmde, F. Lesh, *Spetsial'nye funktsii (Formuly, grafiki, tablitsy)* [Special functions (formulas, graphs, tables)] (Nauka Publ., Moscow, 1977)

Chapter 6
Herz's Task

6.1 Deformation of Adjoining Bodies

Let us consider two elastic bodies contacting before deformation in some point O (Fig. 6.1). Assume this point as the origin of a rectangular system of Cartesian coordinates $Oxyz$ and direct the axis Oz along the common normal line to the surfaces of the adjoining bodies in the point O, and assume the direction of the internal normal line to the surface of body 1 as a positive direction of this axis.

Represent the surfaces of the adjoining bodies 1 and 2 using the functions:

$$z_1 = F_1(x, y), \quad z_2 = F_2(x, y), \quad (F_1(x, y) \geqslant F_2(x, y)). \tag{6.1}$$

Hereinafter, the indexes 1 and 2 mean the first and the second body, respectively. In what follows, it is also suggested that the functions F_1 and F_2 are sufficiently smooth in the point O and can be represented by a formal power series in its vicinity.

After compressing, the equations of deformed surfaces will be

$$\tilde{z}_1 = F_1(x, y) + w_1^o, \quad \tilde{z}_2 = F_2(x, y) + w_2^o, \tag{6.2}$$

where w_1^o and w_2^o designate the movement components of body surface points in the direction of the axis Oz.

Let S mean the contact area of the adjoining bodies after deformation. We have

$$\tilde{z}_1 = \tilde{z}_2 \text{ for } z_1, z_2 \in S; \quad \tilde{z}_1 > \tilde{z}_2 \text{ for } z_1, z_2 \notin S. \tag{6.3}$$

Using the representations of the functions z_1 and z_2 under formulas (6.2), the conditions (6.3) can be written as follows

$$w_1^o - w_2^o = F_2 - F_1 \text{ in the region } S, \tag{6.4}$$

V. Molotnikov, A. Molotnikova, *Theory of Elasticity and Plasticity*,
https://doi.org/10.1007/978-3-030-66622-4_6

Fig. 6.1 Contact of adjoining
bodies

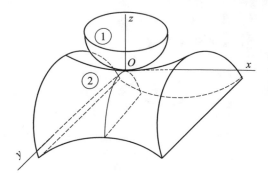

$$w_1^o - w_2^o > F_2 - F_1 \text{ outside the region } S. \tag{6.5}$$

6.2   Primary Assumptions

In what follows in this section, it will be suggested that the adjoining surfaces are
ideally smooth so there are no friction forces in the contact area. Furthermore, we
will assume that the size of the contact area S is small as compared to the minimal
curvature radius of the adjoining surfaces in the point of their initial contact.

With these assumptions, the movements of points of the area S towards the axis
Oz (e. g. the movements w_1^o and w_2^o of the contact area points) can be regarded as
the movements $w(x, y, 0)$ of the elastic half-space boundaries $z \geqslant 0$ subject to the
pressure $p(x, y)$ in the point S, which is yet indefinite. As shown above (see p. 49),
these movements are determined using formula (5.20). By applying formula (5.20),
we will obtain for the first body occupying the half-space $z \geqslant 0$,

$$w_1(x, y, 0) = \frac{1 - v_1^2}{\pi E_1} \iint\limits_{(S)} \frac{p(\xi, \eta)d\xi d\eta}{\sqrt{(x - \xi)^2 + (y - \eta)^2}} + A_1, \tag{6.6}$$

for the second body located in the half-space $z \leqslant 0$,

$$w_2(x, y, 0) = -\frac{1 - v_2^2}{\pi E_2} \iint\limits_{(S)} \frac{p(\xi, \eta)d\xi d\eta}{\sqrt{(x - \xi)^2 + (y - \eta)^2}} + A_2. \tag{6.7}$$

The constant values A_1 and A_2 designate shifts of infinitely distanced points of the
body, e. g. shifts in the direction of the axis Oz of body 1 and body 2 as a rigid link.

Taking this into account, formula (6.4) can be written as follows

$$(k_1 + k_2) \iint\limits_{(S)} \frac{p(\xi, \eta)d\xi d\eta}{\sqrt{(x - \xi)^2 + (y - \eta)^2}} = F_2(x, y) - F_1(x, y) + \delta, \qquad (6.8)$$

where

$$k_1 = \frac{1 - \nu_1^2}{\pi E_1}, \quad k_2 = \frac{1 - \nu_2^2}{\pi E_2}, \quad \delta = A_2 - A_1. \qquad (6.9)$$

We shall note that the constant value δ physically means the approach of infinitely distanced points of the adjoining bodies as a result of their compression.

6.3 Axisymmetric Hertz Problem

Let us consider a case when adjoining bodies are limited by smooth rotation surfaces and the contact point O lies on the rotation axis. Let us take this point as the origin of the rectangular system of Cartesian coordinates. Let us place the axes Ox and Oy in the plane tangential to these bodies in the point O and direct the axis Oz inwards body 1, (Fig. 6.2).

In the considered case, we can represent

$$F_1(x, y) = F_1[r(x, y)], \quad F_2(x, y) = F_2[r(x, y)]; \quad r = \sqrt{x^2 + y^2}.$$

We have

$$F_1[0] = F_2[0] = 0; \quad \left.\frac{dF_1}{dr}\right|_{r=0} = \left.\frac{dF_2}{dr}\right|_{r=0} = 0. \qquad (6.10)$$

Fig. 6.2 Contact of axisymmetric bodies

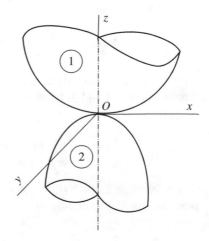

Hence the following can be represented within a sufficiently small vicinity of the reference point

$$F_1[r] \approx \frac{r^2}{2\rho_1}, \quad F_2[r] \approx -\frac{r^2}{2\rho_2}, \tag{6.11}$$

where

$$\frac{1}{\rho_1} = \frac{d^2 F_1}{dr^2}\bigg|_{r=0}, \quad \frac{1}{\rho_2} = \frac{d^2 F_2}{dr^2}\bigg|_{r=0},$$

whereas ρ_1 and ρ_2 are the curvature radii of the forming surfaces F_1 and F_2 in the point $r = 0$. These curvature radii are positive if the body in the contact point is convex and are negative if the body is concave. Taking into account the comments from formulas (6.11), it follows that

$$F_1[r] - F_2[r] = \frac{1}{2}\left(\frac{1}{\rho_1} + \frac{1}{\rho_2}\right)(x^2 + y^2). \tag{6.12}$$

Then formula (6.8) will look like

$$(k_1 + k_2)\Re(x, y) = -\frac{1}{2}\left(\frac{1}{\rho_1} + \frac{1}{\rho_2}\right)(x^2 + y^2) + \delta, \tag{6.13}$$

where $\Re(x, y)$ is the integral operator defined by formula (5.21). A single continuous solution of Eq. (6.13) can be represented as

$$p(x, y) = c\sqrt{a^2 - x^2 - y^2}, \tag{6.14}$$

where c and a are the constant values to be found. To do it, let us use formula (5.29); we obtain

$$\Re(x, y) = c\frac{\pi^2}{4}(2a^2 - x^2 - y^2), \quad (x^2 + y^2 \leqslant a^2).$$

By substituting the last equation to Eq. (6.13) and equaling the coefficients with the same degrees of x and y, we obtain

$$c = \frac{2}{\pi^2}\frac{\frac{1}{\rho_1} + \frac{1}{\rho_2}}{k_1 + k_2}, \quad \delta = \frac{\pi^2}{2}(k_1 + k_2)ca^2. \tag{6.15}$$

The constant value a (the radius of the circular contact area) is determined [1] from the equilibrium condition

$$c \int_0^a \sqrt{a^2 - r^2} \cdot 2\pi r \, dr = P, \tag{6.16}$$

where P is the compressing force.

The condition (6.16) and the first of formulas (6.15) give

$$a = \frac{1}{2} \sqrt[3]{\frac{6\pi \rho_1 \rho_2 (k_1 + k_2) P}{\rho_1 + \rho_2}}. \tag{6.17}$$

By substituting a and c into formula (6.14), we find contact stresses

$$p[r] = \frac{3P}{2\pi a^2} \sqrt{1 - \frac{r^2}{a^2}}, \quad (r \leqslant a). \tag{6.18}$$

The approach of the adjoining bodies is obtained from formulas (6.15) and (6.17):

$$\delta = \frac{1}{2} \sqrt[3]{\frac{9\pi^2 (\rho_1 + \rho_2)(k_1 + k_2) P^2}{2\rho_1 \rho_2}}. \tag{6.19}$$

6.4 Compression of Orthogonal Cylinders

6.4.1 Simplest Case

Two cylinders whose radii are $(R_1 = R_2 = R)$ and whose axes are mutually perpendicular are subject to mutual compression by the force P.

Let us designate the contact point of the cylinders as O and assume it to be the origin of the rectangular coordinate system $Oxyz$. Let us direct the axis Oz inwards the cylinder (1) along the common normal line to the adjoining bodies and axes Ox and Oy along the formed first and second cylinders going through the point O (Fig. 6.3).

In the selected coordinate system, the equations of cylinder surfaces can be represented as follows

$$F_1(x, y) = \frac{1}{2R} y^2, \quad F_2(x, y) = -\frac{1}{2R} x^2. \tag{6.20}$$

In the case under consideration, Eq. (6.8) looks as follows

$$(k_1 + k_2)\Re(x, y) = -\frac{1}{2R}(x^2 + y^2) + \delta. \tag{6.21}$$

Fig. 6.3 Simplest case of
cylinder compression

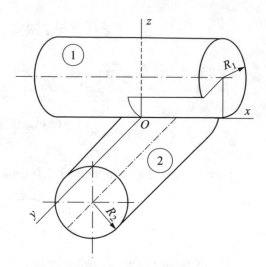

This equation coincides with formula (6.13) if we assume that $\rho_1 = \rho_2 = 2R$. Hence it follows that when compressing opposite cylinders of equal radii, pressure on the contact area is distributed according to the following law

$$p[r] = \frac{3P}{2\pi a^3}\sqrt{a^2 - r^2}, \quad (r \leqslant a),$$

$$a = \frac{1}{2}\sqrt[3]{6\pi PR(k_1 + k_2)}, \tag{6.22}$$

$$\delta = \frac{1}{2}\sqrt[3]{\frac{9\pi^2 P^2(k_1 + k_2)^2}{2R}}.$$

In both considered cases of axisymmetric strain, the maximum pressure in the contact area center will be

$$p_{max} = \frac{3P}{2\pi a^2} = 1,5\frac{P}{F}, \tag{6.23}$$

where F is the contact area. Simple calculations show that for equal compressing forces, the contact area formed when compressing the cylinders is larger approximately by 59% than the contact area for the case of compression of balls escribed within these cylinders of the same material.

6.4.2 Primary Case

Let us consider two cylinders whose radii are R_1 and R_2, $(R_2 \geqslant R_1)$. Let us designate the contact points as O. When selecting the coordinate system shown in Fig. 6.3, the equations of surfaces of these cylinders near the contact point will be

$$F_1(x, y) = \frac{1}{2R_1}y^2, \quad F_2(x, y,) = -\frac{1}{2R_2}x^2.$$

Formula (6.8) in the considered case gives

$$(k_1 + k_2)\Re(x, y) = \delta - Mx^2 - Ny^2, \tag{6.24}$$

where

$$M = \frac{1}{2R_2}, \quad N = \frac{1}{2R_1}. \tag{6.25}$$

A continuous solution of Eq. (6.24) will be found as follows

$$p(x, y) = c\sqrt{1 - \frac{x^2}{a^2} - \frac{y^2}{b^2}}, \tag{6.26}$$

where the constant values a, b, and c are to be determined. Assume for the purpose of certainty that

$$M < N. \tag{6.27}$$

In what follows, we will show that in the case of (6.27) $a > b$. To determine the integral operator $\Re$, one can use formula (5.38). By substituting this formula to Eq. (6.24) and equaling the coefficients with the same degrees of x and y, we obtain

$$\delta = \pi bc(k_1 + k_2)K(e),$$

$$M = \pi bc(k_1 + k_2)\frac{K(e) - E(e)}{a^2 e^2}, \tag{6.28}$$

$$N = \pi bc(k_1 + k_2)\frac{E(e) - (1 - e^2)K(e)}{a^2 e^2(1 - e^2)},$$

where the elliptical integrals $E(e)$ and $K(e)$ are determined earlier by formulas (5.39). For the convenience of calculations of formula (6.28), it is reasonable to make it as follows:

$$\frac{M}{N} = \frac{(1 - e^2)[K(e) - E(e)]}{E(e) - (1 - e^2)K(e)},$$

$$M + N = \pi bc(k_1 + k_2)\frac{E(e)}{(1 - e^2)a^2}. \tag{6.29}$$

The first of these formulas allows making a dependency between the eccentricity (e) and ratio $\frac{M}{N}$.

The resultant of forces with which one of the bodies presses the other will be

$$P = \iint\limits_{(S)} p(x, y)dxdy.$$

The integral on the right represents a volume of the ellipsoid half whose semi-axes equal a, b, c. Hence we have

$$c = \frac{3P}{2\pi ab}. \tag{6.30}$$

From the second Eq. (6.29), it follows that

$$a = \sqrt[3]{\frac{3(k_1 + k_2)E(e)}{2(M + N)(1 - e^2)}P}, \tag{6.31}$$

after which we can find $b = a\sqrt{1 - e^2}$.

The obtained results allow calculating pressure distribution in the contact zone and the approach of bodies caused by deformation. We have

$$p(x, y) = \frac{3P}{2\pi ab}\sqrt{1 - \frac{x^2}{a^2} - \frac{y^2}{b^2}}, \quad \delta = \frac{3}{2}\frac{k_1 + k_2}{a}K(e)P. \tag{6.32}$$

Frequently, it is more convenient to have the following representation of the results found:

$$p_{max} = c_p\sqrt[3]{\frac{(M + N)^2}{(k_1 + k_2)^2}P}, \quad a = c_a\sqrt[3]{\frac{(k_1 + k_2)P}{M + N}}, \tag{6.33}$$

$$\delta = c_\delta\sqrt[3]{(M + N)(k_1 + k_2)^2 P^2},$$

where

$$c_p = \frac{1}{\pi(1 - e^2)} \sqrt[3]{\frac{3}{2} \frac{(1 - e^2)^2}{E^2(e)}},$$

$$c_a = \sqrt[3]{\frac{3}{2} \frac{E(e)}{1 - e^2}}, \quad c_\delta = K(e) \sqrt[3]{\frac{9}{4} \frac{(1 - e^2)}{E(e)}}.$$

6.5 Compression of Barrel-Shaped Bodies

6.5.1 Rotation Bodies with Parallel Axes

In the considered case, equations of surfaces in the vicinity of the contact point O (Fig. 6.4) will be

$$z_1 = F_1(x, y) = \frac{x^2}{2R_1} + \frac{y^2}{2\rho_1}, \quad z_2 = F_2(x, y) = -\frac{x^2}{2R_2} - \frac{y^2}{2\rho_2}.$$

The problem lies in Eq. (6.24) where we must assume

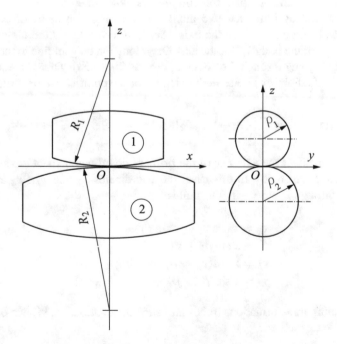

Fig. 6.4 Contact of barrel-shaped bodies with parallel axes

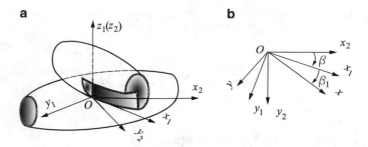

Fig. 6.5 Contact of barrel-shaped bodies with intersecting axes

$$M = \frac{1}{2R_1} + \frac{1}{2\rho_2}, \quad N = \frac{1}{2\rho_1} + \frac{1}{2R_2}. \tag{6.34}$$

6.5.2 Case of Intersecting Axes

The contact of barrel-shaped rotation bodies with intersecting axes is shown in Fig. 6.5a. Let us introduce two auxiliary rectangular systems of coordinates $Ox_1y_1z_1$ and $Ox_2y_2z_2$, where the axis Ox_1 is directed along the tangent line to the generatrix of the body 1, and the axis Ox_2 along the tangent line to the meridian of the body 2 going through the contact point O (Fig. 6.5a). Then the equations of the surfaces of adjoining bodies in the vicinity of the point O can be written as

$$F_1(x_1, y_1) = \frac{x_1^2}{2R_1} + \frac{y_1^2}{2\rho_1}, \quad F_2(x_2, y_2) = -\frac{x_2^2}{2R_2} - \frac{y_2^2}{2\rho_2}. \tag{6.35}$$

Now let us direct the axis Ox of the primary coordinate axis at the angle β_1 to the axis Ox_1, (Fig. 6.5b). In this case, the coordinates x_1 and y_1, x_2 and y_2 are expressed through x and y using formulas

$$\begin{aligned}
x_1 &= x \cos \beta_1 - y \sin \beta_1, \\
y_1 &= x \sin \beta_1 + y \cos \beta_1, \\
x_2 &= x \cos(\beta_1 + \beta) - y \sin(\beta_1 + \beta), \\
y_2 &= x \sin(\beta_1 + \beta) + y \cos(\beta_2 + \beta).
\end{aligned} \tag{6.36}$$

By substituting transformations (6.36) into surface equations (6.35), we obtain

$$F_1(x_1, y_1) - F_2(x_2, y_2) = \left[\frac{\cos^2 \beta_1}{R_1} + \frac{\sin^2 \beta_1}{\rho_1} + \frac{\cos^2(\beta_1 - \beta)}{R_2} + \right.$$
$$+ \left. \frac{\sin^2(\beta_1 + \beta)}{\rho_2}\right]\frac{x^2}{2} - \left[\left(\frac{1}{R_1} - \frac{1}{\rho_1}\right)\sin 2\beta_1 + \left(\frac{1}{R_2} - \frac{1}{\rho_2}\right) \times\right.$$
$$\left. \times \sin 2(\beta_1 + \beta)\right]\frac{xy}{2} + \left[\frac{\sin^2 \beta_1}{R_1} + \frac{\cos^2 \beta_1}{\rho_2} + \right.$$
$$\left. + \frac{\sin^2(\beta_1 + \beta)}{R_2} + \frac{\cos^2(\beta_1 + \beta)}{\rho_2}\right]\frac{y^2}{2}. \tag{6.37}$$

Now let us select the angle β_1 so that the expression [2] in square brackets before the multiplier xy turns zero, e. g.

$$\left(\frac{1}{R_1} - \frac{1}{\rho_1}\right)\sin 2\beta_1 + \left(\frac{1}{R_2} - \frac{1}{\rho_2}\right)\sin 2(\beta_1 + \beta) = 0.$$

To do it, we shall assume

$$tg2\beta_1 = \frac{\left(\dfrac{1}{R_2} - \dfrac{1}{\rho_2}\right)\sin 2\beta}{\left(\dfrac{1}{R_1} - \dfrac{1}{\rho_1}\right) + \left(\dfrac{1}{R_2} - \dfrac{1}{\rho + 2}\right)\cos 2\beta}. \tag{6.38}$$

In this manner, if we direct the axis Ox in the rectangular coordinate system $Oxyz$ at the angle β_1 to the axis Ox_1, Eq. (6.37) will look as follows:

$$F_1(x_1, y_1) - F_2(x_2, y_2) = Mx^2 + Ne^2, \tag{6.39}$$

where

$$M = \frac{1}{2}\left[\frac{1}{R_1}\cos^2 \beta_1 + \frac{1}{\rho_1}\sin^2 \beta_1 + \right.$$
$$\left. + \frac{1}{R_2}\cos^2(\beta_1 + \beta) + \frac{1}{\rho_2}\sin^2(\beta_1 + \beta)\right],$$
$$N = \frac{1}{2}\left[\frac{1}{R_1}\sin^2 \beta_1 + \frac{1}{\rho_1}\cos^2 \beta_1 + \right.$$
$$\left. + \frac{1}{R_2}\sin^2(\beta_1 + \beta) + \frac{1}{\rho_2}\cos^2(\beta_1 + \beta)\right]. \tag{6.40}$$

As above (p. 59), assume that P is the force with which the bodies press each other. Then Eq. (6.8) to determine pressure $p(x, y)$ in the contact area will look as follows:

$$(k_1 + k_2)\Re(x, y) = \delta - Mx^2 - Ny^2, \tag{6.41}$$

where the operator $\Re(x, y)$ is defined by formulas (5.23)–(5.24).

The obtained equation is identical to (6.24). Therefore, we will find the continuous solution of Eq. (6.41) using formulas (6.32), (6.33) at the values M, N, and β_1 determined using formulas (6.38)–(6.40).

6.6   Elongated Contact Area

Assume that $M \ll N$. With this assumption, the eccentricity of the elliptical contact area

$$e = \sqrt{1 - \left(\frac{a}{b}\right)^2}$$

will be close to one, and the value of the elliptical integral $K(e)$ unrestrictedly increases at $e \to 1$.

In these conditions, the ellipsis eccentricity e can be found by transforming formula (6.29). We notice that

$$(1 - e^2)K(e) = \int_0^{\pi/2} \frac{1 - e}{\sqrt{1 - e^2 \sin^2 \varphi}} \, d\varphi, \tag{6.42}$$

and·taking into account that

$$\lim_{e \to 1} \frac{1 - e^2}{\sqrt{1 - e^2 \sin^2 \varphi}} = 0, \tag{6.43}$$

from the first formula (6.29) we obtain

$$\frac{M}{N} \approx (1 - e^2)K(e), \quad (e \approx 1). \tag{6.44}$$

It is known [3] that at $e \approx 1$ can be represented as

$$K(e) \approx \ln \frac{4}{1 - e^2} = \ln \frac{4a^2}{b^2}, \tag{6.45}$$

it is accounted that due to formula (6.43)

$$1 - e^2 = \frac{b^2}{a^2}.$$

Taking into account these comments, formula (6.44) may look as follows:

$$\frac{M}{N} = \frac{2}{\eta^2} \ln 2\eta, \qquad (6.46)$$

where $\eta = a/b$.

After determining η from formula (6.31), we find the large semi-axis of the elliptical area of contact

$$a = \frac{3(k_1 + k_2)}{2N} \eta^2. \qquad (6.47)$$

It is used that $1 - e^2 = b^2/a^2 = 1/\eta$, $E(e) \approx 1$??? $e \approx 1$.

Using formula (6.32), we can define

$$p_{max} = \frac{3P}{2\pi a^2} \eta, \quad \delta = 3\frac{k_1 + k_2}{a} P \ln 2\eta. \qquad (6.48)$$

The last formula (6.48) can be represented as follows

$$\delta = 6(k_1 + k_2) P_1 \ln \frac{2a}{b}, \quad (b = \eta a), \qquad (6.49)$$

where $P_1 = \frac{P}{2a}$ is the load per unit of length of the large axis of the elliptical area of contact.

Formula (6.49) shows that for two elastic half-spaces, the approach (δ) of their infinitely distanced points unrestrictedly grows as a grows at the final P_1.

6.7 Compression of Parallel Cylinders

Assume that two elastic cylinders with parallel axes are subject to mutual pressing across generatrixes. Let us direct the axis Ox along the common generatrix of the adjoining bodies. As before (p. 61), let us designate the curvature radii of these cylinders as R_1 and R_2. In formula (6.8), we must assume

$$F_1 = \frac{1}{2R_1} y^2, \quad F_2 = -\frac{1}{2R_2} y^2.$$

Then for the considered case in Eq. (6.24), we must assume

$$N = \frac{1}{2R_1} + \frac{1}{2R_2}, \quad M = 0.$$

Assuming in the last formula (6.28) $b^2 = a^2(1 - e^2)$, $(e \approx 1)$, we find that

$$\frac{1}{2R_1} + \frac{1}{2R_2} = \pi(k_1 + k_2)\frac{c}{b}. \tag{6.50}$$

By designating the load per unit of generatrix length as P_1, we will obtain

$$P_1 = \int_{-b}^{b} p(y)dy = 2c \int_{0}^{b} \sqrt{1 - \frac{y^2}{b^2}}\, dy, \ ?.?. \ c = \frac{P_1}{\pi b}.$$

Substituting this result into formula (6.51) gives

$$b = \sqrt{\frac{2P_1 R_1 R_2 (k_1 + k_2)}{R_1 + R_2}}, \tag{6.51}$$

$$c = p_{max} = \frac{1}{\pi}\sqrt{\frac{P_1(R_1 + R_2)}{2R_1 R_2(k_1 + k_2)}}. \tag{6.52}$$

References

1. M. Leonov, M. *Osnovy mekhaniki uprugogo tela* (Fundamentals of elastic body mechanics). (Izd-vo AS Kirg. SSR, Frunze, 1963)
2. I. Vorovich, V. Aleksandrov, *Mekhanika kontaktnykh vzaimodeistvii. Pod redaktsiei Vorovicha I.I., Aleksandrova V.M.* (Mechanics of contact interactions. Ed. by Vorovich I.I., Alexandrov V.M.). (Fizmatlit Publ., Moscow, 2001)
3. E. Yanke, F. Ehmde, F. Lesh, *Spetsial'nye funktsii (Formuly, grafiki, tablitsy)* (Special functions (formulas, graphs, tables)). (Nauka Publ., Moscow, 1977)

Chapter 7
Stressed State in a Body Point

7.1 Principal Stresses

Let us select an arbitrary system of Cartesian coordinates $Oxyz$. Let the stress components relative to the selected coordinate systems be known in some point of the body. Let us consider the problem of finding such areas that have no tangential stress. As we know (for example, see [3, p. 67]), such areas are called *principal*, and the values of normal stresses on these areas are called *principal stresses*.

The cosines of angles between the normal line to the principal area and coordinate axes can be found from Eqs. (2.14), to which another condition must be added

$$\cos^2 \widehat{nx} + \cos^2 \widehat{ny} + \cos^2 \widehat{nz} = 1. \tag{7.1}$$

Equations (2.2) and (7.1) form a system of 4 equations relative to unknown cosines and stresses σ. Since cosines in Eq. (7.1) cannot simultaneously turn to zero, the system determinant (2.2) must turn to zero, e. g.

$$\begin{vmatrix} \sigma_x - \sigma & \tau_{xy} & \tau_{xz} \\ \tau_{yx} & \sigma_y - \sigma & \tau_{yz} \\ \tau_{zx} & \tau_{zy} & \sigma_z - \sigma \end{vmatrix} = 0. \tag{7.2}$$

From Eq. (7.2), normal stresses (σ) are determined in principal directions, e. g. principal stresses. By opening the determinant, the last equation can be made to look as follows:

$$\sigma^3 - I_1\sigma^2 + I_2\sigma - I_3 = 0, \tag{7.3}$$

V. Molotnikov, A. Molotnikova, *Theory of Elasticity and Plasticity*,
https://doi.org/10.1007/978-3-030-66622-4_7

where

$$I_1 = \sigma_x + \sigma_y + \sigma_z,$$

$$I_2 = \sigma_x \sigma_y + \sigma_y \sigma_z + \sigma_z \sigma_x - \tau_{xy}^2 - \tau_{yz}^2 - \tau_{zx}^2,$$

$$I_3 = \sigma_x \sigma_y \sigma_z + 2\tau_{xy}\tau_{yz}\tau_{zx} - \sigma_x \tau_{yz}^2 - \sigma_y \tau_{zx}^2 - \sigma_z \tau_{xy}^2. \tag{7.4}$$

By connecting this body at the defined load with any other system of coordinates, we will change the stress coordinates $\sigma_x, \ldots, \tau_{zx}$. However, principal stresses will not depend on the selection of the coordinate system. Consequently, the coefficients of Eq. (7.3) that define principal stresses keep constant values not depending on the selection of coordinate axes. Such values are called *invariants*.

Equation (7.3) has at least one real root. If we know it, we can determine the principal direction from Eqs. (2.2) and (7.1). Consequently, the direction of the area where only normal stress acts can be found. Assume that we selected the axis Oz to be coinciding with this direction. According to the Bezout theorem, Eq. (7.3) can be represented as

$$(\sigma - \sigma_z)[\sigma^2 - \sigma(\sigma_x - \sigma_y) - \sigma_x \sigma_y + \tau_{xy}] = 0. \tag{7.5}$$

By opening this equation and grouping the members at σ and σ^2,, we will obtain a certain view of Eq. (7.3) where $\tau_{zx} = \tau_{zy} = 0$, since the axis Oz is principal. In this case, two other principal stresses are defined by the formulas

$$\sigma_{1,2} = \frac{\sigma_x + \sigma_y}{2} \pm \sqrt{\left(\frac{\sigma_x - \sigma_y}{2}\right)^2 + \tau_{xy}^2}, \quad (\sigma_3 = \sigma_z). \tag{7.6}$$

The angle between principal stresses and the axis Ox is determined from the first Eq. (2.2) when substituting σ under formula (7.6), e. g.

$$\operatorname{tg}\widehat{nx} = \frac{1}{2\tau_{xy}}\left[\sigma_y - \sigma_x \pm \sqrt{(\sigma_y - \sigma_x)^2 + 4\tau_{xy}^2}\right]. \tag{7.7}$$

The roots of Eq. (7.7) define two mutually orthogonal directions $\widehat{n_1 x}$ and $\widehat{n_2 x}$.

7.2 Maximum Stresses

Let us consider some body subject to a homogeneous stressed state. Let us mentally separate a pyramid from the body whose faces coincide with principal areas and the basis is arbitrarily inclined (Fig. 7.1).

By writing the areas of faces F_i, $(i = 1, 2, 3)$, as respective projections of the basis S, we obtain

Fig. 7.1 To analysis of
stresses

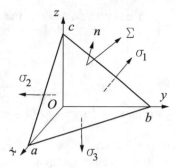

$$F_1 = S\cos\alpha, \quad F_2 = S\cos\beta, \quad F_3 = S\cos\gamma, \tag{7.8}$$

where α, β ? γ mean angles between the normal line n (Fig. 7.1) to the considered
area and respective principal directions.

Let us calculate the square of resultants of all forces distributed over the pyramid
faces. It can be represented by the formula

$$(\sigma_1 F_1)^2 + (\sigma_2 F_2)^2 + (\sigma_3 F_3)^2, \tag{7.9}$$

or as $(\Sigma S)^2$, where Σ is *full stress* on the area S. We will obtain

$$\Sigma = \sqrt{\sigma_1^2 \cos^2\alpha + \sigma_2^2 \cos^2\beta + \sigma_3^2 \cos^2\gamma}. \tag{7.10}$$

In what follows, for certainty reasons, we will assume that

$$\sigma_1 > \sigma_2 > \sigma_3. \tag{7.11}$$

In these conditions, from Eq. (7.10) we have

$$\Sigma^2 \leqslant \sigma_1^2(\cos^2\alpha + \cos^2\beta + \cos^2\gamma) = \sigma_1^2,$$
$$\Sigma^2 \geqslant \sigma_3^2(\cos^2\alpha + \cos^2\beta + \cos^2\gamma) = \sigma_3^2.$$

Hence it follows that principal stresses determine the maximum and minimal
values of full stress.

Normal stress σ in the section abc can be obtained if making up the sum of
projections of all forces applied to that pyramid to the normal line to its basis. We
obtain

$$\sigma_1 F_1 \cos\alpha + \sigma_2 F_2 \cos\beta + \sigma_3 F_3 \cos\gamma = \sigma S;$$

hence we have

$$\sigma = \sigma_1 \cos^2\alpha + \sigma_2 \cos^2\beta + \sigma_3 \cos^2\gamma. \tag{7.12}$$

Then tangential stress in the section abc will be

$$\tau = \sqrt{\Sigma^2 - \sigma^2}$$

or

$$\tau^2 = \sigma_1^2 u + \sigma_2^2 v + \sigma_3(1 - u - v) - [\sigma_1 u + \sigma_2 v + \sigma_3(1 - u - v)]^2, \qquad (7.13)$$

where $u = \cos^2 \alpha, \quad v = \cos^2 \beta$.

To determine the maximum value of τ^2, let us differentiate both parts of this equation upon variables u and v. We obtain

$$\frac{\partial \tau^2}{\partial u} = \sigma_1^2 - \sigma_3^2 - 2[\sigma_1 u + \sigma_2 v + \sigma_3(1 - u - v)](\sigma_1 - \sigma_3),$$
$$\frac{\partial \tau^2}{\partial v} = \sigma_2^2 - \sigma_3^2 - 2[\sigma_1 u + \sigma_2 v + \sigma_3(1 - u - v)](\sigma_2 - \sigma_3). \qquad (7.14)$$

Assume that for some values of the parameters $u = u_1$ and $v = v_1$, the first parts of formulas (7.14) turn to zero. Assuming that all three principal stresses differ from each other in magnitude, in this case we can write the condition u_1 and v_1 as follows:

$$2[\sigma_1 u_1 + \sigma_2 v_1 + \sigma_3(1 - u_1 - v_1)] = \sigma_1 - \sigma_3,$$
$$2[\sigma_1 u_1 + \sigma_2 v_1 + \sigma_3(1 - u - v)] = \sigma_2 - \sigma_3.$$

The same expression is found in left parts of last equations, and different values are given in right parts. The obtained contradiction proves that assumption on the presence of stationary points in the function τ^2 is not true. Hence it follows that *the function τ gets the maximum and minimal values at the boundary of the domain of values $(u; v)$ for which this function makes sense.*

Recalling that the variables u and v are the squares of cosines of some angles, we have the following condition:

$$0 \leqslant u \leqslant 1, \quad 0 \leqslant v \leqslant 1;$$
$$0 \leqslant u + v \leqslant 1.$$

The last formula is a consequence of the condition (7.1).

For the boundary values $u = 0; 1$ or $v = 0; 1$ or $u + v = 0; 1$, we will find that one of the angles α, β or γ becomes straight, e. g. the pyramid basis plane (Fig. 7.1) turns and becomes parallel to one of the principal directions. Having this in mind, assume that the basis of the considered pyramid is parallel to the intermediate principal stress. In this case, $v = 0$, and formula (7.15) gives

$$|\tau| = \frac{\sigma_1 - \sigma_3}{2} |\sin 2\alpha| ; \qquad (7.15)$$

hence, if $\alpha = \pi/4$ we obtain

$$|\tau|_{max} = \frac{\sigma_1 - \sigma_3}{2}. \tag{7.16}$$

Provided that (7.11), formula (7.16) defines the maximum tangential stress.

7.3 Intensity of Stresses

It is known [4] that the experimental values of tangential stresses are called *principal tangential stresses:*

$$\tau_{13} = \frac{\sigma_1 - \sigma_3}{2}, \quad \tau_{21} = \frac{\sigma_2 - \sigma_1}{2}, \quad \tau_{23} = \frac{\sigma_2 - \sigma_3}{2}. \tag{7.17}$$

Octahedral tangential stress (τ_o) is the tangential stress on the area similarly inclined to principal directions. For this area, $u = v = \dfrac{1}{\sqrt{3}}$ and formula (7.15) gives:

$$\tau_o^2 = \frac{2}{9}(\sigma_1^2 + \sigma_2^2 + \sigma_3^2 - \sigma_1\sigma_2 - \sigma_2\sigma_3 - \sigma_3\sigma_1).$$

The last formula can be represented in a different way:

$$\tau_o = \frac{1}{3}\sqrt{(\sigma_1 - \sigma_2)^2 + (\sigma_2 - \sigma_3)^2 + (\sigma_3 - \sigma_1)^2}, \tag{7.18}$$

whereof one can make sure by opening brackets under the radical sign in the last formula.

We will show that octahedral tangential stress is proportional to the mean square of three principal tangential stresses. Indeed, when using formulas (7.17), we can re-write formula (7.18) as follows:

$$\tau_o = \frac{2}{3}\sqrt{\tau_{12}^2 + \tau_{23}^2 + \tau_{31}^2}. \tag{7.19}$$

Octahedral tangential stress is sometimes also referred to as the *intensity of tangential stresses.* Along with this value, its proportional value $\sigma_?$ is often introduced, which is defined by the formula

$$\sigma_? = \sqrt{2(\tau_{12}^2 + \tau_{23}^2 + \tau_{31}^2)}. \tag{7.20}$$

The intensity of (tangential) stresses with the accuracy up to the constant multiplier equals a mean square of tangential stresses defined on the sphere (infinitely small radius in the case of a non-homogeneous stressed state).[1] This means that *The intensity of (tangential) stresses is proportional to the mean square of tangential stresses for all possible areas going through this point of the body.*

7.4   Some Properties of Tangential Stresses

By using formula (7.17), we can obtain

$$\tau_{12} + \tau_{23} = \tau_{31}. \tag{7.21}$$

Similar to the numbering of principal stresses ($\sigma_1 > \sigma_2 > \sigma_3$), (p. 71), let us designate the smallest of principal tangential stresses as τ_1, the middle one as τ_2 and the maximum one as τ_3. Then formula (7.21) will look like [1]

$$\tau_3 = \tau_1 + \tau_2. \tag{7.22}$$

Since we assumed that $\tau_2 > \tau_1$, it follows from formula (7.22) that

$$\tau_2 > \frac{1}{2}\tau_3. \tag{7.23}$$

Let us introduce designations

$$y = \frac{\tau_o}{\tau_3}, \quad x = \frac{\tau_2}{\tau_3}.$$

Formula (7.22) shows that $\tau_1 = \tau_3 - \tau_2$. By substituting this result into formula (7.19), we obtain

$$y(x) = \frac{2}{3}\sqrt{(1-x)^2 + x^2 + 1}. \tag{7.24}$$

We use formula (7.24 to analyze the change in the intensity of tangential stresses τ_o as the tangential stress τ_2 changes, which we will conditionally call the middle stress [2]. Taking into account the condition (7.23), we note that the values of the function $y(x)$ are interesting only at $0,5 \leqslant x \leqslant 1$. Atthe ends of this segment, we will have

---

[1] This property was proved by V.V. Novozhilov.

$$y\left(\frac{1}{2}\right) = \frac{2}{3}\sqrt{0.5^2 + 0.5^2 + 1} \approx 0.82; \quad y(1) = \frac{2}{3}\sqrt{2} \approx 0.94,$$

as the argument x grows, the function y grows monotonously ($y' > 0$ when $\frac{1}{2} < x < 1$). Consequently, as the middle tangential stress grows, the intensity of tangential stresses grows within

$$0.82 \leqslant \frac{\tau_o}{\tau_{max}} \leqslant 0.94. \tag{7.25}$$

The determination of the maximum tangential stresses and the intensity of (tangential) stress is required to formulate the conditions for the occurrence of plastic strain described below (Chap. 13).

References

1. M. Leonov, *Osnovy mekhaniki uprugogo tela* [Fundamentals of elastic body mechanics] (Izd-vo AS Kirg. SSR, Frunze, 1963)
2. M. Leonov, *Mekhanika deformatsii i razrusheniya* [Deformation and fracture mechanics] (Izd-vo Ilim Publ., Frunze, 1981)
3. V. Molotnikov, *Kurs soprotivleniya materialov* [The course of strength of materials] (SPb., Moscow/Lan' Publ., Krasnodar, 2006)
4. V. Molotnikov, *Mekhanika konstruktsii* [Mechanics of structures] (SPb., Moscow/Lan' Publ., Krasnodar, 2012)

Chapter 8
Linear Elastic Systems

8.1 General Comments

Let some system of material points be in equilibrium under the action of a defined load. Due to equilibrium, the resultant of all forces applied to each point of the system equals zero. Let us give infinitely short shifts to system points and keep forces between them unchanged. The work of the defined forces applied to any point of the system will be zero. Hence it follows that for the entire system of material points, the work of the defined forces in equilibrium equals zero. This means that the sum of works of internal and external forces equals zero, e.g.

$$A + W = 0, \qquad (8.1)$$

where A is the work of external forces and W is the work of internal forces.

In the case of an absolute solid body, displacements occur without deformation, distances between particles do not change, and internal forces perform no work. In this case, formula (8.1) shows the principle of possible displacements [3, p. 201].

Let us represent the work of internal forces as a sum of products of parameters (generalized forces P_i) defining the load and some coefficients depending on the type and magnitude of displacements. These coefficients are referred to as generalized displacements (u_i). In a similar way, the work of internal forces at infinitely short displacements of a loaded system can be represented as follows:

$$W = -\Sigma P_i u_i. \qquad (8.2)$$

The work (8.2) is called [2] virtual. It differs from the actual work of forces P_i performed during loading.

V. Molotnikov, A. Molotnikova, *Theory of Elasticity and Plasticity*,
https://doi.org/10.1007/978-3-030-66622-4_8

8.2   Linear System

Let us consider another system loaded by the system of forces P_1, P_2, ..., P_n. Assume that during loading, inertia forces occurring as a result of displacements of system points caused by deformation are negligibly low. This loading is called static. If the magnitude of displacements of system points is increased by λ times under the action of forces λP_1, λP_2, ..., λP_n, the elastic system is called *linear.*

Let us designate the elastic displacement corresponding to the force P_i from the load P_1, P_2, ..., P_n as u_i. Then the displacements at loading λP_1, λP_2, ..., λP_n will be λu_1, λu_2, ..., λu_n. Let us give the parameter λ an infinitely low increment $d\lambda$ and count the work (dA) of last forces on real displacements $u_1 d\lambda$, $u_2 d\lambda$, ..., $u_n d\lambda$:

$$dA = P_1 u_1 \lambda d\lambda + P_2 u_2 \lambda d\lambda + \ldots + P_n u_n \lambda d\lambda.$$

By summing the work when changing λ from zero to one, we obtain

$$A = (P - 1u_1 + P_2 u_2 + \ldots + P_n u_n) \int\limits_0^1 \lambda d\lambda,$$

or

$$A = \frac{1}{2}(P_1 u_1 + P_2 u_2 + \ldots + P_n u_n). \tag{8.3}$$

In this manner, *the work of statically applied external forces on displacements caused by deformations of a linear elastic system equals the half-sum of the products of the final values of each force and the values of respective displacements.* The work of internal forces of an elastic body taken with a reverse sign is called *potential energy of elastic deformations.* According to formula (8.1), numerically it equals the work of external forces. For this reason, the potential energy of elastic deformation and the work of external forces will be designated as A.

As an example, let us count the potential energy of a round rod that elongates by the force N, is bent by the moment M_u, and is twisted by the moment M_k (Fig. 8.1). In the case of elastic deformations, the rod will be elongated by Nl/EF, $(F = \pi d^2/4)$, the end section will rotate in the bending plane by $\Delta\alpha = \dfrac{M_u l}{E I_x}$ $\left(I_x = \dfrac{\pi d^4}{64}\right)$, and the rod will be twisted in the end section by the angle $\Theta l = \dfrac{M_k l}{G I_p}$ $\left(I_p = \dfrac{\pi d^4}{32}\right)$. When considering the forces N, M_k and M_u as external ones relative to the rod under consideration, we will find

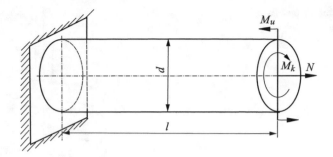

Fig. 8.1 For calculation of the potential energy of elastic strain of a round rod

$$A = \frac{l}{2} \left(\frac{N^2}{EF} + \frac{M_u^2}{EI_x} + \frac{M_k^2}{GI_p} \right). \tag{8.4}$$

Taking into account that $E = 2(1 + v)G$, formula (8.4) will look as follows:

$$A = \frac{l}{2E} \left[\frac{4N^2}{\pi d^2} + \frac{64}{\pi d^4} \left(M_u^2 + (1 + v)M_k^2 \right) \right]. \tag{8.5}$$

8.3 Potential Energy of a Helical Spring

Let us consider a cylindrical helical spring shown in Fig. 8.2. Let us dissect the spring turn by the axial plane. From the condition of equilibrium of the dissected part of the spring, let us find the moment of internal forces relative to the axis going through the center of gravity of the turn section in perpendicular to the specified axial plane:

$$M = PR,$$

where R is the average spring radius.

A component of this moment relative to the axis perpendicular to the cross-section plane represents a torque whose magnitude will be

$$M_k = PR \cos \alpha, \tag{8.6}$$

where α is the inclination level of the helical line of the spring.

Let us calculate the bending moment using the formula:

$$M_u = \sqrt{M^2 - M_k^2} = PR \sin \alpha. \tag{8.7}$$

Fig. 8.2 Deformation of a
helical spring

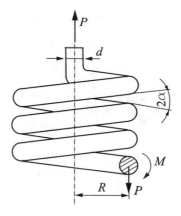

Let us calculate the normal force N as a projection of the force P onto the normal line to the cross-section:

$$N = P \sin \alpha. \tag{8.8}$$

Let us assume that the curvature radius of the spring turn is large as compared to its transverse dimension. While neglecting the transverse force, we can represent the potential energy of spring deformation using formula (8.5), e.g.

$$A = \frac{Pl}{2E} \left[\frac{4 \sin^2 \alpha}{\pi d^2} + \frac{64 R^2}{\pi d^4} (1 + \nu \cos^2 \alpha) \right], \tag{8.9}$$

where l is the spring rod length that, in case of n turns, can be written as follows:

$$l = 2\pi R n \sec \alpha.$$

Formula (8.9) can be represented as follows in a different way:

$$A = \frac{P^2 Rn}{Ed^2} \left[\frac{4 \sin^2 \alpha}{\cos \alpha} + 64 \left(\frac{R}{d} \right)^2 \left(\frac{1}{\cos \alpha} + \nu \cos \alpha \right) \right]. \tag{8.10}$$

8.4   Principle of Mutuality of Works

Let us consider two various states of any linear elastic system loaded by two various loads. Let us designate the loads and internal forces and system displacements in these two states using indexes 1 and 2. Let us represent that the initially non-loaded system is exposed to load 1. The system gets some deformations from load 1, and the load has made work that we will designate as A_{11}. Let us keep load 1 and gradually load the system with load 2. This load will cause additional deformation of the system. Let us call the work of load 2 on displacements caused by additional deformation after applying load 2 as A_{22}.

However, in the case of additional deformation, the work will be made not only by load 2 but also by load 2 since the points of application of forces of system 1 in the case of additional deformation will get additional displacements. Load 1 on these displacements remains constant and will make the work that we will designate as A_{12}. As a result, the work of external forces is expressed by the sum $A_{11} + A_{22} + A_{12}$. In these addends, the first index indicates what load does the work and the second one shows the forces that caused displacements of this load.

Let us change the order of loading. At first, let us apply load 2 to the non-loaded system that will do some work A_{22}. on the displacements caused by it. Let us keep load 2 and will gradually apply load 1. As in the first loading case, load 1 will cause additional deformation of the system and will do work A_{11}. on the displacements caused by it. During this time, load 2 remaining constant will do additional work A_{21} on displacements caused by the application of load 1. As a result, the work of external forces in the second method of loading is expressed by the sum $A_{22} + A_{11} + A_{21}$.

The total work of external forces in each of these cases of loading equals the work of internal forces taken with a reverse sign. This work is defined by the final state of the system and does not depend on the loading sequence. Since the final states of the system in the two considered cases of loading are the same, this means that

$$A_{12} = A_{21}. \tag{8.11}$$

Thus, the following *principle of mutuality of works* (Betty) is proved [1].

Theorem 8.1 *The work of forces of the first system on displacements caused by the second system equals the work of the second system on displacements caused by the first system of forces.*

Let us substitute into the result (8.11) $A_{12} = P_1 \delta_{12} P_2$, where P_1 is the generalized force [4] of the first system and δ_{12} is the generalized displacement of force P_1 in the direction of its action caused by $P_2 = 1$. In a similar fashion, let us substitute $A_{21} = P_2 \delta_{21} P_1$, where δ_{21} is the generalized displacement of force P_2 in the direction of its action caused by $P_1 = 1$. From the condition (8.11), it follows that

$$\delta_{12} = \delta_{21}. \tag{8.12}$$

Equation (8.12) expresses (Maxwell) *the principle of mutuality of displacements* that we formulate as follows.

Theorem 8.2 *The displacement of the point of application of the first force in the direction of its action caused by the second single force equals the displacement* *of the point of application of the second force in the direction of action of the first single force.*

8.5   Castigliano's Theorem

We will consider an elastic system loaded by an arbitrary system of forces $P_1, P_2, \ldots, P_n$ and secured in such a manner that its shifts as a rigid whole are prevented (Fig. 8.3).

Due to deformation, an arbitrary point A will take a new position A'. The segment $\overline{AA'}$ is called full displacement of the point A. Let us take an arbitrary axis l going through the point A, and let us project the point A' upon it. As a result, we obtain a point A'' on the axis l. The segment $\overline{AA}''$ is called *displacement of the point A in the direction l.*

In this manner, if, for example, $\overline{BB'}$ is full displacement of the point (B) by applying the force P_1, then $\delta_1 = \overline{BB''}$ is displacement of the point of application of the force P_1 in the direction of its action.

Let us designate the potential energy of system deformation by the forces $P_1, P_2, \ldots, P_n$ as U.

Let us give one force, for example, the force P_n, an infinitely small increment dP_n. The potential energy will also get an increment and will be

$$U + dU = U + \frac{\partial U}{\partial P_n} dP_n. \tag{8.13}$$

Let us change the order of force application. At first, let us apply only the force dP_n. Due to system deformation, the point of application of this force in the direction of its action will get some displacement that we designate as $d\delta_n$. The work of the force dP_n on the specified displacement based on formula (8.3) will be $dP_n d\delta_n/2$. Now we apply the entire system of external forces $P_1, P_2, \ldots, P_n$. In the case of no force dP_n, the potential energy of the system would be equal to U. However, since there is force dP_n, this force will make additional work

Fig. 8.3 To the Castigliano's theorem

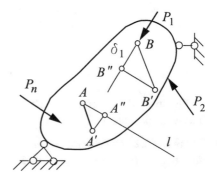

on displacement δn of the point of application of the force P_n in its direction, whereas the displacement δ_n is caused by the entire system of external forces at $dP_n = const$. Consequently, the additional work of the force dP_n will be equal to the product $dP_n\delta_n$. As a result, the potential energy of elastic deformation of the system under the second method of loading will be expressed by the sum:

$$\frac{1}{2}dP_n d\delta_n + U + dP_n\delta_n. \tag{8.14}$$

Since the final state of the system under the first and second methods of loading is the same, we can equate the sum (8.14) to the right part of formula (8.13). We obtain

$$U + \frac{\partial U}{\partial P_n}dP_n = \frac{1}{2}dP_n d\delta_n + U + dP_n\delta_n.$$

By neglecting the first addend in the right part of this equation as an infinitely small and of highest order, we will finally obtain

$$\delta_n = \frac{\partial U}{\partial P_n}. \tag{8.15}$$

In this manner, the following is proved (Castigliano).

Theorem *A partial derivative of the potential energy of the system is equal in force to the displacement of the force application point in the direction of its action.*

8.6   Specific Potential Energy of Elastic Deformation

Let the ribs dl_1, dl_2, dl_3 of an elementary parallelepiped separated from an elastic body be parallel to principal directions. Let us designate their relative deformations as ε_1, ε_2, ε_3. Elongations of ribs caused by deformation will be

$$\Delta(dl_1) = \varepsilon_1 dl_1; \quad \Delta(dl_2) = \varepsilon_2 dl_2; \quad \Delta(dl_3) = \varepsilon_3 dl_3. \tag{8.16}$$

In relation to the considered element, stresses are external loads. During deformation, their work dA on displacements (8.16) goes into the potential energy dU of elastic deformation. Let us calculate this work:

$$dA = dU = \frac{1}{2}\sigma_1 dl_2 dl_3 \Delta(dl_1) + \frac{1}{2}\sigma_2 dl_1 dl_3 \Delta(dl_2) + \frac{1}{2}\sigma_3 dl_1 dl_2 \Delta(dl_3),$$

or, taking into account formulas (8.16),

$$dU = \frac{1}{2}(\sigma_1\varepsilon_1 + \sigma_2\varepsilon_2 + \sigma_3\varepsilon_3).$$

By dividing both parts of this equation by the initial volume of the element dV, we get the specific potential energy of elastic deformation:

$$u = \frac{dU}{dV} = \frac{1}{2}(\sigma_1\varepsilon_1 + \sigma_2\varepsilon_2 + \sigma_3\varepsilon_3). \tag{8.17}$$

The last formula can be represented in a different way. By substituting deformations through stresses according to generalized Hooke's law (1.11), we obtain

$$u = \frac{1}{2E}[\sigma_1^2 + \sigma_2^2 + \sigma_3^2 - 2\nu(\sigma_1\sigma_2 + \sigma_2\sigma_3 + \sigma_3\sigma_1)]. \tag{8.18}$$

Let us represent principal stresses as follows:

$$\sigma_1 = \sigma_0 + \sigma_1'; \quad \sigma_2 = \sigma_0 + \sigma_2'; \quad \sigma_3 = \sigma_0 + \sigma_3', \tag{8.19}$$

where σ_0 is hydrostatic stress defined by formula (1.14). Equations (8.19) and (1.14) show that

$$\sigma_1' + \sigma_2' + \sigma_3' = 0. \tag{8.20}$$

Let all principal stresses be equal between each other and equal to σ_0. By designating specific potential energy in this case as $u_{??}$, we have obtained from formula (8.18) as follows:

$$u_{\text{vol}} = \frac{3(1 - 2\nu)}{2E}\sigma_0^2 = \frac{1 - 2\nu}{6E}(\sigma_1 + \sigma_2 + \sigma_3)^2. \tag{8.21}$$

This value is called *specific potential energy of volume change.* By subtracting it from full specific energy, we find *specific potential energy of volume change:*

$$u_{\text{f}} = u - u_{\text{vol}} = \frac{1 + \nu}{6E}[(\sigma_1 - \sigma_2)^2 + (\sigma_2 - \sigma_3)^2 + (\sigma_3 - \sigma_1)^2]. \tag{8.22}$$

By comparing formulas (8.22) and (7.18), we note that the specific potential energy of volume change with accuracy up to the constant multiplier coincides with the square of octahedral tangential stress.

For the case of plane stressed state ($\sigma_3 = 0$), formulas (8.18), (8.21), and (8.22) will look as follows:

$$u = \frac{1}{2E}(\sigma_1^2 + \sigma_2^2 - 2v\sigma_1\sigma_2);$$

$$u_{\text{vol}} = \frac{1 - 2v}{6E}(\sigma_1 + \sigma_2)^2; \tag{8.23}$$

$$u_{\text{f}} = \frac{1 + v}{3E}(\sigma_1^2 + \sigma_2^2 - \sigma_1\sigma_2).$$

In arbitrary axes x, y, formulas (8.23) will be

$$u = \frac{1}{2E}(\sigma_x^2 + \sigma_y^2 - 2v\sigma_x\sigma_y) + \frac{\tau^2}{2G};$$

$$u_{\text{vol}} = \frac{1 - 2v}{6E}(\sigma_x + \sigma_y)^2; \tag{8.24}$$

$$u_{\text{f}} = \frac{1 + v}{3E}(\sigma_x^2 + \sigma_y^2 - \sigma_x\sigma_y) + \frac{\tau^2}{2G},$$

where the shear modulus G is related to the Young modulus E and Poisson coefficient v by the dependency (1.10).

References

1. M. Leonov, *Osnovy mekhaniki uprugogo tela* (Fundamentals of elastic body mechanics). (Frunze, Izd-vo AS Kirg. SSR, 1963)
2. V. Molotnikov, *Osnovy teoreticheskoi mekhaniki* (Fundamentals of theoretical mechanics). (Feniks Publ., Rostov on Don, 2004)
3. V. Molotnikov, *Mekhanika konstruktsii* (Mechanics of structures). (SPb., Moscow, Krasnodar, Lan' Publ., 2012)
4. E. Nikolai, *Teoreticheskaya mezanika. Ch. 2. Dinamika. Izd. 13-e.* (Theoretical mezanika. Part 2. Dynamics. Prod. The 13th.). (Fizmatlit Publ., Moscow, 1938)

Chapter 9
Plane Problem of Elasticity Theory

9.1 Functions of Stresses

Previously (p. 16), it was established that the dependency between relative elongations and normal stresses in the case of plane strain ($\varepsilon_z = 0$) could be obtained from the respective dependencies in the case of a plane stressed state ($\sigma_z = 0$) by substituting the Young modulus E and Poisson coefficient v with E^* and v^*, respectively, using formulas (1.21). Therefore, both these stress–strain states are combined by the concept of "plane problem" not always mentioning what exact state is considered.

Let us consider plane strain ($\varepsilon_z = 0$). In this case, equilibrium equations:

$$\frac{\partial \sigma_x}{\partial x} + \frac{\partial \tau_{xy}}{\partial y} = 0,$$
$$\frac{\partial \sigma_y}{\partial y} + \frac{\partial \tau_{yx}}{\partial x} = 0 \tag{9.1}$$

can be met if we consider that

$$\sigma_x = \frac{\partial^2 U}{\partial y^2}, \quad \sigma_y = \frac{\partial^2 U}{\partial x^2}, \quad \tau_{xy} = \frac{\partial^2 U}{\partial x \partial y}, \tag{9.2}$$

where U is an arbitrary function differentiated for the sufficient number of times, which is called (Airy, [5]) *the stress function.*

Let us also use the fact that volumetric expansion in an elastic body (1.13) is a harmonic function (see p. 31). This means that in the case of a plane stressed state ($\sigma_z = 0$), the following equation is true:

$$\left(\frac{\partial^2}{\partial x^2} + \frac{\partial^2}{\partial y^2} \right) (\sigma_x + \sigma_y) = 0. \tag{9.3}$$

© The Author(s), under exclusive license to Springer Nature Switzerland AG 2021
V. Molotnikov, A. Molotnikova, *Theory of Elasticity and Plasticity*,
https://doi.org/10.1007/978-3-030-66622-4_9

The condition (9.2) equals the requirements of inadmissibility of structural distortions in the material during deformation at these stresses. With this assumption, it follows from the two last formulas:

$$\left(\frac{\partial^2}{\partial x^2} + \frac{\partial^2}{\partial y^2}\right)^2 U = 0. \tag{9.4}$$

Consequently, for an elastic plane, *the stress function is bi-harmonic.*

9.1.1 Example 1: Concentrated Force in the Wedge Apex

A non-limited wedge is exposed to the concentrated force P applied to its apex at the angle of α to the wedge axis (Fig. 9.1). Let us use the polar system of coordinates (r, ϑ).

Let us set the Airy function as

$$U = Ar\vartheta \sin \vartheta + Br\vartheta \cos \vartheta, \tag{9.5}$$

where A and B—constant and ϑ is the polar angle counted from the wedge axis. The function identically satisfies the bi-harmonic equation:

$$\left(\frac{\partial^2}{\partial r^2} + \frac{1}{r}\frac{\partial}{\partial r} + \frac{1}{r^2}\frac{\partial^2}{\partial \vartheta^2}\right)^2 U = 0. \tag{9.6}$$

Components of stresses

Fig. 9.1 Plane wedge strain

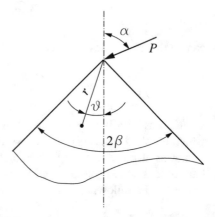

$$\sigma_r = \frac{1}{r^2}\frac{\partial^2 U}{\partial \vartheta^2} + \frac{1}{r}\frac{\partial U}{\partial r},$$

$$\sigma_\vartheta = \frac{\partial^2 U}{\partial r^2}, \tag{9.7}$$

$$\tau_{r\vartheta} = -\frac{\partial}{\partial r}\left(\frac{1}{r}\frac{\partial U}{\partial \vartheta}\right)$$

in this case will be

$$\sigma_r = \frac{1}{r}(A\cos\vartheta - B\sin\vartheta), \quad \sigma_\vartheta = \tau_{r\vartheta} = 0.$$

To determine constant values, let us use the equilibrium equations of the dissected part of the wedge:

$$P\cos\alpha + \int\limits_{-\beta}^{\beta} \sigma_r r \cos\vartheta\, d\vartheta = 0,$$

$$P\sin\alpha + \int\limits_{-\beta}^{\beta} \sigma_r r \sin\vartheta\, d\vartheta = 0.$$

Let us find as follows from the last conditions:

$$A = \frac{2P\cos\alpha}{2\beta + \sin 2\beta}, \quad B = \frac{2P\sin\alpha}{2\beta - \sin 2\beta}.$$

For the stress σ_r, we will finally obtain

$$\sigma_r = -\frac{2P}{r}\left(\frac{\cos\alpha\cos\vartheta}{2\beta + \sin 2\beta} + \frac{\sin\alpha\sin\vartheta}{2\beta - \sin 2\beta}\right). \tag{9.8}$$

9.1.2 Example 2: Wedge Bending by Uniform Pressure

Let a uniform pressure q be applied perpendicular to one of the faces of an unlimited wedge (Fig. 9.2). The stress function in this case will be

$$U = -\frac{qr^2}{2(\beta - \mathrm{tg}\,\beta)}\left(\beta - \vartheta + \sin\vartheta\cos\vartheta - \cos^2\vartheta\,\mathrm{tg}\,\beta\right). \tag{9.9}$$

By substituting this function into Eq. (9.6), we make sure that it is bi-harmonic. Let us find components of the stress tensor:

Fig. 9.2 Wedge bending by
uniform pressure

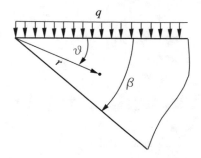

$$\sigma_r = \frac{q}{\beta - \operatorname{tg}\beta} \left(\vartheta - \beta + \operatorname{tg}\beta \sin^2 \vartheta + \sin \vartheta \cos \vartheta \right),$$

$$\sigma_\vartheta = \frac{q}{\beta - \operatorname{tg}\beta} \left(\vartheta - \beta + \operatorname{tg}\beta \cos^2 \vartheta - \sin \vartheta \cos \vartheta \right), \qquad (9.10)$$

$$\tau_{r\vartheta} = \frac{q \sin \vartheta}{\beta - \operatorname{tg}\beta} \left(\operatorname{tg}\beta \cos \vartheta - \sin \vartheta \right).$$

9.2   Complex Representation of a Bi-Harmonic Function

Any bi-harmonic function (U) under the Goursat formula can be expressed through
two analytical functions (φ and Ω) of a complex variable ($z = x + iy$):

$$U = Re\left[\bar{z}\varphi(z) + \Omega \right], \qquad (9.11)$$

where Re is the real part of the complex function enclosed by brackets and a dash
above means a complex conjugate value ($\bar{z} = x - iy$).

Let us consider an arbitrary arch AB in the area occupied by an elastic body; let
S be the arch length counted in the positive direction, which is direction from A to
B. Let us select the direction of the normal line to this arch to the right relative to
the viewer moving from A to B as a positive direction and designate components
of the force acting on the element (ds) of the arch from the external normal line as
follows:

$$X_n ds, \quad Y_n ds \qquad (9.12)$$

will have

$$X_n = \frac{\partial^2 U}{\partial y^2} \cos \widehat{nx} - \frac{\partial^2 U}{\partial x \partial y} \cos \widehat{ny},$$

$$Y_n = \frac{\partial^2 U}{\partial x^2} \cos \widehat{ny} - \frac{\partial^2 U}{\partial x \partial y} \cos \widehat{nx}.$$

These formulas can be written in a more compact way:

$$X_n = \frac{d}{ds}\frac{\partial U}{\partial y}, \quad Y_n = -\frac{d}{ds}\frac{\partial U}{\partial x},$$

or

$$(X_n + i Y_n)\, ds = id\left[\varphi(z) + z\overline{\varphi'(z)} + \psi(z)\right], \tag{9.13}$$

where

$$\psi(z) = \frac{d\Omega}{dz}. \tag{9.14}$$

Let the arch AB represent a body outline. The last formulas give boundary conditions using which we can define two unknown homomorphic functions φ and ψ. By selecting the element ds to be oriented along the axis Oy, we will obtain

$$X_x + i Y_x = \varphi'(z) + \overline{\varphi'(z)} - z\varphi''(z) - \overline{\psi'(z)}.$$

By selecting ds to be parallel to axis Ox, we will have

$$Y_y - i X_y = \varphi'(z) + \overline{\varphi'(z)} + z\overline{\varphi''(z)} + \overline{\psi'(z)}.$$

The last formulas can be represented in a different way:

$$\begin{aligned}
X_x + Y_y &= 2\left[\Phi(z) + \overline{\Phi(z)}\right], \\
Y_y - X_x + 2i X_y &= 2\left[\overline{z}\Phi'(z) + \Psi(z)\right],
\end{aligned} \tag{9.15}$$

where

$$\Phi(z) = \varphi'(z), \quad \Psi(z) = \psi'(z), \tag{9.16}$$

due to designations (9.12), X_x, Y_y, and X_y ($= Y_x$) represent σ_x, σ_y, and τ_{xy}, respectively.

9.3 Kolosov Displacement Integral

To define displacements to required directions, we should express deformations through displacements and use Hooke's law. Not to trouble the readers with the described calculations, let us give the final formula to define displacements (u, v):

$$2G(u + iv) = k\varphi(z) - z\overline{\varphi'(z)} - \overline{\psi(z)}, \tag{9.17}$$

where G is the shift modulus, φ and ψ are holomorphic functions of the complex variable, and k is the material constant, which is defined as follows:

$$k = \begin{vmatrix} 3 - 4v & \text{for } \varepsilon_z = 0, \\ \dfrac{3 - v}{1 + v} & \text{for } \sigma_z = 0. \end{vmatrix} \tag{9.18}$$

In a general case, finding holomorphic functions φ and ψ from boundary conditions represents a complicated problem that has been circumstantially studied by N. I. Muskhelishvili and is described in his monograph [9]. Therefore, the functions of φ, ψ, and Φ, Ψ are usually referred to as *the Muskhelishvili functions*.

9.4 Action of Concentrated Force

By integrating the stress (9.13) over the closed outline S in a clockwise manner, we obtain the main vector $(X + iY)$ of forces acting in the positive direction of the normal line to the outline:

$$i(X + iY) = \left[\varphi(z) + z\overline{\varphi'(z)} + \overline{\psi(z)} \right]. \tag{9.19}$$

In this case, square brackets mean an increment of the expression within when the point z runs through the closed outline S.

If the main vector of forces applied within S is not zero, the functions of $\varphi(z)$ and $\psi(z)$ are not uniquely defined.

Assume that the concentrated force (X, Y) acts in the infinite plane in the point $z = 0$ located inside the outline S. In this case, Muskhelishvili functions are discrete, but the requirement to the continuity of displacements implies a continuity (unambiguity) condition on the following combination (9.17) of these functions:

$$k\varphi(z) - z\varphi'(z) - \psi(z). \tag{9.20}$$

By using expression (9.20) in formula (9.19), we obtain

$$[\varphi(z)] = i\frac{X + iY}{1 + k}, \tag{9.21}$$

where square brackets are used in the above-mentioned sense.

When going along the outline S clockwise, the function $\ln z$ gives an increment $(-2\pi i)$, so

$$\varphi(z) = A \ln z, \tag{9.22}$$

where

$$A = -\frac{X + iY}{2\pi(1 + k)}.$$

Taking into account the continuity of expression (9.20), we also find

$$\psi(z) = B \ln z, \tag{9.23}$$

where

$$B = k\frac{X - iY}{2\pi(1 + k)}.$$

By differentiating the found functions $\varphi(z)$ and $\psi(z)$, we obtain

$$\Phi(z) = \frac{A}{z}, \quad \Psi(z) = \frac{B}{z}. \tag{9.24}$$

By using transformations [9] of the functions Φ and Ψ when adopting new coordinate axes (9.22) and (9.23), we will obtain the Muskhelishvili functions under the action of a concentrated force applied in an arbitrary point z_0 of an infinite plane:

$$\Phi(z) = \frac{A}{z - z_0}, \quad \Psi(z) = \frac{B}{z - z_0} + \frac{Az_0}{(z - z_0)^2}. \tag{9.25}$$

9.5   Solution of the First Principal Problem for a Circle

Let us place the reference point in the center of the circle and designate its radius as R. Let us consider elastic equilibrium of the area $|z| \leqslant R$ loaded at the boundary L by the self-equilibrated system of normal and tangential stresses

$$\sigma_r = N = f_1(\theta), \quad \tau_{r\theta} = T = f_2(\theta) \text{ for } |z| = R. \tag{9.26}$$

To simplify writing, let us solve the considered task for a single circle, for which we substitute the following designations:

$$z = R\zeta; \quad \zeta = \rho e^{i\theta} = \rho\sigma; \quad \sigma = e^{i\theta}; \quad d\sigma = ie^{i\theta}d\theta = i\sigma d\theta, \tag{9.27}$$

where ζ is an affix of a point inside a single circle and σ is a point belonging to the single circumference γ. Boundary conditions (9.26) on the outline γ are represented [6] as

$$\Phi(\sigma) + \overline{\Phi}\left(\frac{1}{\sigma}\right) - \frac{1}{\sigma}\overline{\Phi}'\left(\frac{1}{\sigma}\right) - \frac{1}{\sigma^2}\overline{\Psi}\left(\frac{1}{\sigma}\right) = f(\theta), \qquad (9.28)$$

where

$$f(\theta) = f_1(\theta) + if_2(\theta) = N + iT.$$

If we switch to conjugate values in formula (9.28), we obtain

$$\overline{\Phi}\left(\frac{1}{\sigma}\right) + \Phi(\sigma) - \sigma\Phi'(\sigma) - \sigma^2\Psi(\sigma) = \overline{f}(\theta). \qquad (9.29)$$

By multiplying both formulas (9.28) by

$$\frac{1}{2\pi i}\frac{d\sigma}{\sigma - \zeta}$$

and integrating along the outline of the single circumference γ, we obtain

$$\frac{1}{2\pi i}\left[\oint_\gamma \frac{\Phi(\sigma)}{\sigma - \zeta}d\sigma + \oint_\gamma \overline{\Phi}\left(\frac{1}{\sigma}\right)\frac{d\sigma}{\sigma - \zeta} - \oint_\gamma \overline{\Phi}'\left(\frac{1}{\sigma}\right)\frac{d\sigma}{\sigma(\sigma - \zeta)}\right.$$

$$\left. - \oint_\gamma \overline{\Psi}\left(\frac{1}{\sigma}\right)\frac{d\sigma}{(\sigma - \zeta)\sigma^2}\right] = \frac{1}{2\pi i}\oint_\gamma \frac{f d\sigma}{\sigma - \zeta}.$$

By using integral formulas of Cauchy and taking into account that $\overline{\Phi}\left(\frac{1}{\zeta}\right)$ is a holomorphic function beyond γ, we have from the last formula

$$\Phi(\zeta) = \frac{1}{2\pi i}\oint_\gamma \frac{(N + iT)d\sigma}{\sigma - \zeta} - a_0, \qquad (9.30)$$

where

$$a_0 = \overline{\Phi}(0) = \frac{1}{4\pi i}\oint_\gamma (N - iT)\frac{d\sigma}{\sigma}. \qquad (9.31)$$

Acting in a similar way with the condition (9.29), we obtain

$$\overline{\Phi}(0) + \Phi(\zeta) - \zeta\Phi'(\zeta) - \zeta^2\Psi(\zeta) = \frac{1}{2\pi i}\oint_\gamma \frac{\overline{f}(\theta)d\sigma}{\sigma - \zeta}. \qquad (9.32)$$

Then from formulas (9.30) and (9.32), we find

$$\Psi(\zeta) = -\frac{1}{2\pi i \zeta^2} \oint_{\gamma} \frac{(N - iT)d\sigma}{\sigma - \zeta} - \frac{\Phi'(\zeta)}{\zeta} + \frac{(\zeta)}{\zeta^2} + \frac{a_0}{\zeta^2}. \tag{9.33}$$

9.6   Annex to the Brazilian Test

Problem Statement Direct uniaxial compression tests of non-metallic brittle materials are related to the technical problem of application of elongating axial forces to the specimen. This problem is solved by the termination of specimen ends in special lugs using epoxy [7] or other adhesive materials, Wood's alloy [11] equipping specimen with such devices makes experiments much more expensive and in fact prevents mass or prompt testing.

For this reasons, an indirect method to determine the material elongation resistance proposed in 1947 by the Brazilian engineer F. Carneiro [10] is widely used in materials science and production. The method consists in that the material elongation strength is determined by a test of a cylindrical specimen for compression along a diametric plane evenly distributed along the generatrix by the load transmitted by a triangle prism rib. Tests in scientific publications were named the Brazilian test or Brazilian method. Regulatory documents (state standards, departmental instructions, etc.) *call the Brazilian test* as the "cleavage test."

Hertz Solution Standards for testing specimens of carbonic materials [4], concretes [1], and some other materials define the diametric compression of a cylindrical specimen under the scheme given in Fig. 9.3a. If we assume the specimen material to be elastic until destruction, for this scheme we have a problem of plane strain of a cylinder loaded by two diametrically opposite forces P (Fig. 9.4a).

It is known [9] that the solution to this problem was given by H. Hertz in 1883. Let us write the Muskhelishvili functions representing the solution to this problem as follows:[1] must be before the third addend in brackets

$$\Phi(\zeta) = -\frac{p}{2\pi R}\left(\frac{1}{1-\zeta} + \frac{1}{1+\zeta} - 1\right),$$

$$\Psi(\zeta) = \frac{p}{2\pi R}\left(\frac{1}{1-\zeta} + \frac{1}{1+\zeta} + \frac{1}{(1-\zeta)^2} + \frac{1}{(1+\zeta)^2}\right), \tag{9.34}$$

where $p = P/l$ is the loads per unit length l of the generatrix of the compressed cylindrical specimen with a radius R.

---

[1]There is a misprint in the formula for the function Ψ on page 298 of the paper [9] that we used to obtain formulas (9.34): the sign «+».

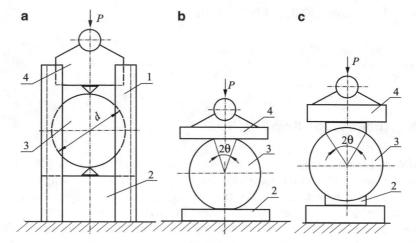

Fig. 9.3 Methods of load application in experiments under the Brazilian method: (**a**) initial contact along the line, (**b**) contact along the rectangular area, (**c**) contact along the cylindrical surface, 1— guides, 2—lower support plate, 3—specimen, 4—upper loading plate

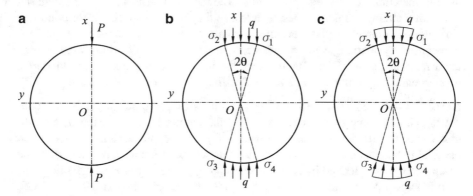

Fig. 9.4 Options of disc loading

Even Loading by the System of Parallel Forces The instructions for engineering surveys in mines [2] when preparing cores for tests by the Brazilian method require that two flat cuts 3...5 mm wide to be made along two diametrically opposite generatrixes on a flat grinding base plate. This implements the method and scheme of loading shown in Figs. 9.3b and 9.4b.

In this manner, this option of loading brings us to the first principal problem of elasticity theory for a circle loaded along symmetric arches $\sigma_1\sigma_2$ and $\sigma_3\sigma_4$ by the system of parallel evenly distributed loads q statically equivalent to the force P:

$$q = \frac{p}{2R\theta},\tag{9.35}$$

where 2θ is the angle tied up by the arch $\sigma_1\sigma_2$ (Fig. 9.4b).

Due to the low width of flat cuts, we can neglect the change in the shape of the specimen cross-section and substitute chords limiting the flat cuts with arches $\sigma_1\sigma_2$ and $\sigma_3\sigma_4$. Then we will use the general solution for the first problem of elasticity theory for a circle represented by formulas (9.30), (9.31), and (9.33). In the considered case [8]:

$$N = \begin{cases} -q\cos\varphi \text{ on } \sigma_1\sigma_2, \\ 0 \text{ on } \sigma_2\sigma_3, \\ q\cos\varphi \text{ on } \sigma_3\sigma_4, \\ 0 \text{ on } \sigma_4\sigma_1; \end{cases} \qquad T = \begin{cases} q\sin\varphi \text{ on } \sigma_1\sigma_2, \\ 0 \text{ on } \sigma_2\sigma_3, \\ -q\sin\varphi \text{ on } \sigma_3\sigma_4, \\ 0 \text{ on } \sigma_4\sigma_1, \end{cases}$$

where φ is a polar angle counted from the positive half-axis Ox in the counterclockwise direction.

Using the last results and having in mind that

$$\cos\varphi + i\sin\varphi = e^{i\varphi} = \sigma, \ \sigma_1 = e^{-i\theta}, \ \sigma_2 = e^{i\theta}, \ \sigma_3 = e^{i(\pi-\theta)}, \ \sigma_4 = e^{i(\pi+\theta)},$$

we can write as

$$N + iT = \begin{cases} -\dfrac{q}{\sigma} \text{ on } \sigma_1\sigma_2, \\ 0 \text{ on } \sigma_2\sigma_3 \text{ and } \sigma_4\sigma_1, \\ \dfrac{q}{\sigma} \text{ on } \sigma_3\sigma_4. \end{cases} \tag{9.36}$$

Adopting conjugate values in formula (9.36), we obtain

$$N - iT = \begin{cases} -q\sigma \text{ on } \sigma_1\sigma_2, \\ 0 \text{ on } \sigma_2\sigma_3 \text{ and } \sigma_4\sigma_1, \\ q\sigma \text{ on } \sigma_3\sigma_4. \end{cases} \tag{9.37}$$

By substituting (9.36) and (9.37) into the solution (9.30), (9.31), and (9.33) and after calculating integrals, we will find

$$a_0 = -\frac{q\sin\theta}{\pi};$$

$$\Phi(\zeta) = -\frac{q}{2\pi i\zeta}\ln\frac{(\sigma_1+\zeta)(\sigma_2-\zeta)}{(\sigma_1-\zeta)(\sigma_2+\zeta)} + \frac{q\sin\theta}{\pi},$$

$$\Psi(\zeta) = -\frac{q}{2\pi i\zeta^2}\left[2(\sigma_1-\sigma_2) + \zeta\ln\frac{(\sigma_1-\zeta)(\sigma_2+\zeta)}{(\sigma_1+\zeta)(\sigma_2-\zeta)} - \right.$$

$$\left. -\frac{2(\sigma_2-\sigma_1)(1+\zeta^2)}{(\sigma_1^2-\zeta^2)(\sigma_2^2-\zeta^2)} + \frac{2}{\zeta}\ln\frac{(\sigma_1+\zeta)(\sigma_2-\zeta)}{(\sigma_1-\zeta)(\sigma_2+\zeta)}\right]. \tag{9.38}$$

Even Radial Load The standard [3] for testing of chemically resistant ceramic materials under the Brazilian method recommends the specimen loading scheme given in position c in Figs. 9.3 and 9.4. In this case,

$$N + iT = N - iT = \begin{cases} -q \text{ on } \sigma_1\sigma_2 \text{ and } \sigma_3\sigma_4, \\ 0 \text{ on } \sigma_2\sigma_3 \text{ and } \sigma_4\sigma_1, \end{cases} \tag{9.39}$$

we assume from the condition of static equivalence of loads at radial distributed load as follows:

$$q = \frac{p}{2R\sin\theta}. \tag{9.40}$$

By substituting the function (9.39) into formulas (9.30), (9.31), and (9.33) and calculating the respective integrals, we obtain Muskhelishvili functions for a circle bearing axisymmetric even load on the arches $\sigma_1\sigma_2$? $\sigma_3\sigma_4$:

$$\Phi(\zeta) = -\frac{q}{2\pi i}\ln\frac{\sigma_2^2 - \zeta^2}{\sigma_1^2 - \zeta^2} + \frac{q\theta}{\pi},$$

$$\Psi(\zeta) = -\frac{q}{2\pi i}\cdot\frac{2(\sigma_1^2 - \sigma_2^2)}{(\sigma_1^2 - \zeta^2)(\sigma_2^2 - \zeta^2)}, \tag{9.41}$$

where q is defined by formula (9.40).

Conclusions Figure 9.5 gives curves of normal stresses in diametrical points located on the axis Ox. The adopted designations are

$$\tilde{X}_x(\zeta) = X_x(\zeta)\cdot\frac{\pi d}{p}; \quad \tilde{Y}_y(\zeta) = Y_y(\zeta)\cdot\frac{\pi d}{p},$$

where the stress components X_x and Y_y are defined by equations (9.15) with known $\Phi(\zeta)$ and $\Psi(\zeta)$. In all positions of Fig. 9.5, a continuous straight line depicts the component $\tilde{Y}_y$ under the action of concentrated forces P, and curves depicted with dots show the component $\tilde{X}_x$ of the same load. Dashed curves depict the component $\tilde{Y}_y$ when loading the disc by the system of parallel forces (positions a and c) or radial loads (positions b and d), and dash-and-dot curves show the component $\tilde{X}_x$ at these loads. For curves in the positions a and b, the arches $\sigma_1\sigma_2$ and $\sigma_3\sigma_4$ are taken equal to 5^0, and in the positions c and d, they are equal to 10^0.

The analysis of the curves represented in Fig. 9.5 leads [8] to the following conclusions:

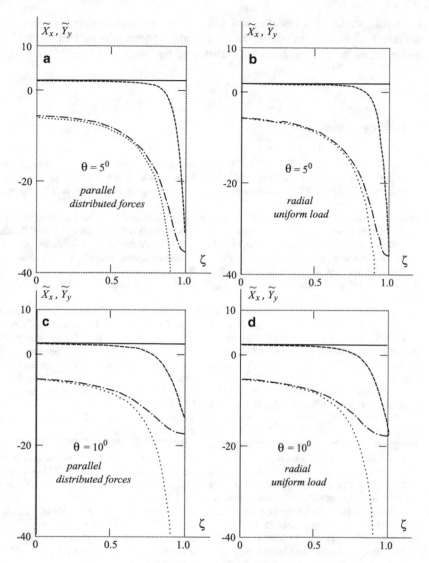

Fig. 9.5 Stress curves for various methods of applying statically equivalent loads

1. Normal stress in the diametrical plane of specimen crushing in the Brazilian method is not the material's ultimate tensile strength and for some brittle bodies can be several times less than the specified ultimate strength.
2. Despite the solid identification of this break-off stress and ultimate tensile strength, it is reasonable (at least in educational materials) to underline the non-identity of these two characteristics of material strength.
3. The method of load application when crushing the specimen has no significant influence on the normal break-off stress during destruction.

4. For strong rocks with low ratios of ultimate tensile and compressive strength, the break-off stress in the destruction plane almost coincides with the equivalent Mohr stress, so the value of this stress can be taken as the ultimate tensile strength.

References

1. *Betony. Metody opredeleniya prochnosti po kontrol'nym obraztsam* (Concretes. Methods for determining the strength of control samples)n.d. *Betony. Metody opredeleniya prochnosti po kontrol'nym obraztsam* (Concretes. Methods for determining the strength of control samples) (n.d.)
2. *Instruktsiya po inzhenernym izyskaniyam v gornykh vyrabotkakh, prednaznachaemykh dlya razmeshcheniya ob"ektov narodnogo khozyaistva* (Instructions for engineering surveys in mine workings intended for the placement of objects of the national economy) 1976 *Instruktsiya po inzhenernym izyskaniyam v gornykh vyrabotkakh, prednaznachaemykh dlya razmeshcheniya ob"ektov narodnogo khozyaistva* (Instructions for engineering surveys in mine workings intended for the placement of objects of the national economy). (Min. of the USSR on Affairs of construction Publ., Moscow, 1976)
3. *Izdeliya khimicheski stoikie i termostoikie keramicheskie. Metod opredeleniya predela prochnosti pri razryve: GOST 473.7–81. Vzamen GOST 473.7–72; vved. 1982–06–01* (Chemical-resistant and heat-resistant ceramic products. The method of determining the tensile strength at break: GOST 473.7–81. Instead of GOST 473.7–72; introduced 1982–06–01)n.d. *Izdeliya khimicheski stoikie i termostoikie keramicheskie. Metod opredeleniya predela prochnosti pri razryve: GOST 473.7–81. Vzamen GOST 473.7–72; vved. 1982–06–01* (Chemical-resistant and heat-resistant ceramic products. The method of determining the tensile strength at break: GOST 473.7–81. Instead of GOST 473.7–72; introduced 1982–06–01) (n.d.)
4. *Izdeliya uglerodnye. Metody opredeleniya predela prochnosti na szhatie, izgib, razryv (diametral'noe szhatie): GOST 23775–79. – Vved. 1981.01.01.* (Carbon products. Methods for determining the ultimate strength for compression, bending, and rupture (Diametric compression): GOST 23775–79. Entered 1981.01.01.)n.d. *Izdeliya uglerodnye. Metody opredeleniya predela prochnosti na szhatie, izgib, razryv (diametral'noe szhatie): GOST 23775–79. Vved. 1981.01.01.* (Carbon products. Methods for determining the ultimate strength for compression, bending, and rupture (Diametric compression): GOST 23775–79. Entered 1981.01.01.) (n.d.)
5. M. Lavrent'ev, B. Shabat, *Metody teorii funktsii kompleksnogo. Izd. 3-e* (Methods of the theory of functions of a complex variable. Ed. 3). (Nauka Publ., Moscow, 1965)
6. A. Lur'e, *Teoriya uprugosti* (Theory of elastic strength). (Nauka Publ., Moscow, 1970)
7. V. Molotnikov, *Metody modelirovaniya v zemledel'cheskoi mekhanike* (Modeling methods in agricultural mechanics). Ph.D. thesis, Sankt-Peterburg, ASFI, 1994
8. V. Molotnikov, A. Molotnikova, *Zamechaniya k brazil'skomu metodu opredeleniya predela prochnosti pri rastyazhenii khrupkikh materialov* (Notes on the Brazilian method for determining the tensile strength of brittle materials). *Vestn. Donsk. gos. tekhn. un-ta* (Bulletin of the don state technical University) 4(79), 30–38 (2014)
9. N. Muskhelishvili, *Nekotorye osnovnye zadachi matematicheskoi teorii uprugosti* (Some main problems of mathematical theory elasticity). (Nauka Publ., Moscow, 1966)
10. T. Rodriguez, C. Navarro, V. Sanchez-Galvez, Splitting tests: an alternative to determine the dynamic tensile strength of ceramic materials. J. Phys. **IV**, 101–106. (1994)
11. A. Spivak, A. Popov, *Razrushenie gornykh porod pri burenii skvazhin: uchebn. dlya vuzov* (Destruction of rocks during well drilling: training. for higher education institutions). (Nedra Publ., Moscow, 1986)

Chapter 10
Mathematical Structural Imperfections

10.1 Mathematical and Physical Theories of Structural Imperfections

In multiply-connected bodies such as a hollow cylinder, many-valued displacements are possible. This circumstance was first noted by Weingarten [20] early in the twentieth century. Timpe [18] studied this phenomenon in detail on flat systems such as annulus. A more general mathematical theory of structural imperfections was later developed by Volterra [19] who introduced the term of "distortions" (distorsioni) for this type of strain, which was later substituted by Love [10] for "dislocations" (dislocation).

In solid body physics, dislocation theory occurred somewhat later than mathematical theory. The founders [7] of the physical theory of dislocations were Orowan [15], Taylor [17], Burgers [21] et al. Physically, a dislocation was represented as a distortion of a correct crystalline lattice of a solid body. These representations were experimentally proved when electronic microscopes appeared [6]. Another method for studying dislocations was growing crystals with defects of a structure (image of a growth spiral in the paraffin crystal at the outlet of a screw-type dislocation is shown in Fig. 10.2a). It is important to note that a distinctive feature of dislocations, unlike other defects in crystals, is the significant distortion of regular atom ordering in the small vicinity of some line piercing the crystal.

Researchers pay the highest attention to one-dimensional (linear) defects whose size in one direction is much more than the lattice parameter and is comparable with it in two other directions. Linear defects include *dislocations* and *disclinations*. The simplest type of linear defects are *boundary* and *screw-type* dislocations.

A boundary dislocation can be represented (Fig. 10.1) as an introduction (or removal) of one half-plane in the crystal lattice. In this case, planes surrounding the defects will not be straight (Fig. 10.1a). They will envelope the boundary of the implemented (or removed) half-plane so that the lattice structure on crystal faces will not be distorted and the defects will not be seen. Figure 10.1b gives [6] an

V. Molotnikov, A. Molotnikova, *Theory of Elasticity and Plasticity*,
https://doi.org/10.1007/978-3-030-66622-4_10

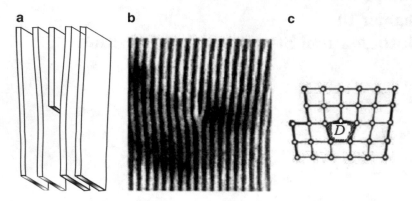

Fig. 10.1 Boundary dislocation: (**a**) introduction "of a spare" atomic half-space; (**b**) electronic microscope image; (**c**) flat scheme of boundary dislocation

electronic microscope image of the dislocation in the crystal, and the position c designates the atom arrangement near the core D.

Let us consider the formation mechanism of a screw-type dislocation. Let us mentally make a cut in the crystal along the plane $ABCD$ (Fig. 10.2a) and move the front right part for one period of the lattice downwards. The step formed during this shift on the upper face does not go across the entire crystal width and ends in the point B. As a result of such operation, the cubic lattice looks as shown in Fig. 10.2b, c.

The shift near the front edge of the crystal was one period, so the upper atomic plant to the right from the point A coincides with the second plane of the left from the point A. Since the cut $ABCD$ reached only the middle of the crystal, the right part of the crystal cannot be fully moved relative to the left part for one period of the lattice, and shift of the right part relative to the left part ends in the point B. The upper atomic plane becomes bent. The same bending of atomic planes occurs in all underlying planes. If the crystal consisted of parallel horizontal atomic layers before the shift, after a non-through shift along the plane $ABCD$ the crystal turns into atomic planes twisted as a helix (helical stairs).

After the crystal shift along the plane $ABCD$ at a distance from the line BC, the lattice remains imperfect, and the imperfection area goes along it near the line BC. The imperfection area along the line BC is comparable with the crystal size, and the perpendicular to the line BC is several lattice periods, so when shifting along the plane $ABCD$ around the line BC, linear imperfection occurs. They say that imperfections around the line BC are structural imperfections or dislocations. Since this dislocation consists in atomic planes twisted into helical stairs, it is called screw-type dislocation. A precise location of atoms in the core of the screw-type dislocation is unknown.

As a thread, the screw-type dislocation can be right-hand or left-hand. The line of the right-hand dislocation from the upper horizon to the lower one must be covered clockwise. If we shift the left part of the crystal downwards along the plane $ABCD$,

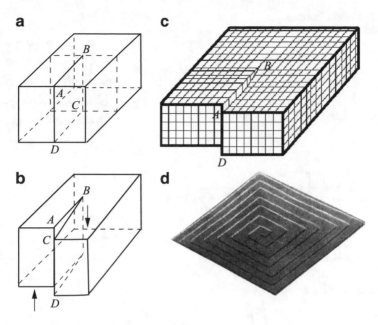

Fig. 10.2 Formation scheme of screw-type dislocation

a left-hand dislocation would form. The right-hand dislocation can be turned into a left-hand one by simply turning the crystal.

Both mathematical and physical theories of dislocations are widely applied in the reinforcement analysis in plastic strain, the study of polygonization, recrystallizing, creep and other important phenomena occurring in metals and alloys.

However, dislocation theory cannot be deemed universal and complete. This is supported by multiple ongoing studies both in Russia and abroad. A sufficiently compressive overview of the latest works in this field can be found in publications by L. M. Zubov [22, 23], theses by S. V. Derezin [1] et al.

10.2 Edge Dislocation in an Infinite Body

Let us place the beginning of the coordinate system $Oxyz$ with the dislocation core (Fig. 10.3) and direct the Ox axis parallel to the burgers vector [5]. The Oy axis is located in the "extra" atomic half-plane, and the third coordinate axis Oz is directed so that the orts of the selected system form the right three (Fig. 10.3).

With this choice of coordinate axes, the displacement components u, v, and w will be

Fig. 10.3 Model of an edge
dislocation

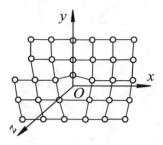

$$u = \frac{b}{2\pi}\left[\arctan\frac{y}{x} + \frac{1}{2(1-v)}\frac{xy}{x^2+y^2}\right],$$

$$v = -\frac{b}{8\pi(1-v)}\left[(1-2v)\ln(x^2+y^2) + \frac{x^2-y^2}{x^2+y^2}\right], \tag{10.1}$$

$$w = 0,$$

or in polar coordinates (r, φ):

$$u = \frac{b}{2\pi}\left[\varphi + \frac{\sin^2\varphi}{4(1-v)}\right],$$

$$v = -\frac{b}{2\pi}\left[\frac{1-2v}{2(1-v)}\ln r + \frac{\cos 2\varphi}{4(1-2v)}\right], \tag{10.2}$$

$$w = 0,$$

where b is the module of the burgers vector and

$$\varphi = \text{arctg}\frac{y}{x}.$$

Since there is no movement in the direction of the dislocation line, there is
a flat deformation. For known displacements by the formulas (2.2)–(2.3), the
deformations ε_x, ε_y, and γ_{xy} are determined, and then by Hooke's law (1.20)–
(1.21)—stress components

$$\sigma_x = -\frac{Gb}{2\pi(1-v)}\frac{y(3x^2+y^2)}{(x^2+y^2)^2},$$

$$\sigma_y = \frac{Gb}{2\pi(1-v)}\frac{y(x^2-y^2)}{(x^2+y^2)^2}, \tag{10.3}$$

$$\tau_{xy} = \frac{Gb}{2\pi(1-v)}\frac{x(x^2-y^2)}{(x^2+y^2)^2}.$$

In the polar coordinates of the formula (10.3) take the form

$$\sigma_x = -\frac{Gb}{2\pi(1-v)}\frac{b}{r}(2+\cos 2\varphi)\sin\varphi,$$

$$\sigma_y = \frac{Gb}{2\pi(1-v)}\frac{\sin\varphi\cos 2\varphi}{r}, \qquad (10.4)$$

$$\tau_{xy} = \frac{Gb}{2\pi(1-v)}\frac{\cos\varphi\cos 2\varphi}{r}.$$

The largest normal stress σ_x acts along the Ox axis and is tensile at $y < 0$ and compressive at $y > 0$. This is also seen in the model of crystal lattice distortions in the vicinity of the dislocation core, shown in Fig. 10.3.

10.3 Mathematical Wedge-Shaped Dislocation

We cut out a wedge from an unbounded elastic plane with apex at the point $(x_0; 0)$. The banks of the formed section are spaced from each other at a distance of $\delta = \varepsilon(x - x_0)$, $(\varepsilon - const)$. Now let us combine the points of the opposite edges of the cut, equidistant from the top of the wedge, and glue the plane material. The resulting structural imperfection is called a [9] *wedge-shaped (wedge) dislocation.* In this case, a plane stress state arises, axisymmetric with respect to the point $(x_0; 0)$.

Let us take this point as the origin of the polar coordinate system (r, α). Using the solution of the Lame problem (p. 18), we write down the radial displacement component

$$u_r = \frac{1-v}{4\pi}\varepsilon r \ln Ar + \frac{B}{r}, \qquad (10.5)$$

where A and B are constants, and in this case $B = 0$ should be set from the condition of bounded displacement at the origin, and the parameter ε is called the power of the wedge dislocation.

The constant A corresponds to uniform axially symmetric tension. Next, we put $A = 1$.

We find the tangential component of the displacement (u_α) from the condition that any circle centered at the origin of coordinates deforms uniformly along its length (axisymmetric stress state). Assuming that there is no rotation of the cut, we find

$$u_\alpha = -\frac{\varepsilon r}{2}\left(1 - \frac{\alpha}{\pi}\right), \quad (0 < \alpha < 2\pi), \qquad (10.6)$$

where

$$\alpha = \operatorname{arctg} \frac{y}{x - x_0}, \quad r = \sqrt{(x - x_0)^2 + y^2}. \tag{10.7}$$

10.4 Mathematical Biclination

Let, as in the previous paragraph, at the point $(x_0, 0)$ of an unbounded elastic plane, there is a wedge dislocation of power ε. We place at the point $(x_0 + \Delta, 0)$ a wedge dislocation of power $-\varepsilon$. Then the components $(\overline{u}, \overline{v})$ of displacement in the direction of the axes Ox and Oy from two dislocations at $\Delta \to 0$ will be

$$\overline{u} = \frac{\partial u}{\partial x} \Delta, \quad \overline{v} = \frac{\partial v}{\partial x} \Delta,$$

where u and v are the components of movement in the direction of the Ox and Oy axes for a wedge dislocation, defined by the formulas

$$u = u_r \cos \alpha - u_\alpha \sin \alpha, \quad v = u_r \sin \alpha + u_\alpha \cos \alpha. \tag{10.8}$$

Denoting further $\varepsilon \Delta = b$, we get

$$\begin{aligned} \overline{u} &= \frac{b}{2\pi} \left[\frac{1 - v}{2} \left(\ln r + \cos^2 \alpha \right) + \sin^2 \alpha \right], \\ \overline{v} &= -\frac{1 + v}{8\pi} b \sin 2\alpha - \frac{b}{2} \left(1 - \frac{\alpha}{\pi} \right). \end{aligned} \tag{10.9}$$

From the formula (10.9), it follows that the displacement component $\overline{v}$ for $\alpha = 2\pi$ exceeds by b its value for $\alpha = 0$, i.e. there is a gap of displacements on the semi-direct $x > 0$, $y = 0$. The latter means that at the point $(0, 0)$ the edge dislocation considered above (p. 103) is created with the burgers vector b parallel to the Oy axis.

10.5 Flat Dislocation of Somigliana

Assume a section is made along some curve L in an unlimited elastic plane. The section edges can be split and material is inserted into the slit, or material can be cut off from the edges of the cut. If the edges of such a slit are than glued by preliminary compressing and shifting them relative to each other, this structural imperfection is referred to as [2, 9, 12] *the flat dislocation of Somigliana*. Let us set the displacement difference on the curve L after gluing as

$$g(l) = g_1(l) + ig_2(l), \quad (i^2 = -1), \tag{10.10}$$

where l is the arch length of the curve L counted from the starting point to the considered point.

$$g_1(l) = u^-(l) - u^+(l), \quad g_2(l) = v^-(l) - v^+(l), \tag{10.11}$$

whereas $u^\pm$ and $v^\pm$ are the ultimate values of the components of displacement along the axes Ox and Oy to the left $(+)$ and right $(-)$ from the curve L when moving along this curve from its start.

The definition of the Somigliana dislocation shows that a ruptural deformation (10.10)–(10.11) can be represented as a result of location of biclination in the initial point $l = 0$ with the Burgers vector $b(0) = g_1(0) + ig_2(0)$ and biclinations distributed along the length of the curve L whose density will be

$$b(l) = g_1'(l) + ig_2'(l), \tag{10.12}$$

where derivatives are taken in the direction of the tangential line to the curve L.

Consequently, it can be deemed that biclinations are distributed on the element dl of the curve L whose total power (total Burgers vector) is the vector with the components

$$X = g_1'(l)dl, \quad Y = g_2'(l)dl. \tag{10.13}$$

Note It can be shown that a wedge dislocation located in some point O is equivalent to the system of biclinations evenly distributed with the density of ε along any half-straight line coming from the point O with Burgers vectors normal to this line. Indeed, by representing the biclination with a pair of wedge dislocations with an opposite sign, we can make a conclusion that structural imperfections inside the area of even distribution of biclinations are mutually destroyed (wedge-like dislocation remains at the beginning of the half-straight line).

10.6 Somigliana Dislocation in Half-Plane

Assume that L is the beam O_1y_1 coming from the point O_1 (Fig. 10.4a) located at the distance H from the half-plane boundary and making an angle α with a positive direction of the axis Ox. Moreover, the following conditions are met indexDislocation!Somigliana!in a half plane

$$-\frac{\pi}{2} < \alpha < \frac{\pi}{2},$$
$$u_1^+ = v_1^+ = u_1^- = 0; \quad v_1^- = \beta, \quad (\beta - const), \tag{10.14}$$

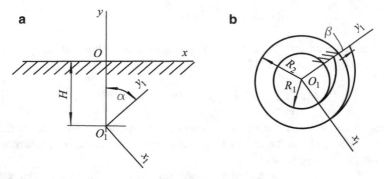

Fig. 10.4 Somigliana dislocation in half-plane (**a**) and in annulus (**b**)

where $u_1^\pm$, $v_1^\pm$ mean the ultimate values of the components of displacement along the axes Ox and Oy to the left $(+)$ and right $(-)$ from the beam $O_1 y_1$ while moving in the positive direction of the axis $O_1 y_1$.

There is a problem: find elastic equilibrium of the half-plane $y \leqslant 0$ (Fig. 10.4b) in the conditions of (10.14)

10.6.1 Functions Φ, Ψ for the Plane with Dislocation

Let us use the solution to the problem for an annulus with an internal radius R_1 and external radius R_2 (Fig. 10.4b) where a cut is made along the axis $O_1 y_1$ and the right edge of the cut is moved relative to the left secured edge by the value of β in the direction indicated by the arrow in Fig. 10.4b. After the shift, cut edges are soldered together and the left edge is released from securing.

Muskhelishvili functions for this ring are given [14] by the formulas[1]

$$\Phi(z_1) = -\frac{(\lambda + \mu)}{2\pi(\lambda + 2\mu)} \left[\frac{2z_1}{R_1^2 + R_2^2} - \frac{1}{z_1} \right],$$

$$\Psi(z_1) = \frac{(\lambda + \mu)}{2\pi(\lambda + 2\mu)} \left[\frac{1}{z_1} + \frac{2R_1^2 R_2^2}{R_1^2 + R_2^2} \cdot \frac{1}{z_1^3} \right], \tag{10.15}$$

$$(z_1 = x_1 + iy_1),$$

---

[1]In the fifth edition of the monograph [14] the formula for the function $\Psi(z_1)$ contains misprints.

where[2]

$$\lambda = \frac{E\nu}{(1+\nu)(1-2\nu)}, \quad \mu = \frac{E}{2(1+\nu)};$$

as previously (p. 9), E is the Young modulus, and ν is the Poisson coefficient.

By making passages to the limit in formula (10.15) at $R_1 \to 0$, $R_2 \to \infty$, we will obtain Muskhelishvili functions for an infinite plane with a dislocation introduced along the axis $O_1 y_1$:

$$\Phi_1(z_1) = \Psi_1(z_1) = \frac{k}{z_1}, \tag{10.16}$$

where the following designation is introduced for short

$$k = \frac{(\lambda + \mu)}{2\pi(\lambda + 2\mu)}. \tag{10.17}$$

Using the known [14] formulas of conversion of Muskhelishvili functions when replacing rectangular coordinate axes, let us write functions (10.16) in the axes xOy (Fig. 10.4a). To do it, let us at first turn the axes $x_1 O_1 y_1$ by the angle α so that the new Y-axis takes a vertical position and is directed from bottom to top. Then we will make a parallel transfer of axes by the value of H (Fig. 10.4a) vertically upwards. As a result, for an infinite plane with dislocation, we obtain

$$\Phi_2(z) = \frac{k}{(z+iH)e^{i\alpha}}, \quad (z = x+iy),$$

$$\Psi_2(z) = \frac{k[(z+2iH)\cos\alpha + iz\sin\alpha]}{(z+iH)^2}. \tag{10.18}$$

10.6.2 Functions Φ, Ψ for a Half-Plane with Dislocation

The components of stresses corresponding to functions (10.18) are designated as $\widetilde{X}_x$, $\widetilde{Y}_y$, and $\widetilde{X}_y$. They are related to functions (10.18) by dependencies (9.15)

$$\widetilde{Y}_y + \widetilde{X}_x = 2\left[\Phi_2(z) + \overline{\Phi_2(z)}\right],$$

$$\widetilde{Y}_y - \widetilde{X}_x + 2i\widetilde{X}_y = 2\left[\bar{z}\Phi_2'(z) + \Psi_2(z)\right], \tag{10.19}$$

where the straight line above still (p. 90) means a complex conjugate value, and the dash means the differentiation operation under the argument z.

---

[2]Considered plane strain.

From formulas (10.18)–(10.19), we obtain the stress field in the infinite half-plane with a straight line dislocation of Somigliana type with a core in the point O_1 and line L coinciding with the beam O_1y_1:

$$\widetilde{X}_x = \frac{k[x\cos\alpha - 3(y+H)\sin\alpha]}{x^2 + (y+H)^2} +$$

$$+k\frac{[x^2 + 3(y+H)^2]x\cos\alpha - [3x^2 - (y+H)^2](y+H)\sin\alpha}{[x^2 + (y+H)^2]^2},$$

$$\widetilde{Y}_y = \frac{k[3x\cos\alpha - (y+H)\sin\alpha]}{x^2 + (y+H)^2} +$$

$$+k\frac{[3(y+H)^2 - x^2]x\cos\alpha + [3x^2 - (y+H)^2](y+H)\sin\alpha}{[x^2 + (y+H)^2]^2},$$

$$\widetilde{X}_y = \frac{k[x\sin\alpha - (y+H)\cos\alpha]}{x^2 + (y+H)^2} +$$

$$+k\frac{[3x^2 - (y+H)^2](y+H)\cos\alpha + [x^2 - 3(y+H)^2]x\sin\alpha}{[x^2 + (y+H)^2]^2}.$$

$$(10.20)$$

Now let us route the normal $(N(x))$ and tangential $(T(x))$ loads along the axis x equal in magnitude to the normal $\widetilde{Y}_y$ and tangential $\widetilde{X}_y$ stresses on the axis x taken with opposite signs. Assuming in the second and third formulas (10.20) $y = 0$ and changing the signs, we obtain

$$N(x) = k\frac{H\sin\alpha - 3x\cos\alpha}{x^2 + H^2} -$$

$$-k\frac{(3H^2 - x^2)x\cos\alpha + (3x^2 - H^2)H\sin\alpha}{(x^2 + H^2)^2},$$

$$T(x) = k\frac{H\cos\alpha - x\sin\alpha}{x^2 + H^2} -$$

$$-k\frac{(3x^2 - H^2)H\cos\alpha + (x^2 - 3H^2)x\sin\alpha}{(x^2 + H^2)^2}.$$

$$(10.21)$$

According to Galin [3], let us designate

$$\omega_1(z) = \frac{1}{2\pi i}\int_{-\infty}^{\infty}\frac{N(\zeta)d\zeta}{\zeta - z}, \quad \omega_2(z) = \frac{1}{2\pi i}\int_{-\infty}^{\infty}\frac{T(\zeta)d\zeta}{\zeta - z}, \quad (z \in S^-), \qquad (10.22)$$

where S^- designates the area located below the axis x, for all points of which $y < 0$.

Muskhelishvili functions $[\Phi_3(z), \Psi_3(z)]$ for a half-plane loaded at the boundary $y = 0$ by forces (10.21) are expressed [14] through Galin functions (10.22)

Fig. 10.5 Integration outline

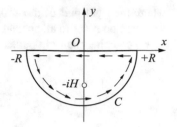

$$\Phi_3(z) = -\omega_1(z) + i\omega_2(z),$$

$$\Psi_3(z) = -2i\omega_2(z) + z[\omega_1'(z) - i\omega_2'(z)].$$

$$(10.23)$$

10.6.3 Calculation of Galin Functions

By analyzing formulas (10.21), one can easily make sure that in integrals of Cauchy type (10.22), the functions $N(z)$ and $T(z)$ are holomorphic everywhere in S^-, except the point $z = -iH$.

Let us consider, for example, integral calculation in the first of the formulas (10.22). Let us make an arch of a semi-circumference in the lower half-plane $y \leqslant 0$ with the center in the origin of the coordinate system xOy and radius $R > H$, (Fig. 10.5). Let us calculate the integral

$$F(z) = \frac{1}{2\pi i} \oint_\Gamma \frac{N(\zeta)d\zeta}{\zeta - z}, \qquad (10.24)$$

where Γ is a closed outline consisting of a segment $[-R; +R]$ of the X-axis and an arch of the semi-circumference C (Fig. 10.5). The outline Γ will be found in the positive direction so that the part of the area S^- confined by them when going over Γ is located to the left all the time. In Fig. 10.5, the travel direction Γ is shown with arrows

Due to the adaptivity of the integral, let us substitute formula (10.24) as follows:

$$F(z) = -\frac{1}{2\pi i} \int_{-R}^{+R} \frac{N(\zeta)d\zeta}{\zeta - z} + \frac{1}{2\pi i} \int_C \frac{N(\zeta)d\zeta}{\zeta - z}. \qquad (10.25)$$

On the other side, according to the theorem of deductions [16]

$$F(z) = \Sigma \, res \, N(a_i), \qquad (10.26)$$

where the right part of the equation symbolically writes the sum of deductions of the function $N(z)$ in all special points a_i of the area limited by the outline Γ.

By equating the right parts of equations (10.25) and (10.26), we will obtain

$$\frac{1}{2\pi i}\int_{-R}^{+R}\frac{N(\zeta)d\zeta}{\zeta-z} = \frac{1}{2\pi i}\int_{C}\frac{N(\zeta)d\zeta}{\zeta-z} - \Sigma\,res\,N(a_i). \tag{10.27}$$

Using the first of the formulas (10.21), it can be easily ensured that for $R \to \infty$, the integral in the right part of formula (10.27) tends to zero. By making the passage to the limit in formula (10.27), taking into account designations (10.22), we will obtain

$$\omega_1(z) = -\Sigma\,res\,N(a_i). \tag{10.28}$$

From formula (10.28) after calculation and summing of deductions, let us find

$$\omega_1(z) = k\frac{(z-2iH)\cos\alpha + H\sin\alpha}{(z-iH)^2}. \tag{10.29}$$

In a similar way, let us find

$$\omega_2(z) = k\frac{H\cos\alpha + z\sin\alpha}{(z-iH)^2}. \tag{10.30}$$

10.6.4   Completion of Problem Solution

Substituting functions (10.29)–(10.30) and the respective derivatives into formulas (10.23) gives Muskhelishvili functions for the half-plane on which boundary distributed loads (10.21) are applied:

$$\Phi_3(z) = k\frac{(3iH - z)\cos\alpha + i(z + iH)\sin\alpha}{(z-iH)^2},$$

$$\Psi_3(z) = k\frac{(3iHz - 2H^2 - z^2)\cos\alpha - (iz^2 + 5Hz)\sin\alpha}{(z-iH)^3}. \tag{10.31}$$

The components of stresses corresponding to these functions are designated by the upper index "0". Using the formulas of type (24.6) and dependencies (10.31), we obtain

$$X_x^0 = \frac{2k\{2x(y-H)V_1 - V_2[x^2 - (y-H)^2]\}}{[x^2 + (y-H)^2]^2} -$$

$$- \frac{2ky\{x[3(y-H)^2 - x^2]V_3 + (y+H)[(y-H)^2 - 3x^2]V_4\}}{[x^2 + (y-H)^2]^3};$$

$$Y_y^0 = -\frac{2k\{V_5[x^2 - (y-H)^2] + 2x(y-H)(y-2H)\cos\alpha\}}{[x^2 + (y-H)^2]^2} +$$

$$+ \frac{2ky\{x[3(y-H)^2 - x^2]V3 + (y+H)[(y-H)^2 - 3x^2]V6\}}{[x^2 + (y-H)^2]^3};$$

$$X_y^0 = -\frac{2k\{(H\cos\alpha + x\sin\alpha)[x^2 - (y-H)^2] + 2xy(y-H)\sin\alpha\}}{[x^2 + (y-H)^2]^2} -$$

$$- \frac{2ky\{(y-H)[(y-H)^2 - 3x^2]V_8 + x[x^2 - 3(y-H)^2]V_7\}}{[x^2 + (y-H)^2]^3}$$

$$(10.32)$$

where

$$
\begin{aligned}
&V_1 = (4H - y)\cos\alpha + 2x\sin\alpha, \quad V_2 = x\cos\alpha + (H + 2y)\sin\alpha, \\
&V_3 = (5H - y)\cos\alpha + x\sin\alpha, \quad V_4 = x\cos\alpha + (y + H)\sin\alpha, \\
&V_5 = x\cos\alpha + H\sin\alpha, \quad V_6 = (y + H)\sin\alpha + x\cos\alpha, \\
&V_7 = x\cos\alpha + (3H + y)\sin\alpha, \quad V_8 = x\sin\alpha + (5H - y)\cos\alpha.
\end{aligned}
\quad (10.33)
$$

By summing the corresponding components of stresses in formulas (10.20) and 10.32), we obtain the solution to the problem:

$$X_x = \tilde{X}_x + X_x^0; \quad Y_y = \tilde{Y}_y + Y_y^0; \quad X_y = \tilde{X}_y + X_y^0. \quad (10.34)$$

10.6.5 Addition to Geomechanics

It is known [4] that one of the manifestations of geodynamic processes in the Earth's crust is the formation of faults called *disjunctive dislocations*. Schematically, such dislocation is shown in Fig. 10.6 where the digits 1 and 2 designate a fixed and shifted crust blocks, respectively. This type of fracture is called [4] upcast in geology. An arrow in Fig. 10.6 indicates the direction of block 2 shift. Stresses caused by such movements of the crust are called tectonic.

If a rock massif in the vicinity of the dislocation is conditionally considered as homogeneous, isotropic, and ideally elastic, it is obvious that the problem solved in this paragraph serves an almost adequate mathematical model of a rock massif with upcast dislocation. In other conditions, the answer to the question about the applicability of the solution (10.34) to the assessment of tectonic stresses in the vicinity of a disjunctive dislocation can be obtained by specifying the model and comparing the results of field and numerical experiments.

Fig. 10.6 Formation scheme
of upcast dislocation

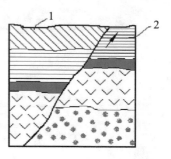

To do this comparison, let us use [13] the information given in the collective work [8]. Figure 10.8 schematically depicts the section of the upper part of the crust borrowed from [8] in the area of the "Kadamjay" antimony mine. Pilot works were carried out at measurement stations marked on the scheme by employees of VNIMI (Saint Petersburg) and IFMGP (Bishkek, Kyrgyzstan) to study the stressed state of rocks enveloping the fault whose "healing" resulted in the ore deposit formation.

The cited work had almost all the information needed for calculations based on formulas (10.34). This information and the nominal values obtained using it for maximum normal tectonic stresses are given in Table 10.1.

Figure 10.7 shows the curves of the maximum (σ_{max}) and minimal (σ_{min}) (in plane xOy, Fig. 10.4a) normal stresses calculated using the formula

$$\sigma_{\substack{max \\ min}} = \frac{X_x + Y_y}{2} \pm \frac{1}{2}\sqrt{(X_x - Y_y)^2 + 4X_y^2},$$

as well as curves of normal stress (σ_z) perpendicular to the plane xOy

$$\sigma_z = \nu(X_x + Y_y),$$

whereas the components X_x, Y_y, X_y are defined by formulas (10.34), (10.20) and (10.32).

Figure 10.7 also gives curves of equivalent stresses (σ_{equ}^V)

for fifth strength [11] (Mohr):

$$\sigma_{equ}^V = \sigma_1 - m\sigma_3,$$

where m is the ratio of tensile and compressive strengths, and σ_1 and σ_3 are the algebraically highest and lowest principal stress, respectively. Fixed values of the x-coordinate x were selected so that the vertical $x = const$ was completely to the left (Fig. 10.7a) or to the right (Fig. 10.7b) from the line of the displacement break. The values of the parameters of calculation formulas were taken from the last line of Table 10.1.

The analysis of the given charts shows [13] the following specifics of distribution of tectonic stresses in the vicinity of the dislocation core. To the left from the break

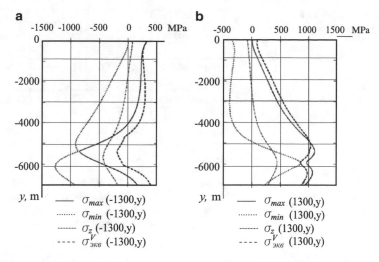

Fig. 10.7 Stress curves on verticals: (**a**) to the left and (**b**) to the right from the dislocation core

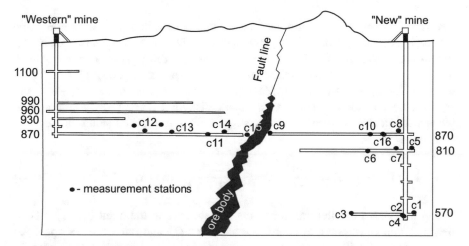

Fig. 10.8 Scheme of the "Kadamjay" mine area

line (Fig. 10.7a) everywhere $\sigma_1 = \sigma_{max}$, except for some vicinity of the dislocation core where $\sigma_1 = \sigma_z$. In a similar way, to the right from the break line (Fig. 10.7b) $\sigma_3 = \sigma_{min}$ everywhere, except for some vicinity of the core where $\sigma_3 = \sigma_z$.

Referring to columns 4 and 5 of Table 10.1, it can be noted that the coincidence of the nominal and experimental values of maximum tectonic stresses is acceptable. Apart from numerical values, there is also satisfactory coincidence of directions of these stresses. By making simple calculations, we can find that in the horizon of 930 m (southern gate), the stress σ_{max} has an azimuth of 163°, whereas the experimental value of this azimuth is 156°, [8, p. 265].

Table 10.1 Studies for the "Kadamjay mine"

			Maximum stress, MPa	
Area, horizon	y, m	t, m	Measured	Nominal
1	2	3	4	5
Horizon 960 m Northern gate	240	200	11.3	7.9
Horizon 930 m Southern gate	330	25	17.1	14.2
Horizon 930 m Mine shaft	380	10	22.0	20.0
"New"	425	700	7.8	11.1

Parameters: $E = 2.2 \cdot 10^4$ MPa; $\nu = 0.24$; $H = 5800$ m $\beta = 700$ m; ratio of ultimate tensile strength and compression $m = 0.25$; $\alpha = \pi/15$; y—depth from day surface; t—depth to the fault in horizontal

10.7   Pair of Fislocations in a Plane

Let us at first consider an unlimited plane where an absolutely rigid infinite stripe δ thick is introduced along the negative half-axis Oy of the system of rectangular coordinates $x_1 Oy$. Muskhelishvili functions [14] for this problem can be obtained using the same ring cutting method as used in the previous paragraph (p. 108) to create a Somigliana dislocation. Making a passage to the limit at $R_1 \rightarrow 0$ and $R_2 \rightarrow \infty$, we obtain the functions of Muskhelishvili $\Phi_1(z)$? $\Psi_1(z)$ for an infinite plane with an absolutely rigid infinite stripe δ thick introduced along the negative half-axis Oy:

$$\Phi_1(z) = -\Psi_1(z) = \frac{K\delta i}{z}, \quad K = \frac{E\delta i}{8\pi(1-\nu^2)} \quad (z = x_1 + iy). \tag{10.35}$$

Let us make a parallel transfer of coordinate axes to the point $(0, -t)$. Let us designate the new system of coordinate axes as xOy and introduce a stripe $(-\delta)$ thick in the point $(0, -t)$ of the new system along the negative Y-half-axis. By the superposition of solutions, let us find Muskhelishvili functions for an unlimited plane where an absolutely rigid infinite stripe δ thick and $2t$ long is introduced at the section $(-t < y < t)$. Muskhelishvili functions for this problem will be

$$\Phi_2(z) = -\frac{2K\delta t}{z^2 + t^2}, \quad \Psi_2(z) = \frac{4K\delta t z^2}{(z^2 + t^2)^2}, \quad (z = x + iy). \tag{10.36}$$

Let us use the known formulas (9.15) associating the stress components σ_x, σ_y, τ_{xy} with Muskhelishvili functions

$$\sigma_x + \sigma_y = 2\left[\Phi(z) + \overline{\Phi(z)}\right],$$
$$\sigma_x - \sigma_y + 2i\tau_{xy} = 2\left[\bar{z}\Phi'(z) + \Psi(z)\right],$$

where the top line means a complex conjugate value and the "dash" symbol means the differentiation operation using the variable z. By substituting into these dependencies functions (10.36) instead of Φ and Ψ, let us find stresses in the points of the axis Ox after separating the real and imaginary parts:

$$\sigma_y(x, 0) = \frac{4K\delta t(x^2 - t^2)}{(x^2 + t^2)^2}, \quad \tau_{xy}(x, 0) = 0. \tag{10.37}$$

10.8 Edge Dislocation in a Half-Plane

Let us then consider a half-plane $y \leqslant 0$, along the boundary of which $(y = 0)$ a normal pressure N is applied, which is equal to the normal stress σ_y according to formula (10.37), but of an opposite sign. For such a half-plane, Muskhelishvili functions are represented [14] as:

$$\Phi_3(z) = \frac{1}{2\pi i} \int_{-\infty}^{\infty} \frac{N d\xi}{\xi - z}, \quad \Psi_3(z) = -z\Phi'(z). \tag{10.38}$$

By substituting into the last formulas, instead of N, the formula

$$N(\xi) = -\frac{4K\delta t(x^2 - t^2)}{(x^2 + t^2)^2},$$

we obtain as follows using formulas (10.38) the after calculation of the integral and differentiation

$$\Phi_3(z) = -\frac{2K\delta t}{(z + it)^2}, \quad \Psi_3(z) = -\frac{4K\delta tz}{(z + it)^3}. \tag{10.39}$$

Muskhelishvili functions for a half-plane $y \leqslant 0$ with the introduced stripe δ wide on the section $-t \leqslant y \leqslant 0$ of the negative half-axis y, by superposing respective functions defined by formulas (10.36) and (10.39), we obtain as follows, e.g.

$$\Phi_4(z) = \Phi_2(z) + \Phi_3(z) = -\frac{4K\delta tz}{(z + it)^2(z - it)},$$

$$\Psi_4(z) = \Psi_2(z) + \Psi_3(z) = \frac{4K\delta zt^2(3iz + t)}{(z + it)^3(z - it)^2}. \tag{10.40}$$

10.9 Half-Plane with a System of Dislocations

Assume that infinitely many parallel rigid stripes of a similar length t and width δ distanced from each other to the same distance a are introduced into the half-plane $y < 0$. Using expression (10.40), the principle of superposition and formulas for conversion of Muskhelishvili functions in the case of parallel transfer of coordinate axes, we will obtain the functions $\Phi(z)$ and $\Psi(z)$ for the specified half-plane as follows:

$$\Phi(z) = -4K\delta t \sum_{k=-\infty}^{k=\infty} \frac{z - ak}{(z - ak + it)^2(z - ak - it)},$$

$$\Psi(z) = 4K\delta t \sum_{k=-\infty}^{k=\infty} \frac{(3it - 2ak)(z - ak)^2 + (t^2 + akti)(z - ak) + akt^2}{(z - ak + it)^3(z - ak - it)^2}.$$

$$(10.41)$$

Using decomposition:

$$\sum_{k=1}^{\infty} \frac{1}{z^2 - k^2} = \frac{1}{2z}\left(\pi \cot \pi z - \frac{1}{z}\right), \quad \sum_{k=-\infty}^{\infty} \frac{1}{(z - k)^2} = \frac{\pi^2}{\sin^2 \pi z};$$

$$\sum_{k=-\infty}^{\infty} \frac{1}{z - k} = \pi \cot \pi z, \quad \sum_{k=-\infty}^{\infty} \frac{1}{(z - k)^3} = \frac{\pi^3 \cos \pi z}{\sin^3 \pi z},$$

formulas (10.41) can be converted to

$$\Phi(z) = -\frac{K\delta}{a}\left[\pi i(\cot \pi z_1 - \cot \pi z_2) + \frac{2\pi^2 t}{a \sin^2 \pi z_1}\right],$$

$$\Psi(z) = \frac{K\delta}{a}\left[\frac{4\pi^3 t(it - az_1)\cos \pi z_1}{a^2 \sin^3 \pi z_1} + \frac{\pi^2(t - z_1 ai)}{a \sin^2 \pi z_1} + \right.$$

$$\left. + \frac{\pi^2(t + z_2 ai)}{a \sin^2 \pi z_2} + 2\pi i(\cot \pi z_1 - \cot \pi z_2)\right],$$

$$(10.42)$$

where $z_1 = \frac{1}{a}(z + it)$, $z_2 = \frac{1}{a}(z - it)$.

In this manner, Muskhelishvili functions for an elastic half-plane wedged by an infinite number of parallel rectangular rigid stripes of the same width δ and length t spaced with even pitch a are built. It is easy to determine stresses using the ratios (10.42), (9.15). If we consider the half-plane as the basis of a pile foundation of a building, and implemented absolutely rigid stripes as piles, the given solution allows finding an additional loading that the single pile can take.

Indeed, with the known normal pressure on side edges of the pile $\sigma_y(\pm\delta/2, y)$ and the friction coefficients f between the pile and soil of basis, additional loading that the single pile can take will be proportional to the value:

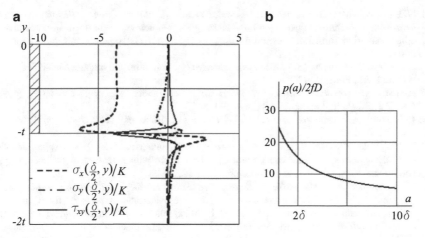

Fig. 10.9 Effects of a single pile on: (**a**) distribution of stresses; (**b**) bearing capacity of basis

$$p(a) = 2f \int_0^t H\left[-\sigma_y\left(\frac{\delta}{2}, \xi\right)\right]\sigma_y\left(\frac{\delta}{2}, \xi\right)d\xi, \quad H(x) = \begin{vmatrix} 1 & \text{for } x > 0, \\ 0 & \text{for } x \leqslant 0. \end{vmatrix}$$

$$(10.43)$$

In Fig. 10.9, using formulas (9.15), (10.42), stress distribution curves (*a*) are built along with the function $p(a)$, (*b*). In the position *b*, *D* designates the integral from formula (10.43).

References

1. S. Derezin, *Sobstvennye napryazheniya v nelineino uprugikh telakh s dislokatsiyami i disklinatsiyami* (Proper stresses in nonlinear elastic bodies with dislocations and disclinations). Cand. philos. sci. diss. Abstr., PhD thesis, Rostov-on-Don, Izdatelstvo RSU Publ., 2011
2. D. Ehshelbi, *Kontinual'naya teoriya dislokatsii* (Continual theory of dislocations). (IL Publ., Moscow, 1963)
3. L. Galin, *Kontaktnye zadachi teorii uprugosti i vyazkouprugosti* (Contact problems of elasticity and viscoelasticity theory). (Nauka Publ., Moscow, 1980)
4. V. Khain, M. Lomize, *Geotektonika s osnovami geodinamiki: ucheb. dlya vuzov. 3-e izd.* (Geotectonics with the basics of geodynamics: studies for universities. The 3rd prod.). (Knizhn. dom Un-t Publ., (KDU), Moscow, 2010)
5. R. Khonikomb, *Plasticheskaya deformatsiya metallov* (Plastic deformation of metals). (Mir Publ., Moscow, 1972), 408 p
6. A. Kosevich, *Osnovy mekhaniki kristallicheskoi reshetki* (Fundamentals of crystal lattice mechanics). (Nauka Publ., Moscow, 1972)
7. A. Kottrell, *Teoriya dislokatsii i plasticheskoe techenie v kristallakh: per. s angl* (Dislocation theory and plastic flow in crystals: TRANS. from English). (GNTIL po chern. i tsv. metallurgii Publ., Moscow, 1958)

8. N. Laverov, *Sovremennaya geodinamika oblastei vnutrikontinental'nogo kollizionnogo goroo-brazovaniya (Tsentral'naya Aziya)* (Modern geodynamics of regions of intracontinental collisional urban formation (Central Asia)/Ed. N.P. Lavyorov). (Nauchnyi mir Publ., Moscow, 2005)

9. M. Leonov, *Mekhanika deformatsii i razrusheniya* (Deformation and fracture mechanics). (Izdvo Ilim Publ., Frunze, 1981)

10. A. Lyav, *Matematicheskaya teoriya uprugosti* (Mathematical theory of elasticity). (ONTI NKTP SSSR Publ., Leningrad, Moscow, 1935)

11. V. Molotnikov, *Mekhanika konstruktsii* (Mechanics of structures). (SPb., Moscow, Krasnodar, Lan' Publ., 2012)

12. V. Molotnikov, A. Molotnikova, *Uprugoe ravnovesie poluploskosti s pryamolinejnoj dislokaciej tipa somiliany* (Elastic equilibrium of a half plane with a rectilinear dislocation type somigliana). Vestn. Donsk. texn. un-ta, no. 1(52) **11**, 21–28. (2011)

13. V. Molotnikov, A. Molotnikova, *K otsenke tektonicheskikh napryazhenii v okrestnosti diz'yunktivnykh dislokatsii v zemnoi kore* [to assess tectonic stresses in the vicinity of disjunctive dislocations in the earth's crust]. bishkek, pp. 310–315, in *Sovremennye problemy mekhaniki sploshnoi sredy* (Modern problems of continuum mechanics: collection of works). (2012)

14. N. Muskhelishvili, *Nekotorye osnovnye zadachi matematicheskoi teorii uprugosti* (Some main problems of mathematical theory elasticity). (Nauka Publ., Moscow, 1966)

15. E. Orowan, Kristallplastizitat. i. Zeitsch. Phys. Bd. **89**(1), 605–613 (1934)

16. V. Smirnov, *Kurs vysshei matematiki. T. 3, ch. 2.: uchebn. posobie* (Course in higher mathematics. T. 3, part 2.: study. grant). (Nauka Publ., Moscow, 1974)

17. G. Taylor, The mechanism of plastic deformation of crystals. Part I. theoretical., Part II. Comparison with observations. Proc. R. Soc. A. V. **145**, 362–404 (1934)

18. A. Timpe, Probleme der Spannungsverteilung in ebenen Systemen einfach gelöst mit Hilfe der Airyschen Funktion. Göttingen, Diss. Leipzig. PhD thesis, Göttingen, Diss. Leipzig, 1905, 348 p

19. V. Volterra, Sur l'équilibre des corps élastiqnes multiplement connexes. Ann. Éc. Norm. Paris. V. 24. Ser. 3, 401–517 (1907)

20. J. Weingarten, Sulle superficie di discontinuite nella teoria della elasticite dei copi solidi. Atti Accad. Naz. Lincei Rend. Cl. Sei. fis., mat., natur. **5**, 57–60 (1901)

21. S. Zherebtsov, G. Salishchev, W. Lojkowski, Strengthening of a ti–6al–4v titanium alloy by means of hydrostatic extrusion and other methods. Mater. Sci. Eng. A. **515**, 43–48 (2009)

22. L. Zubov, *Nonlinear Theory of Dislocations and Disclinations in Elastic Bodies* (Springer, Berlin, 1997)

23. L. Zubov, Uravneniya karmana dlya uprugoi plastinki s dislokatsiyami i disklinatsiyami (Karman equations for an elastic plate with dislocations and disclinations). DAN SSSR [Reports of the USSR Academy of Sciences] **412**(3), 343–346 (2007)

Chapter 11
The Beginning of the Theory of Stability of Equilibrium

11.1 Stability and Instability

Consider some mechanical system that is in equilibrium. Imagine that the points of this system were given infinitesimal speeds. The system will start a movement called *perturbed*. If with an unlimited increase in time, the deviations of the system points from the equilibrium position remain small for all perturbed movements, then the equilibrium of the system is called s t a b l e.

In the event that the deviations of the system from the equilibrium position decrease indefinitely, the equilibrium is called a s y m p t o t i c a l l y   s t a b l e. Otherwise, the equilibrium is called u n s t a b l e.

11.2 Work and Classification of Forces

Let some constant force be applied at an arbitrary point on the body. Let, further, the body make infinitely small movements. If a force moves with its point of application, then they say that the work of a given force is equal to the product of the magnitude of the force and the projection of its movement in the direction of the force. If the displacements of the force and the points of its application do not coincide, the work should be defined as the product of the modulus of force and the corresponding projection of the displacement of the initial point of its application. With this in mind, we consider the work of a force applied to a completely rigid body subject to the following conditions:

(a) the force does not change the line of its action;
(b) the force is associated with the same point of application and when the body moves, it is always perpendicular to some line invariably connected with the body (*tracking force*).

© The Author(s), under exclusive license to Springer Nature Switzerland AG 2021   121
V. Molotnikov, A. Molotnikova, *Theory of Elasticity and Plasticity*,
https://doi.org/10.1007/978-3-030-66622-4_11

If the body in question moves translationally (any straight line segment, rigidly connected with the moving body, remains parallel to its original position), then the work will be equal to the scalar product of force and displacement. If the body rotates around a certain center located on the line of action of the force, then the displacements of the points will be perpendicular to this line and the work of the force is zero.

It is known [4] that any plane motion of a solid can be represented as the sum of the translational motion with a displacement vector equal to the displacement vector of the body at an arbitrary given point, called *pole*, and of rotation around the specified pole.

We show that the work of force in the case of *a* or *b* depends on the order in which the body moves from one position to another. Let, for example, the body be a bar $ABCD$ (Fig. 11.1), subject to the action of the tracking force F. We move the body of $ABCD$ from position I to position II in two ways. In the first case, as a pole, we take some point O lying on the line of action of the force before moving the body (Fig. 11.1a).

Turn the rod counterclockwise by the angle $\pi/2$. The force F will occupy the position F', not doing any work. Then we give the body a vertical movement to position II. On this displacement, the work of the force F' is also zero. Thus, in the first method of moving the body from position I to position II, the total work is zero.

In the second way of moving, we select the point K (Fig. 11.1b) on the line of action of the force F as a pole and turn the rod around the selected pole by the angle $\pi/2$ counterclockwise. In this case, the force F will move to the position F', without doing any work. The body will occupy the position of $A'B'C'D'$ according to Fig. 11.1b. Now move the body vertically up so that the displacement vector of its pole is $\overrightarrow{KK'}$. Then move the body horizontally to the left by $\overrightarrow{K'K''}$. In the last move, the force F will do the work $A = F \cdot K'K''$ and take position II, the same as at the position a of Fig. 11.1.

Fig. 11.1 To the unconservative nature of the tracking force

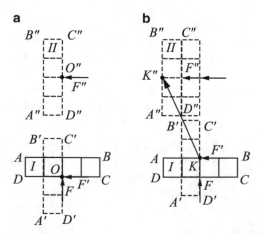

Thus, it turns out that in the considered example, the work of force performed during the movement of the body depends on the method in which the movement is performed. Forces with the latter property are called n o n - c o n s e r v a t i v e. Forces whose work is determined only by the initial and final position of the system are called c o n s e r v a t i v e. By definition, the potential (potential energy) of such forces is the work done by them on the movements of the body from the final position to the initial. In this case, the initial position can be arbitrarily selected. It follows that the potential energy of the system is determined with an accuracy up to a constant.

In addition to conservative and non-conservative forces in real mechanical systems, there are always forces whose work during actual motion is negative (for example, friction forces). Such forces are called d i s s i p a t i v e. They are a special case of non-conservative forces.

11.3 Stability with Conservative and Dissipative Forces

Lagrange theorem[1] *A system subject to the action of only conservative and dissipative forces will be stable if in the equilibrium position the potential energy of conservative forces has a minimum.*

Proof The position of the system will be determined by the generalized coordinates q_1, q_2, ..., q_n, where n is the number of degrees of freedom of the system. We assume that in the equilibrium position $q_1 = q_2 = \ldots = q_n = 0$. The potential of the conservative forces $\Pi(q_1, q_2, \ldots, q_n)$ in the equilibrium position is also assumed to be zero, i.e. $\Pi(0, 0, \ldots, 0) = 0$.

Suppose that the potential of conservative forces is monotonously increasing, but this increase cannot be unlimited. After the potential reaches a certain value Π_1, it can decrease. There may be several such values of Π_1. The smaller of them is denoted by E and is called *potential barrier*. The range of generalized coordinates in the vicinity of the equilibrium under study, in which the potential of conservative forces does not reach the potential barrier E, is called the minimum region of the function Π, or *potential well*. Under the above assumptions, the function Pi vanishes only at the point $(q_1, q_2, \ldots, q_n = 0)$, if the system does not go out of the potential well.

We derive the system from the considered equilibrium position by communicating to its elements such velocities at which the kinetic energy (T) does not reach the potential barrier E. From the kinetic energy theorem [4] in this case, it follows that the potential energy will always be less than E, i.e. the system cannot go beyond the potential barrier and will be in a potential well. With an unlimited decrease in the

---

[1]The wording given here is borrowed from [3] and is somewhat different from the generally accepted ones.

initial velocities of the perturbed motion, the potential energy will tend to zero, and the position of the system will differ infinitely little from the equilibrium or coincide with it, which proves the Lagrange theorem.

The Lagrange theorem defines only a sufficient, but, generally speaking, not a necessary condition for stability. In particular, in systems with dissipative forces such as Coulomb friction, the equilibrium can be stable in the absence of a minimum of potential energy.

11.4   Lyapunov–Chetaev Theorem

Suppose that there is a certain equilibrium position of a conservative system in which its potential energy is not minimal. The following statement is true.

Theorem *If the absence of a minimum of potential energy is determined by the lowest order terms that are in the expansion of the potential energy in a power series in generalized coordinates, then the equilibrium state under consideration is u n s t a b l e.*

The proof of this theorem is very complicated. In order not to bother the reader with mathematical calculations, here we give only an explanation of the theorem. Let, for example, the potential energy of the system have a decomposition

$$\Pi(q_1, q_2, \ldots, q_n) = \frac{1}{2} \sum_{i=1}^{n} a_i q_i^2 + \varepsilon(q_1, q_2, \ldots, q_n), \tag{11.1}$$

where $\varepsilon(q_1, q_2, \ldots, q_n)$ is a small value of the third or higher order, when the values of the generalized coordinates $(q_1, q_2, \ldots, q_n)$ are small.

For $a_i > 0 \ \forall \ i \in 1..n$, the function Π has a minimum at the point $(q_1..q_n = 0)$ and the equilibrium is stable. If at least one of the coefficients a_i is negative, then the equilibrium is unstable.[2] In the case when some of the coefficients a_i are positive, and the rest are equal to zero and the potential energy of the system does not turn to a minimum, the question of the stability or instability of the system remains open. However, if all the coefficients a_i are equal to zero and the expansion (11.1) actually begins with third-order small values, then the minimum potential energy is no longer possible and the equilibrium is unstable.[3]

---

[2]This result belongs to Lyapunov.

[3]The proof and generalization of the latter statement was given by Chetaev.

11.5   Instability in the First Approximation

Let the equations of perturbed motion near the considered equilibrium position $(q_1, q_2, \ldots, q_n = 0)$ contain only terms that are linear with respect to generalized coordinates, their velocities and accelerations. Such equations are called *first approximation equations*. If they have only bounded solutions, then the considered equilibrium state is called *stable in the first approximation*.

Theorem *If only conservative forces act on the system, then its stability is guaranteed by stability as a first approximation*

An interested reader will find a proof of the theorem in a course on the theory of oscillations (see, for example, [2]). It is based on the use of the main (normal) coordinates, at which the kinetic and potential energy for the perturbed motion in the first approximation are represented in the form:

$$T = \frac{1}{2} \sum_{i=1}^{n} m_i \dot{q}_i^2, \quad \Pi = \frac{1}{2} \sum_{i=1}^{n} a_i q_i^2. \tag{11.2}$$

The corresponding equations of perturbed motion in the case of the action of only conservative forces will have the form

$$m_i \ddot{q}_i + a_i q_i = 0 \ (m_i > 0). \tag{11.3}$$

Equations (11.3) can have bounded (periodic) solutions only for positive a_i. On the other hand, if all the coefficients are positive, then the potential energy (11.2) has a minimum. The latter means that the condition for the boundedness of the solution of Eq. (11.3) is a sufficient condition for stability under a conservative load.

11.6   Critical Load

Let all the forces applied at various points of the elastic system arise as a result of their monotonic growth from zero values. At sufficiently small loads, the deformation of the studied system is uniquely determined by the load. Therefore, for simplicity, we can assume that the load arose by the growth of all forces in proportion to the same parameter (H). We call H the load parameter.

Let us denote by H_k such a value of the load parameter that for $H < H_k$, the equilibrium of the system is stable, and for an infinitesimal excess of the value H_k, the equilibrium is either unstable or impossible. The load corresponding to H_k is called c r i t i c a l.

The critical load is determined either by a direct study of perturbed motion (dynamic method) or by a study of potential energy using the Lagrange and Lyapunov-Chetaev theorems (energy method), or by a static method involving the so-called Euler forces. The latter method is familiar to the reader from the course of resistance of materials.

The problem of determining the critical load is often replaced by a simpler problem of investigating first approximation equations or conditions for the existence of extreme values of potential energy (in the presence of a force potential) or a state of indifferent equilibrium, etc. Despite the fact that this substitution is not always legal, these particular methods of determining the critical load are widely used in many computer programs for analyzing structures.

11.7   The Theorem on Stability by the First Approximation

In the case of a system with a finite (n) number of degrees of freedom, the first approximation equations of perturbed motion are a system of ordinary homogeneous differential equations with constant coefficients depending on the load parameter. Representing the generalized coordinates in this case as

$$q_i = A_i \exp \lambda t \quad (i = 1, 2, \ldots, n), \tag{11.4}$$

we obtain from the first approximation equations a known (characteristic) equation for determining the c h a r a c t e r i s t i c   i n d i c a t o r λ:

$$p_0 \lambda^{2n} + p_1 \lambda^{2n-1} + \ldots + p_{2n} = 0 \quad (p_0 > 0), \tag{11.5}$$

where $p_0, \ldots, p_{2n}$ are coefficients that depend in a known way on the system and the value of the load parameter.

For these systems, Lyapunov proved the following theorems.

Theorem 11.1 *If the real parts of all the roots of Eq. (11.5) are negative, then the considered equilibrium state will be stable (asymptotically).*

Theorem 11.2 *If among the roots of Eq. (11.5) there is at least one with a positive real part, then the considered equilibrium state is unstable.*

Note The study of (non)stability by the first approximation, generally speaking, cannot give a solution to the problem of (non)stability only in the case when the real parts of the roots of the characteristic equation have zero and negative values.

11.8   The Raus–Hurwitz criterion

In order for all the roots of Eq. (11.5) to have negative real parts, it is necessary and sufficient to perform the inequalities

$$\Delta_1 > 0, \ \Delta_2 > 0, \ \ldots, \ \Delta_{2n-1} > 0, \tag{11.6}$$

where

$$\Delta_1 = p_1, \quad \Delta_2 = \begin{vmatrix} p_1 & p_0 \\ p_3 & p_2 \end{vmatrix}, \quad \Delta_3 = \begin{vmatrix} p_1 & p_0 & 0 \\ p_3 & p_2 & p_1 \\ p_5 & p_4 & p_3 \end{vmatrix},$$

$$\Delta_{2n} = \begin{vmatrix} p_1 & p_0 & 0 & 0 & 0 & 0 & \ldots & 0 \\ p_3 & p_2 & p_1 & p_0 & 9 & 0 & \ldots & 0 \\ p_5 & p_4 & p_3 & p_2 & p_1 & p_0 & \cdots & 0 \\ \ldots & \ldots & \multicolumn{5}{c}{\ldots\ldots\ldots\ldots\ldots} & 0 \\ \ldots & \ldots & \multicolumn{5}{c}{\ldots\ldots\ldots\ldots\ldots} & 0 \\ p_{4n-1} & p_{4n-2} & \multicolumn{5}{c}{\cdots\cdots\cdots\cdots\cdots\cdots} & p_{2n} \end{vmatrix},$$

moreover, p_i should be replaced with zero if $i > 2n$. For example, for a system with two degrees of freedom ($n = 2$), i.e. for the equation

$$p_0 \lambda^4 + p_1 \lambda^3 + p_2 \lambda^2 + p_3 \lambda + p_4 = 0$$

the determinant Δ_{2n} takes the form

$$\Delta_4 = \begin{vmatrix} p_1 & p_0 & 0 & 0 \\ p_3 & p_2 & p_1 & p_0 \\ 0 & p_4 & p_3 & p_2 \\ 0 & 0 & 0 & p_4 \end{vmatrix} = p_4 \begin{vmatrix} p_1 & p_0 & 0 \\ p_3 & p_2 & p_1 \\ 0 & p_4 & p_3 \end{vmatrix} = p_4 \cdot \Delta_3.$$

It follows from the above explanation that

$$\Delta_2 = p_{2n} \Delta_{2n-1}. \tag{11.7}$$

11.9 Main Types of Stability Loss

Let us consider a system that is asymptotically stable at a fairly low load. As the load parameter increases, the coefficients of the characteristic equation for p_i ($i = 0, 1, \ldots, 2n$), and hence the values Δ_j ($j = 2, \ldots, 2n$) usually change. We will consider these changes continuous. In this case, the loss of stability can occur either when at least one of the roots λ_i of Eq. (11.5), passing through zero, becomes positive, or when two complex-conjugate roots with negative real parts turn into purely imaginary ones, and then their real parts become positive. It is known [1] that in the first case, the coefficient p_{2n} vanishes, and in the second—the value Δ_{2n-1}. It follows that if instability occurs with the growth of the load parameter, then under these conditions, the critical value of the parameter coincides with the smallest root of one of the two equations

$$p_{2n} = 0, \quad \Delta_{2n-1} = 0, \tag{11.8}$$

or, which is the same thing, with the smallest root of the equation $\Delta_{2n} = 0$.

Keeping in mind that for $p_{2n} = 0$, Eq. (11.5) has a zero root ($\lambda = 0$), which corresponds to arbitrary constant values of the generalized coordinates (11.4), we have in this case a state of indifferent equilibrium. If the load is critical, we will say that the elastic system loses *Euler stability*.

From the above, it follows that for $\Delta_{2n-1} = 0$, at least two characteristic numbers are purely imaginary ($\lambda_{1,2} = \pm\omega i$). Since the real parts of the remaining roots of Eq. (11.5) are negative, the system sets the periodic motion. This type of stability loss is called s e l f - o s c i l l a t i n g.

11.10   Methods for Determining Critical Load

The smallest load at which the total potential energy of the elastic system under the influence of forces loses a minimum in the state of equilibrium under study is called e n e r g y - c r i t i c a l. The determination of this load is the e n e r g y m e t h o d task of studying the stability of elastic systems.

The value of the load parameter, at which infinitely close forms of equilibrium are possible, is called *critical in Euler's sense*, and the corresponding load is *Euler's*.[4]

The value of the load parameter, at an arbitrarily small excess of which there is no form of equilibrium infinitely close to the studied one, is called the *ultimate*, and the corresponding load is called the u l t i m a t e d. Determining Euler and limit loads is the task of a static method for studying the stability of elastic systems.

In the presence of non-conservative forces in an elastic system, the concept of energy load, generally speaking, loses its meaning. Below (p. 131) we show that the Euler load in such systems can be critical under certain conditions.

In some special cases, the Euler and energy-critical loads in conservative systems may differ. For example, when the Euler value of the load parameter $H = H_e$ does not change the sign of any of the coefficients $a_n(H)$ in the expression (11.2), the potential energy may not lose its minimum at $H = H_e$, i.e. the Euler load may be lower than the energy-critical one. The Euler load may not exist when the coefficient $a_n(H)$ changes the sign after going to infinity, and the energy-critical load exists.

It follows from the above, in particular, that the presence of the potential of forces acting on an elastic system is neither necessary nor sufficient for the existence of an Euler load or its coincidence with the critical load.

---

[4]This load is also sometimes called the bifurcation or branching load of equilibrium forms.

11.11 The Perturbed Motion of the Compressed Rod

Let the resistance force of the external environment be proportional to the speed. The equation of perturbed motion of a longitudinally bent rod (Fig. 11.2) is obtained by adding inertia forces to the acting forces. Assuming that the intensity (p) of the transverse load is equal to

$$p = -b\frac{\partial v}{\partial t} - m\frac{\partial^2 v}{\partial t^2}, \tag{11.9}$$

where b is a certain coefficient of friction, m is the linear mass of a unit of length of the rod; these values will be considered variables along the length of the rod.

The equation of the bending of a beam exposed to a given transverse load $p(z)$ and the compressive force H has the form [3]

$$D\frac{d^4 v}{dz^4} + H\frac{d^2 v}{dz^2} = p(z), \tag{11.10}$$

where for a rectangular-section rod with a height of h

$$D = \frac{Eh^2}{12(1 - v^2)};$$

here E and v, respectively, are Young's modulus of elasticity and Poisson's ratio of the bar material.

If we substitute the expression pz into Eq. (11.10) by formula (11.9), and then replace the ordinary derivatives with partial ones, then we obtain

$$D\frac{\partial^4 v}{\partial z^4} + H\frac{\partial^2 v}{\partial z^2} + b(z)\frac{\partial v}{\partial z} + m(z)\frac{\partial^2 v}{\partial z^2} = 0. \tag{11.11}$$

We will look for a partial solution of this equation in the form

$$v = f(z)T(t). \tag{11.12}$$

In this case, Eq. (11.11) takes the form

$$(Df^{IV} + Hf^{II})T + [m(z)\ddot{T} + b(z)\dot{T}]f = 0 \tag{11.13}$$

Fig. 11.2 Stability in longitudinal and transverse bending

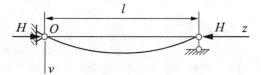

or

$$\frac{Df^{IV}(z) + H^{II}f(z)}{m(z)f(z)} = -\frac{\ddot{T}(t) + \dfrac{b(z)}{m(z)}\dot{T}(t)}{T(t)}. \tag{11.14}$$

If $b(z)/m(z) = const$, i.e. if the coefficient of friction is proportional to the linear mass of the bar, then the right part of the latter equality will not depend on the variable z. Since here the left part does not depend on the variable t, it follows that the right and left parts of Eq. (11.14) must be equal to a constant, which we call e i g e n v a l u e λ, i.e.

$$Df^{IV}(z) + Hf^{II}(z) - \lambda mf(z) = 0, \tag{11.15}$$

$$\ddot{T}(t) + 2c\dot{T}(t) + \lambda T(t) = 0, \tag{11.16}$$

where

$$c = \frac{1}{2}b(z)/m(z) \tag{11.17}$$

is assumed to be a constant called a r e d u c e d c o e f f i c i e n t o f f r i c -
t i o n.

In the case of the rod shown in Fig. 11.2, we have $v(0) = v(l) = 0$,

$$\left.\frac{d^2v}{dz^2}\right|_{z=0} = \left.\frac{d^2v}{dz^2}\right|_{z=l} = 0.$$

We will satisfy these conditions by assuming

$$f = \sin\frac{\pi n}{l}z \quad (n = 1, 2, \ldots, \infty),$$

and from Eq. (11.15) for $m = const$, we find $\lambda = \lambda_n$, where

$$\lambda_n = \frac{\pi^2 n^2}{ml^2}\left(n^2 D\frac{\pi^2}{l^2} - H\right) \quad (n = 1, 2, \ldots). \tag{11.18}$$

Note that

$$\lambda_{min} = \lambda_1 = \frac{\pi^2}{ml^2}\left(D\frac{\pi^2}{l^2} - H\right). \tag{11.19}$$

If all eigenvalues are positive ($\lambda_1 > 0$), then the solutions of Eq. (11.16) will fade over time at $c > 0$ (or they will be periodic in the absence of friction when $c = 0$). If λ_1 becomes negative, which is possible only when the compressive force exceeds the

Euler force, then Eq. (11.16) for $\lambda = \lambda_1$ will have a solution that grows indefinitely over time according to the potential law. For the case under consideration, the Euler force is critical.

When $\lambda = 0$, Eq. (11.15) becomes independent of the nature of the mass distribution and passes into the static equation, from which the Euler forces are determined. Thus, the Euler load can be defined as the smallest load, at which the eigenvalue of the corresponding dynamic problem vanishes.

11.12   Stability Under Non-conservative Load (Example)

The study of equilibrium stability in the general case of non-conservative loads is a complex mathematical problem. Therefore, examples of this type of problem in publications are quite rare. This section only provides an illustration of the effect of non-conservative loads on the oscillation and stability of the simplest model of a cantilever rod compressed by spaced masses.

11.12.1   Equations of Perturbed Motion

We will consider an elastic rod of length l with two separated masses at the free end, compressed by the gravity G of the end masses and the tracking force H (Fig. 11.3a). Ignoring the mass of the rod, we consider here a system with two degrees of freedom.

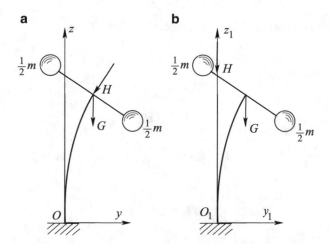

Fig. 11.3 The simplest non-conservative system

Let us make up the equations of small vibrations of the rod near the rectilinear form of equilibrium. For generalized coordinates, we take the deflection $v(l)$ and the angle of rotation $\varphi = \dfrac{dv}{dz}\Big|_{z=l}$ of the free end.

Let us assume that in the system under consideration, there are small friction forces proportional to the speeds. Under these conditions, small fluctuations of the system are described by the equations

$$m\frac{d^2v}{dt^2} = P_0 - h_1\frac{dv}{dt}, \quad I\frac{d^2\varphi}{dt^2} = M_0 - h_2\frac{d\varphi}{dt}, \tag{11.20}$$

where P_0 is the force, M_0 is the moment acting on the concentrated masses from the rod side; h_1 and h_2 are some small positive parameters; $I = m\rho^2$ is the central moment of inertia of the end masses, the distance between which is equal to 2ρ.

The values P_0 and M_0 depend in a certain way on the movement and angle of rotation of the free end of the rod, namely [3, p. 66]

$$v = (P_0 - G\varphi)a_{11} + M_0 a_{12}, \\ \varphi = (P_0 - G\varphi)a_{21} + M_0 a_{22}, \tag{11.21}$$

where

$$a_{11} = \frac{1}{k(G+H)}(kl\cos kl - \sin kl), \quad a_{12} = \frac{1}{G+H}(1 - \cos kl - kl\sin kl), \\ a_{21} = \frac{1}{G+H}(\cos kl - 1), \quad a_{22} = -\frac{1}{G+H}k\sin kl, \\ k = \sqrt{\frac{G+H}{D}}. \tag{11.22}$$

Solving the system (11.22) with respect to P_0 and M_0, we get

$$P_0 = c_{11}v - c_{12}\varphi, \quad M_0 == -c_{21}v - c_{22}\varphi, \tag{11.23}$$

where

$$c_{11} = \frac{G+H}{\Gamma}k\sin kl, \quad c_{12} = \frac{G+H}{\Gamma}(\cos kl - 1 + \eta\Gamma), \\ c_{21} = \frac{G+H}{\Gamma}(\cos kl - 1), \quad c_{22} = \frac{G+H}{k\Gamma}(\sin kl - kl\cos kl); \tag{11.24}$$

$$\Gamma = 2 - 2\cos kl - kl\sin kl, \quad \frac{H}{G+H}. \tag{11.25}$$

Taking into account the results (11.23), Eqs. (11.20) of small vibrations of the system can be written as:

$$m\frac{d^2v}{dt^2} + h_1\frac{dv}{dt} + c_{11}v + c_{12}\varphi = 0,$$

$$I\frac{d^2\varphi}{dt^2} + h_2\frac{d\varphi}{dt} + c_{21}v + c_{22}\varphi = 0.$$

(11.26)

We look for solutions to Eqs. (11.26) in the form

$$v = A\exp\lambda t, \quad \varphi = B\exp\lambda t,$$

(11.27)

where A, B are constants, and λ is a characteristic indicator.

Substituting the solutions (11.27) into Eqs. (11.26) gives a system of homogeneous algebraic equations with respect to A and B. The condition for the existence of a non-zero solution of this system is that its determinant is equal to zero

$$\begin{vmatrix} m\lambda_2 + b_1\lambda + c_{11} & c_{12} \\ c_{21} & I\lambda^2 + h_2\lambda + c_{22} \end{vmatrix} = 0.$$

By revealing this determinant, we obtain the characteristic equation:

$$p_0\lambda^4 + p_1\lambda^3 + p_2\lambda^2 + p_3\lambda + p_4 = 0,$$

(11.28)

where

$$p_0 = Im, \quad p_1 = h_2m(\mu\rho^2 + 1),$$

$$p_2 = m(\rho^2c_{11} + c_{22}) + b_1b_2, \quad p_3 = h_2(\mu c_{22} + c_{11}),$$

(11.29)

$$p_4 = c_{11}c_{22} - c_{12}c_{21}, \quad \mu = h_1/h_2.$$

In this case, in the coefficient p_2, the product of small quantities b_1, b_2 will be ignored in the future (unless otherwise specified).

In conclusion of this point, we note the following. The oscillations of the system shown in Fig. 11.3b are described under similar assumptions by the following equations

$$m\frac{d^2v_1}{dt^2} + b_1\frac{dv_1}{dt} + c_{11}v_1 + c_{21}\varphi_1 = 0,$$

$$I\frac{d^2\varphi_1}{dt^2} + b_2\frac{d\varphi_1}{dt} + c_{12}v_1 + c_{22}\varphi = 0,$$

(11.30)

moreover, the coefficients c_{ij} $(i, j = 1, 2)$ are determined by formulas (11.24, 11.25). The systems of Eqs. (11.26) and (11.30) differ from each other only in the arrangement of the coefficients c_{12} and c_{21}. These values are included in the stability conditions only as a product of $(c_{12}c_{21})$. It follows that the systems in Fig. 11.3a, b are equivalent with respect to the stability of their equilibrium.

11.12.2 Area of Valid Stability

Below, we will investigate the system of Eq. (11.26). To determine the critical load (see p. 125) it is enough to examine the coefficient p_4 and the value

$$\Delta\mu \equiv p_3(p_1 p_2 - p_0 p_3) - p_1^2 p_4. \tag{11.31}$$

It is easy to show that the function $\Gamma(kl)$ defined by formula (11.25) on the interval $(0, 2\pi)$ has no zeros. It follows that the functions (11.24), as well as $p_2(kl)$, $p_3(kl)$, $p_4(kl)$ and $\Delta\mu(kl)$, are continuous when $0 < kl < 2\pi$. Let us limit ourselves to the specified interval for now.

Using formulas (11.24) and (11.25), the coefficient p_4 can be converted as follows:

$$p_4 = \frac{(G+H)^2}{\Gamma(\gamma)}[\eta + (1-\eta)\cos\gamma] \ (\gamma = kl). \tag{11.32}$$

This shows that p_4 can change the sign for those values of γ that nullify either the expression in the square brackets or the denominator of the right side of the equality (11.32). The denominator $\Gamma(\gamma)$ of formula (11.32) changes the sign for $\gamma > 0$ for the first time when $\gamma_1 = 2\pi$. Equating the numerator to zero, we get

$$\cos\gamma = -\frac{\eta}{1-\eta}. \tag{11.33}$$

The equality (11.33) is possible if and only if

$$\left|-\frac{\eta}{1-\eta}\right| \leqslant 1,$$

i.e. for $\eta \leqslant \frac{1}{2}$. Setting the values $\eta \left(0 < \eta < \frac{1}{2}\right)$ from Eq. (11.33), we find the smallest values of $\gamma_0(\eta)$, which changes the sign of the expression in the square brackets of formula (11.32). The visualization of the calculation results is shown in Fig. 11.4.

The figure shows that the value p_4 for the values η that satisfy the condition $0 \leqslant \eta < \frac{1}{2}$ becomes negative when $\gamma > \gamma_0(\eta)$ and the inequality $\gamma_0(0) < \gamma_0(\beta) <$ $\gamma_0 \left(\frac{1}{2}\right)$ is right; for the values $\eta \geqslant \frac{1}{2}$, the coefficient p_4 goes to negative values

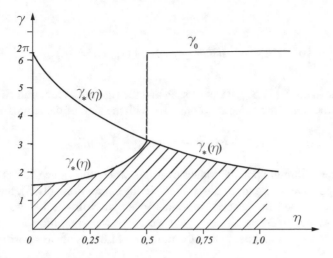

Fig. 11.4 Research on the stability of a non-conservative system

when γ exceeds 2π. Note that in the interval $\left(0, \dfrac{1}{2}\right)$ for η, the value p_4 passes from positive to negative values through zero (indifferent equilibrium), and p_4 reaches zero without changing the sign, at $\gamma = \pi$, when $\eta = \dfrac{1}{2}$; for $\eta \geqslant \dfrac{1}{2}$, the coefficient p_4 changes the sign, turning to infinity.

11.12.3 *Investigation of the Value* $\Delta\mu$, *(Formula (11.31))*

Using formulas (11.29), $\Delta\mu$ can be converted to the form:

$$\Delta\mu = mh_2^2(\mu\rho^2 + 1)L(f, \rho^2, c_{ij}), \tag{11.34}$$

where

$$L(f, \rho^2, c_{ij}) = (\rho^2 c_{11} - c_{22})f(\mu) + c_{12}c_{21}, \tag{11.35}$$

$$f(\mu) = \frac{\mu}{(\mu\rho^2 + 1)}^2. \tag{11.36}$$

Substituting in (11.35) formula (11.24), we get

$$L(f, \rho^2, c_{ij}) == \frac{(G + H)^2}{\Gamma^2}[fS^2 + (\cos\gamma - 1)(\cos\gamma - 1 - \eta\Gamma)], \tag{11.37}$$

where

$$S = \left[a^2 \gamma \sin \gamma - \frac{1}{\gamma} (\sin \gamma - \gamma \cos \gamma) \right] l, \quad \left(a = \frac{\rho}{l} \right).$$ (11.38)

From the identity (11.37), it can be seen that the sign of the function $L(f, \rho^2, c_{ij})$ or, what is the same, the value $\Delta \mu$ coincides with the sign of the expression

$$f S^2 + (\cos \gamma - 1)(\cos \gamma - 1 + \eta \Gamma).$$ (11.39)

Define the minimum positive values of $\gamma_*(\eta)$, where the value (11.39) or (11.31) goes from positive to negative values. Since in the expression (11.39), the first term is non-negative, and the second for small γ, as we will see below, is positive, the values we are interested in are $\gamma_*(\eta)$ we will have when the value $f S^2$ is the smallest. It is easy to make sure that the function (11.36) is always positive, and

$$f_{min} < f(\mu) \leqslant f_{max},$$ (11.40)

where

$$f_{min} = \lim_{\mu \to 0} f(\mu) = 0; \quad f_{max} = f \Big|_{\mu = \frac{1}{\rho^2}} = \frac{1}{4\rho^2}.$$

Since the relation (μ) of unknown friction coefficients can take any value from the range $(0, \infty)$, then the function $f(\mu)$ can be arbitrarily small. Since the value $S(\gamma)$ (11.38) is bounded at $\gamma > 0$, then in the expression (11.39), the value $f S^2$ can also be infinitesimal. Hence, $\gamma_*(\eta)$ are the smallest positive roots of the equation

$$(\cos \gamma - 1)[\cos \gamma - 1 + \eta \Gamma(\gamma)] = 0.$$

The first multiplier can only reach zero, but it does not change the sign. Equating the second multiplicand to zero and taking into account (11.25), we get

$$\cos \gamma - 1 + \eta(2 - 2\cos \gamma - \gamma \sin \gamma) = 0.$$

The latter equation is easily converted to the form

$$-2 \sin \frac{\gamma}{2} \left[\sin \frac{\gamma}{2} + \eta \left(\gamma \cos \frac{\gamma}{2} - 2 \sin \frac{\gamma}{2} \right) \right] = 0.$$ (11.41)

The smallest positive root of the equation $\sin \frac{\gamma}{2} = 0$ is $\gamma_1 = 2\pi$ and does not depend on η. Equating to zero the multiplier in (11.41) in the square brackets, we get

$$\sin \frac{\gamma}{2} + \eta \left(\gamma \cos \frac{\gamma}{2} - 2 \sin \frac{\gamma}{2} \right) = 0.$$ (11.42)

Hence, for $\eta = \dfrac{1}{2}$, we find that $\gamma_* \left(\dfrac{1}{2} \right) = \pi$. Let $\eta \neq \dfrac{1}{2}$; then Eq. (11.42) can be converted:

$$\text{tg}\,\frac{\gamma}{2} = \frac{\dfrac{\gamma}{2}}{1 - \dfrac{1}{2\eta}}. \tag{11.43}$$

Setting η, one can find the corresponding value of $\gamma_0(\eta)$ from formula (11.43). These values are shown graphically in Fig. 11.4.

From the above arguments, it follows that the equilibrium of the system under consideration is always stable in the area that is covered with shading in Fig. 11.4. Let us call it the *reliable stability* area. In addition, one can say the following:

1. when the parameter η is in the range $\left(0, \dfrac{1}{2} \right)$ or, equivalently, $H < G$ ($H \geqslant 0$, $G > 0$), the loss of stability occurs statically, and the critical load coincides with the Euler load. The critical parameter $\gamma_0(\eta)$ increases with an increase in the non-conservative component of the load, which indicates the stabilizing influence of H;

2. for values η that satisfy the condition $\dfrac{1}{2} \leqslant \eta \leqslant 1$, in other words, $H \geqslant G$ ($H > 0$, $G \geqslant 0$), the loss of stability is self-oscillating; for $\eta > \dfrac{1}{2}$, the Euler load does not exist. The critical parameter $\gamma_*(\eta)$ decreases with an increase in the non-conservative component of the load. This means that the tracking force can have a destabilizing effect.

11.12.4 Investigation of the Effect of Friction

Now we will show that in this problem, taking into account small friction can sometimes have a significant impact on the stability of the system equilibrium.

Let us first assume that there is no friction ($b_1 = b_2 = 0$). Then instead of Eqs. (11.26), we have the system

$$m\frac{d^2 v}{dt^2} + c_{11} v + c_{12} \varphi = 0,$$
$$I\frac{d^2 \varphi}{dt^2} + c_{21} v + c_{22} \varphi = 0. \tag{11.44}$$

The characteristic equation of the system (11.44) has the form

$$p_0 \omega^4 + p_2 \omega^2 + p_4 = 0, \tag{11.45}$$

and the coefficients p_0, p_2, p_4 are still determined by formulas (11.29).

The zero solution of the system (11.44) is stable by the first approximation if and only if all the coefficients of Eq. (11.45) and its discriminant

$$\Delta_0 = p_2^2 - 4p_0 p_4 \tag{11.46}$$

are positive because all the roots of Eq. (11.45) are purely imaginary.

Reasoning similarly, as in the presence of friction, it is not difficult to establish that a trivial solution of the system (11.44) can become unstable due to violation of one of the inequalities

$$p_4 > 0, \quad \Delta_0 > 0. \tag{11.47}$$

If the first one is violated, the system under study loses stability statically, passing through a state of indifferent equilibrium at $p_4 = 0$; if the second one is violated, the loss of stability is self-oscillating.

As shown above (see p. 134), in the presence of friction, the necessary and sufficient stability conditions for the zero solution of the system (11.26) had the form:

$$p_4 > 0, \quad \Delta_\mu > 0. \tag{11.48}$$

Comparing the conditions (11.47) and (11.48), we see that in the case of s t a t i c stability loss, small friction does not affect the critical parameters.

Let the stability loss be self-oscillating $\left(\dfrac{1}{2} \leqslant \eta \leqslant 1\right)$. Compare the conditions $\Delta_0 > 0$ and $\Delta_\mu > 0$. Substituting the values of the coefficients (11.29) and (11.46), in the absence of friction, we will have the following stability condition

$$\frac{1}{4\rho^2}(\rho^2 c_{11} - c_{22})^2 + c_{12}c_{21} > 0. \tag{11.49}$$

In the presence of friction, the stability condition according to formula (11.35) has the form

$$(\rho^2 c_{11} - c_{22})^2 \cdot f(\mu) + c_{12}c_{21} > 0. \tag{11.50}$$

Since the left side of the latter inequality depends on the ratio of small friction coefficients μ, at $b_1 \to 0$ and $b_2 \to 0$, the condition (11.50) does not go, generally speaking, to the condition (11.49). In other words, the critical parameters determined with low friction are usually different from the critical parameters found without it.

The question arises: how significant can this difference be?

Obviously, the inequality (11.50) goes to (11.49) when $f = \dfrac{1}{4\rho^2}$. As already noted, this value is the maximum and is reached when

$$b_2 = b_1 \rho^2. \tag{11.51}$$

The expressions (11.37) and (11.38) show that the left side of the inequality (11.50) depends on the parameters γ, μ, a and η. Denote the smallest positive values of the parameter γ, for which the condition (11.50) is violated, by $\gamma_\mu^0(a, \eta)$, and those of them that correspond to R_{max} will be denoted by $\gamma_*^0(a, \eta)$.

Since the function $f(\mu)$ is positive, it is obvious that at the given parameters a and η, the value of $\gamma_*^0(a, \eta)$ exceeds, as a rule, the corresponding value of $\gamma_\mu^0(a, \eta)$. Therefore, at $b_2 = b_1\rho^2$, or at $\mu = \dfrac{1}{\rho^2}$, the stability loss usually occurs later than at other values of μ. Therefore, if the coefficients of small friction satisfy the relation (11.51), then friction will be called b e s t.

The minimum values of $\gamma_\mu^0(a, \eta)$, as previously defined, are $\gamma_*(\eta)$ and are reached at $f = 0$ when $\mu \to 0$ or $\mu \to \infty$. In this case, friction is called w o r s t.

Figure 11.5 shows graphs of the values γ_* and $\gamma_*^0(a)$ for the extreme values of the parameter $\eta = \dfrac{1}{2}$ and $\eta = 1$. Note that the values $\gamma_\mu^0\left(a, \dfrac{1}{2}\right)$ fill the area bounded by the straight line $\gamma_* = \pi$ and the curve $\gamma_*^0\left(a, \dfrac{1}{2}\right)$; the value area $\gamma_\mu^0(a, 1)$ in Fig. 11.5 is covered with hatching. It can be shown that for any value $\eta = \eta_0$ that belongs to the interval $\left(\dfrac{1}{2}, 1\right)$, curves $\gamma_\mu^0(a, \eta_0)$ fill some similar area.

From the above and Fig. 11.5, the following follows. Low friction in the system under consideration can cause destabilization; the critical load value obtained with any small difference from the best friction is usually less than the value obtained without taking into account friction. This discrepancy can be significant;

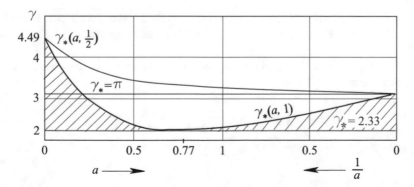

Fig. 11.5 Investigation of the effect of friction on stability

for example, for $\eta = 1$ and small values of the parameter a, the worst-case friction reduces the critical load by almost 4 times.

Since the ratio μ of small friction coefficients is uncertain, the critical parameters have to be determined from the assumption that the friction is the worst. In this way, the area of reliable stability was determined above (see p. 134), where the equilibrium of the system is stable under arbitrary small friction.

11.12.5   The influence of the spacing of the End Masses

Let us show that the effect of friction on the value of the critical load depends significantly on the separation of the end masses, set by the parameter a. For this purpose, let us consider the case when $\eta = 1$. At the same time (see Fig. 11.5), the straight line $\gamma_* = 2.33$ touches the curve $\gamma_*^0(a, 1)$ at the point $a_0 = 0.77$. Obviously, for the values $a = 0.77$, the critical load is minimal (γ_0) and its value does not depend on the relation (μ) of low friction coefficients.

Let $\gamma_0 \neq 1$. It can be shown that when η is continuously reduced from 1 to $\frac{1}{2}$, the point $a_0(\eta)$ will move along the numeric axis from the value $a_0(1) = 0.77$ to $a_0\left(\frac{1}{2}\right) = \infty$, respectively. Each point $a_0(\eta_0)$ gives the minimum of the corresponding curve $\gamma_*^0(a, \eta_0)$ and has similar properties as $a_0(1) = 0.77$. At such points of the minimum, there is an equality

$$\rho^2 c_{11} = c_{22}. \tag{11.52}$$

In this case, the conditions (11.49) and (11.50) obviously coincide for any μ; the critical load is minimal and there is no destabilization.

In the area of reliable stability (Fig. 11.4), the equilibrium of the system is stable for any values of the parameter a. Obviously, if the separation of the end masses is unfavorable $(a = a_0)$, then the loss of stability can occur directly above the curve $\gamma_*(\eta)$.

The significant effect of low friction on critical parameters in self-oscillating loss of stability becomes apparent if we consider the general condition (11.31) in more detail.

In fact, let the loss of stability occur due to a violation of the condition (11.31), which can be represented as

$$p_1^2 \left[\frac{p_3}{p_1} p_2 - \left(\frac{p_3}{p_1}\right)^2 p_0 - p_4 \right] > 0. \tag{11.53}$$

All coefficients of the characteristic equation (11.28) in this case are positive.[5]
Hence, the area of equilibrium stability is determined by the inequality

$$p_0 \left(\frac{p_2}{p_1} \right)^2 - p_2 \left(\frac{p_3}{p_1} \right) + p_4 < 0, \tag{11.54}$$

or

$$\frac{p_2 - \sqrt{\Delta}}{2p_0} < \frac{p_3}{p_1} < \frac{p_2 + \sqrt{\Delta}}{2p_0}, \tag{11.55}$$

where

$$\Delta = p_2^2 - 4p_0 p_4. \tag{11.56}$$

Note that the value of $\dfrac{p_3}{p_1}$ depends on the ratio (μ) of small friction coefficients.
In the absence of friction, the characteristic equation goes into the following:

$$p_0 \lambda^4 + p_2^* \lambda^2 + p_4 = 0, \tag{11.57}$$

and also the equality is true:

$$p_2 = p_2^* + \varepsilon^2 \quad (\varepsilon^2 = b_1 b_2). \tag{11.58}$$

Based on Eq. (11.57), we find the boundary of instability of the equilibrium by
equating its discriminant Δ_0 to zero:

$$\Delta_0 = p_2^{*2} - 4p_0 p_4. \tag{11.59}$$

Comparing formulas (11.56) and (11.57), we get

$$\Delta = \Delta_0 + \varepsilon^2 (2p_2^* + \varepsilon^2). \tag{11.60}$$

Let the friction be so small that the value of ε^2 in equality (11.58) can be
neglected. Then, with the accuracy of the values of the second order of smallness
$p_2 = p_2^*$ and $\Delta = \Delta_0$, the following statements arise from the conditions (11.55).

1. Low friction, as a rule, causes destabilization, and its value depends on the
   parameter μ.
2. Destabilization is absent when and only when the system parameters satisfy the
   condition

---

[5]Here it is assumed that the product of small values of b_1 and b_2 is stored in the coefficient p_2.

$$\frac{p_3}{p_1} = \frac{p_2}{2p_0}.$$ (11.61)

In fact, if the equality (11.61) is fulfilled, then, as can be seen from formula (11.55), stability occurs as long as the value (11.56) is positive, so that the boundaries of the stability ($\Delta = 0$) and instability ($\Delta_0 = 0$) regions in this case coincide. On the contrary, when there is no destabilization, the system parameters must obviously meet the condition (11.61).

Note that small friction can stabilize an unstable equilibrium, while taking into account the quantities of the second order of smallness (ε^2) is significant. Indeed, the presence of the second term in the right part of formula (11.60) may slightly expand the stability area (11.55). The latter becomes important in cases when the system without taking into account friction is unstable or is located on the boundary of the instability region ($\Delta_0 \leqslant 0$).

References

1. N. Chetaev, *Ustoichivost' dvizheniya* (Dynamic stability). (Nauka Publ., Moscow, 1965)
2. M. Il'in, K. Kolesnikov, S. Yu.S., *Teoriya kolebanii: ucheb. dlya vuzov* (Theory of vibrations: textbook. for universities). (MGTU im. N.Eh. Baumana Publ., Moscow, 2003)
3. M. Leonov, *Osnovy mekhaniki uprugogo tela* (Fundamentals of elastic body mechanics). (Izd-vo AS Kirg. SSR, Frunze, 1963)
4. V. Molotnikov, *Osnovy teoreticheskoi mekhaniki* (Fundamentals of theoretical mechanics). (Feniks Publ., Rostov on Don, 2004)

Part II
Principal Variants of Mathematical Plasticity Theory

Chapter 12
Origin and Development of Plasticity Theory

12.1 Primary Definitions[1]

As said above (p. 10), the difference in distance between different particles of a body before and after loading characterizes permanent deformation. Experiments show that a property to acquire permanent deformations is observed in specific conditions in all real solid bodies. In many cases, permanent deformations are almost as high as elastic ones or even exceed them.

The studies have shown that for some bodies (steel, non-ferrous metals, their alloys, etc.), permanent deformations are accompanied by a residual change in body volume. We will call them plastic deformations. For elastic bodies (cast irons, soils, rocks, etc.), a residual change in the volume is observed. In this case, one always separates that part of permanent deformation that is not related to the change in body volume. Let us call this part of permanent deformation purely plastic or simply plastic deformation. The difference between full non-elastic deformation and purely plastic deformation is called softening deformation. Let us call negative softening deformation as compaction.

Experiments found that this component of non-elastic deformation accompanied plastic deformation in some conditions. In what follows, we will use the term "complex deformation" to denote such solid body deformation when it contains all three previously defined types of deformation, e.g. plastic and softening deformation.

---

[1]This chapter can be omitted during first reading.

© The Author(s), under exclusive license to Springer Nature Switzerland AG 2021   145
V. Molotnikov, A. Molotnikova, *Theory of Elasticity and Plasticity*,
https://doi.org/10.1007/978-3-030-66622-4_12

12.2   The Subject and Tasks of the Theory of Plasticity

The above shows that the study of solid body complex deformation should include the mechanics of plastic deformation, in other words, *the mathematical theory of plasticity*. The mathematical theory of plasticity is a section of mechanics of a continuous medium, a science of deforming plastic bodies.

The task of plasticity theory is to describe the load behavior of materials getting irreversible deformations provided that the dependency between stresses and strains contains no time. This is true in the first approximation for metals and their alloys usually used in engineering at normal temperatures. Therefore, plasticity theory, in fact, describes the behavior of steels, brass, aluminum alloys, and some other materials and alloys at normal and low temperatures. At high temperatures under load, creep occurs in materials and the link between stresses and strains greatly depends upon time. Creep theory describes these phenomena and we will not highlight them, except cases when the creep is found even at normal temperatures.

The mathematical theory of plasticity takes experimental data of observations over the macroscopic behavior of a solid body as initial positions. A systematic description of the mechanical properties of metal will be found by the reader in the third part of the Oding's book [59]. Among domestic studies of metal behavior in the case of elastic and non-elastic deformations, the classical monograph by Y. B. Fridman must be mentioned [13]. However, the mathematical theory of plasticity is not directly intended to physically describe the plasticity properties based on ideas of the real structure of materials. As in elasticity theory, a real solid body is replaced by its continuous ideal model, to which various material properties, obtained from an experiment with macro-samples, are attributed in an idealized state. It is suggested that the plastic resistance of a continuous homogeneous medium reproduces the behavior of a real solid body in an integral form.

The aim of plasticity theory is twofold: first of all, building sufficiently precise relations between stresses and strains aligned with the experiment as close as possible; secondly, developing mathematical methods of solving boundary problems, important for all applications.

In this manner, actually, two inter-related problems must be solved: (1) finding the determinant law and formulation of the closed boundary problem; (2) solution of the boundary problem in cases representing applied interest. The first problem is solved by a method traditional for the mechanics of deformable media. Experimental data is generalized and inputted into the theory as principles based on which a possible type of the determinant law is set (determinant equation). Determinant equations of plasticity theory have a long and complicated history. The development history of plasticity theory starting with the primary works by Saint-Venant and Levy (M. Levy, 1871) can be tracked using Russian translations of original articles published in the translation collection [65] "Plasticity Theory" (edited by Yu.N. Rabotnov). The collection includes 28 articles by Saint-Venant, Levy, R. von Mises, L. Prandtl, H. Hencky, A. Reuss, and W. Prager. These works reflect the origin and development of mathematical plasticity theory and allow getting familiar with its primary concepts, methods, and results, whose originality and distinction allowed the collection editor in 1948 to assert as follows: "This theory that is often called plasticity theory (in a narrow sense) can't be deemed final; however, the studies of recent years have definitely explored some principal laws that permit regarding many results as truly valid."

The below history of the origin and early stage of development of plasticity theory is mostly based on the materials of that collection.

As indicated above, the subject of research of plasticity theory is currently metals and their alloys, though it can be applied to such materials as rocks, soils, ice, etc .

The tasks of plasticity theory are rather diverse. All static and dynamic tasks of elasticity must be set for areas of plastic deformations. The thing is that in many structures of modern engineering, such deformations are inevitable, permitted, or desirable. Taking into account plastic deformations and creep, the problems of rigidity, stability, and strength must be solved, and the issues of the synthesis of structures of minimal weight, etc. must be raised. Plasticity theory must give methods to solve many problems of metalworking by pressure and permit raising the issues of quality.

12.3   Early Development Stages of Plasticity Theory

Plastic properties of various materials have been known for a long time and were studied by Coulomb (C. A. Coulomb, 1776) in the 1770s. An overview of these studies can be found in a book by S. P. Timoschenko [74]. It is notable that in the report of the French Academy of Science in 1773, Coulomb stated that the destruction of a prismatic specimen in compression occurred by sliding of one of its parts relative to the other one on a plane making an angle of 45° with the compression direction. Coulomb also found that sliding occurred when the tangential stress in the sliding plane reached a magnitude sufficient to overcome shear resistance along this plane.

After Coulomb, there were almost no experiments in plastic deformation for a long time, which was one of the reasons for the retarded birth of plasticity theory. The insufficiency of experimental capability has been typical of plasticity theory since its birth until now. The difference is that in the eighteenth–nineteenth centuries, there were few experiments, then there were too many experiments, with their results frequently not matching each other, or the same results being interpreted differently. This situation caused multiple versions of plasticity theory to appear, which sometimes contradicted each other.

Almost a 100 years after Coulomb's experiments, H. Tresca carried out systematic research of the plastic yield of metals and used Coulomb's idea to formulate a historically first condition of yield.

We have already said (p. 146) that the development history of plasticity theory starting with the fundamental works by Saint-Venant and M. Levy (1871) can be very well traced using the original articles [65] by its authors where one can find the bibliographic descriptions of the originals.

Using Tresca's results, B. Saint-Venant (1870) created the first mathematic plasticity theory. Apart from Coulomb's idea and the yield condition of Tresca, Saint-Venant built his theory based on the idea of the coinciding directions of the maximum shear speed and maximum tangential stress, as well as an idea of the non-compressibility of a body in plastic yielding. Saint-Venant's theory was created for the case of plane plastic deformation and basically was a generalization of the viscous liquid movement equation of Navier–Stokes.

M. Levy (1871) generalized Saint-Venant's results by writing the equation of the 3D-problem of the ideal plasticity theory. In the suggested ratios, the first three equations represented an equation of solid body equilibrium. The fourth equation expressed Tresca's yield condition and the fifth ratio is the mathematical expression of the body non-compressibility condition in plastic yield. These ratios were then supplemented by the condition of scalar linear dependency between the strain tensor and the stress speed tensor. In a particular case of plastic yield, Levy's ratios go to Saint-Venant's equation.

The detailed analysis of the Saint-Venant–Levy ratios can be found in an article by S. G. Mihklin [49].

12.4 Development of Plasticity Theory in the Twentieth Century

The end of the nineteenth century brought almost no new development to plasticity theory. In the early twentieth century, plasticity problems started to attract the attention of large scientists. Here we will describe only the most significant studies of that period.

In 1909, a paper by A. Haar and T. Karman appeared, which made an effort to obtain theory equations using a variation principle. It was stated that stresses minimized some functionality in an elastic–plastic condition of the body.

Later, Haar and Karman formulated the so-called full plasticity theory that D. D. Ivlev called fundamental for the entire theory of ideal plasticity. D. D. Ivlev believed that the condition of full plasticity allowed formulating the general theory of ideal plasticity with a unified mathematical tool of statically definable equations of a hyperbolic type corresponding to the shift nature of ideally plastic deformation.

However, it should be noted this point of view is not accepted by everyone. For example, A. A. Vakulenko and L. M. Kachanov believe that arguments of physical nature for the full plasticity scheme "... are dictated by the tempting simplicity of mathematical analysis rather than the essence of the issue" [26, p. 100]. A similar evaluation of the full plasticity condition is also given by R. Hill [30, p. 320–321] who calls the Haar–Karman condition as "artificial and unreal."

Together with this, other authors (see, for example, [26]) note that the solutions obtained under the scheme of full plasticity can be interesting.

In 1913, Richard Edler von Mises generalized and partially simplified the Saint-Venant–Levy ratios. The theory was based on [50, 65] the following ideas.

1. Any body is elastic if stresses are sufficiently low.
2. When reaching the elasticity limit, the body shows the properties of a non-compressible viscous liquid.
   Bringing this idea to a mathematical form, Mises comes to the Levi equations.
3. During plastic deformations, stresses remain the same as at the elastic limit. The concept of the elastic limit is defined as follows by Mises.
   Assume that σ_1, σ_2, σ_3 are principal stresses and

$$\tau_1 = \frac{\sigma_2 - \sigma_3}{2}, \quad \tau_2 = \frac{\sigma_3 - \sigma_1}{2}, \quad \tau_3 = \frac{\sigma_1 - \sigma_2}{2}.$$

Then

$$\tau_1 + \tau_2 + \tau_3 = 0. \tag{12.1}$$

By taking τ_1, τ_2, τ_3 as the coordinates of the 3D space point, Mises formulates the fourth hypothesis.

4. In plane (12.1), the elastic limit is depicted by a closed circle, for which the reference point is an internal point.

It can be easily shown that the Saint-Venant hypothesis is a particular case of the Mises fourth hypothesis. Indeed, since the maximum shear stress equals the biggest half-difference of principal stresses, according to Saint-Venant

$$|\tau_1| \leqslant K, \quad |\tau_2| \leqslant K, \quad |\tau_3| \leqslant K. \tag{12.2}$$

In the coordinates τ_1, τ_2, τ_3, equations-inequations (12.2) define a cube whose crossing with the plane (12.1) gives a regular hexagon. This hexagon is the closed outline described by Mises in the 4th hypothesis.

Mises then replaces the Saint-Venant hypothesis with another one, a simpler one, namely, he defines the elastic limit by crossing of the plane (12.1) with the sphere

$$\tau_1^2 + \tau_2^2 + \tau_3^2 = 2K^2.$$

We note that Mises does not give any assumptions to substantiate hypothesis No. 4 except for the simplicity of ratios resulting from it. It is also notable that in a plain case, the Mises hypothesis coincides with the Saint-Venant hypothesis.

L. Prandtl [62] considered a plain problem. Similarly to Mises, he replaces the Saint Venant hypothesis with a more general one, believing that the maximum shear stress is some function (f) of hydrostatic pressure.

$$\frac{\sigma_1 - \sigma_2}{2} = f\left[\frac{\sigma_1 + \sigma_2}{2}\right].$$

Assuming linear approximation for f in the first approximation, the author believes that

$$\frac{\sigma_2 - \sigma_1}{2} = K - a\frac{\sigma_1 + \sigma_2}{2}.$$

The further intensive development of plasticity theory took place in this period in the papers by Hencky [16], (for Russian translation, see [14]), Prandtl [62, 63], Reuss [66, 67], Hill [30], Nadai [56], Odquist [60], etc.

12.5   Soviet Period of Plasticity Theory Development

After the October Revolution of 1917, the development of mechanics in Soviet Russia and then in the Soviet Union was defined for many decades by healthy traditions of domestic science in general and by those scientists who inherited these traditions. They made an invaluable contribution to the upbringing and establishment of the first Soviet generations of scientists [81].

Starting from 1935, the USSR becomes the center of research activity in plasticity theory and retains leadership for more than 10 years. In chronological order, we will indicate the papers by S. G. Mikhlin [48] and S. L. Sobolev [73].

Fundamental papers in plasticity theory in the USSR appeared in 1936. They are associated with the names of A. A. Ilyushin and S. A. Khristianovich [32, 33]. After the war, more than two hundred papers were published in the periodicals of the Academy of Sciences. Their overview is given in the article by A. A. Vakulenko and L. M. Kachanov [26, p. 79–118], and in the seventh chapter of the book by J. Goodier and F. Hodge [15]. These papers were accompanied by developments of efficient experimental methods for researching the plasticity of materials in a complex stressed state basically in the case of simple loading.

The development of plasticity theory in the USSR was not isolated. Many studies were associated with developments of foreign scientists, were noted, admitted, and rarely criticized by Western colleagues.

At the brink of the 1950s, Prager developed the concept of a limit loading surface of strain-hardening materials and established a link of increments of plastic deformations with this surface in general, and also created the general Melan–Prager theory. This area was further developed by A. Yu. Ishlinsky, V. V. Novozhilov, Yu. I. Kadashevich [58], and other Soviet scientists.

In 1944, A. Yu. Ishlinsky studied the axisymmetric task of plasticity theory [21] suggesting the fulfillment of the full plasticity condition of Haar–Karnan by proving the static definability and hyperbolicity of principal equations. Using the numerical method, the authors of the same paper solved the problem of pressing a solid ball into an ideally plastic medium. The solution of A. Yu. Ishlinsky caused critical comments of R. Hill and others [36, 68, 75]. In particular, Hill believed that "... such calculations had a low and no value, since the Haar–Karman hypothesis for metals is physically unreal and it introduces an error of unknown magnitude." Hill based his objections on the impossibility to define the distribution of stresses within the Levi–Mises theory, which would satisfy the condition of full plasticity due to the overdetermination of the system of kinematic ratios. Hill's objection was later eliminated by Ishlinsky by substituting [22] the Levi–Mises law with the generalized flow law associated with the Coulomb–Tresca plasticity condition.

In these years, A. A. Ilyushin introduced the notions of director tensors, simple and complex loading, and theoretically proved the identity of primary plasticity theories existing at that time in the case of simple loading using the single general theory of low elastic–plastic deformations. Searching for ways to create the general plasticity theory in complex loading led A. A. Ilyushin to introducing linear coordinate Euclidian 5D spaces into the theory of plasticity, introducing the concepts of process image, isotropy postulate, and the principle of delay.

The need to check the isotropy postulate in complex trajectories of deformation led to the creation of automated test stations at MSU, the Institute of Mechanics of the Academy of Sciences of the USSR, and the Institute of Mechanics of the National Academy of Sciences of Ukraine. Systematic experiments for checking the isotropy postulate and the principle of delay were conducted by V. S. Lensky.

A huge contribution to the development of the new area in plasticity theory called the theory of processes was made by V. S. Lensky [38, 39], V. G. Zubchaninov [81], V. V. Moskvitin [54], and others. A significant contribution to the development of ideal plasticity theory and limit states was made by D. D. Ivlev, S. L. Khristianovich, A. Yu. Ishlinsky, V. V. Sokolovsky, E. I. Shemyakin, V. D. Klushnikov, etc.

In the 1990s, V. G. Zubchaninov [83, 84] developed a general theory of determinant ratios of the process theory. He guided (with the participation of A. A. Ilyushin) the project and built an SN-EVM automated test station used to conduct important systematic tests at complex loading of samples of structural materials.

Another area that originated in the West [6] and later actively developed [33, 34, 40, 42, 44, 46], [43, 53] in the USSR was the sliding concept in plasticity theory. Researchers intended to build determinant ratios which are true at arbitrary loading beyond the yield strength. The most successful results in the development of the sliding concept were obtained by M. Ya. Leonov and his school. An undeniable advantage of their results is the rehabilitation of the sliding concept by eliminating the flaw found by Cicala [8] and Iosimura [20] in the Batdorf and Budiansky model.

12.6 Russian Mechanics in the Post-Soviet Period

12.6.1 General Situation and Dangerous Trends

It is commonly admitted that mechanics is a driver of technical progress. It is undeniable that mathematics developed in the context of mechanics problems (Newton, Lagrange, Lyapunov, Pontryagin, Muskhelishvili). As early as the nineteenth century, mechanics was the queen of physical and mathematical sciences, and its brightest decoration was elasticity theory.

Mechanics is the scientific basis of most (if not all) areas of engineering: industry, construction, transport (land, water, air, space), agricultural equipment, and military equipment. No one doubts that all existing types of armament are the products of mechanics despite the high role of computer and informational technologies in its development. Therefore, it is clear that creating modern hi-tech types of armament without progress in mechanics is impossible.

The state has always paid high attention to the development of mechanics in Russia. It started before the Revolution and was continued after it. Soviet decrees in 1918 established the Physical-Technical Institute headed by A. F. Ioffe and the Central Aerohydrodynamic Institute headed by N. E. Zhukovsky and S. A. Chaplynin. This was the start of creating a wide network of institutes in natural science and engineering in the country.

In the 1930s–1970s, by virtue of multiple works and the pool of mechanical scientists living in those times, the USSR held a strong position in global mechanics. All universities of the country created mechanical and mathematical faculties, and almost all technical institutes established mechanical departments. The USSR Academy of Science entered IUTAM (International Union of Theoretical and Applied Mechanics) and was active in it, while Soviet scientists were always among the management of that union. A methodological council for the strength of materials, constructional mechanics, plasticity and elasticity theory worked in the Ministry of Higher and Secondary Vocational Education, which held crowded meetings of heads of mechanics departments once in 5 years. Mechanics was unequivocally admitted a fundamental science, a development basis for the economy and defense sector.

The development of many areas in mechanics was frequently dictated by the rapid development of engineering and technologies. As early as the 1930s, the engineering agenda included the problems of continuous movement of media where the medium changes its mechanical properties or even aggregate state under the action of high or rapidly changing temperatures. The solution of the most complicated problems of mechanics was supported by a reasonable strategy, state attention, and care.

Leadership in space and nuclear areas allowed creating the defensive shield of the country in the 1960s–1970s. The situation arose certain concerns across the ocean. Correctly understood reasons for lagging induced foreign leaders to respond immediately. For more than 10 years in the 1970s, the USA opened about

400 universities in physical-technical, technological, and mathematical areas! New universities invited (and still do) leading scientists and talented specialists from all over the world.

The position of mechanics in new Russia and the attitude toward it at the state level radically changed. During the first 10 years, the authorities discontinued the system of scientific institutes of the defense industry that actively worked over many fundamental problems of mechanics. Multiple highly qualified engineers and scientists who previously worked in the organizations of the Ministry of Medium Machine Building and then in the Ministry of Atomic Energy and Industry along with well-equipped laboratories became non-demanded. The further elimination of mechanics is described in detail in the lecture [7] of the Corresponding Member of the USSR Academy of Sciences at the Department of Technical Sciences (Mechanics), the Academician at the Department of Energy, Engineering, Mechanics and Control Processes (theoretical and applied mechanics, engineering and machine science), Secretary Academician of the Department, Professor G. G. Cherny. The below information is taken from this lecture read by G. G. Cherny at the Moscow Polytechnical Museum in 2008.

By the decision of authorities with the acquiescence of the Russian Academy of Sciences (RAS) and the entire scientific and technical community, mechanics was excluded from the list of fundamental sciences, and developments in the field of mechanics were excluded from the list of key technologies of federal value. The Russian Ministry of Education and Science conducts no contests for works in mechanics problems. After changing the RAS structure, the Department of Mechanics, Engineering, and Control Processes became the part of the Department of Energy, Engineering, Mechanics, and Control Processes. The leading role of mechanics in the previous department in the new structure of the department was lost.

The number of mechanical departments at technical universities greatly decreased due to their aggregation into departments of applied mechanics and modeling. Recently, the term "mechanics" has been eliminated from the names of university faculties and institutes. The Mechanics Institute at Saint Petersburg University was actually eliminated. At Ulyanovsk State University, the Mechanical and Mathematical Faculty became the Faculty of Mathematics and Computer Technologies. There are multiple examples of this kind. Even at Moscow State University, the "Mechanics" qualification that existed for more than 70 years has transformed into "Mechanics and Mathematical Modeling" under the pressure of the Ministry of Education and Science, and the qualification passport has almost no description of mechanics, since it concerns only mathematical modeling. The curriculum for specialists in "Applied Mechanics" at either classical or technical universities ("Dynamics and Strength of Machines", "Designing, Engineering and Production of Aircraft," etc.) has no such discipline as "Plasticity Theory." The discipline of "Experimental Methods of Studying Structural Strength" was allocated with 72 h at the Space Faculty of Moscow Aviation Institute, while 400 h were allocated for physical culture.

At provincial universities, the situation is even worse. If educational programs for training engineers of all mechanical qualifications in the Soviet times suggested a 3-semester course of theoretical mechanics, a 2-semester course of the strength of materials, a 2-semester course of the theory of mechanisms and machines, a 1–2-semester course of parts of machines and hoisting gear, and a 2-semester course of hydraulics, hydraulic machines, and hydraulic drive, now, instead of that depositary of fundamental knowledge, only a 36-h (or even 18-h) course of applied mechanics is adopted without any designing works, theses, projects, laboratory practicals. The word "calculation" in mathematic modeling for centuries used by engineers and designers in designing is replaced by the word "analysis." Software products of such "analysis" are frequently created by groups of high-class programmers individually without participation of competent specialists in the field of mechanics. An example is the CosmosDesign STAR application by SRAC in the SolidWorks package [2], where in the rush for universality and multi-functionality, the program includes ". . . stability analysis beyond elastic limit" based on the model of isotropic hardening that does not exist [2, p. 143] (quote: ". . . von Mises Plasticity isotropic"—a plastic body model with isotropic hardening and Mises plasticity condition). It does not take into account the dependency of plastic properties of the material on the loading trajectory, especially in its angular points that occur in the case of flexure–torsion loss of stability beyond the elastic limit.

The results of such "analysis" can be easily found: the collapse of the Moscow water park roofing (Yasenevo). The first conclusions regarding the reasons for this catastrophe were terrorism, poor quality of cement, and lack of columns supporting the roofing. Only after the second building designed by the same authors collapsed (building of the Basmanny Market in Moscow), the reasons became obvious. The investigation was summarized by the project head: ". . . software . . ."used for strength analysis is to be blamed. It is annoying that even in the presence of such late confessions, the criticism of these software applications (see, for example, [52]) is either left unnoticed or suppressed. In the existing situation, the twenty-first century gradually turns for Russia into the age of cruel man-made catastrophes, natural and environmental disasters. We frequently hear messages on falling rockets, airplanes, explosions at environmental hazardous industrial facilities, collapses of new buildings. Among the reasons are designing errors, first of all, related to the insufficient knowledge of mechanics laws and their incorrect use (lack of qualification).

A trend of the previous 10 years in the derogation of the place and role of mechanics in some fundamental sciences is completely unjustified, extremely harmful, and may cause heavy consequences. This trend will have detrimental consequences for fostering the economy of the country and especially for preserving and strengthening the state's defensive capacity. These consequences are already seen due to recent economic sanctions imposed on Russia by foreign partners.

12.6.2   Plasticity Theory in Russia in the Post-Soviet Period

The first half of the period after the USSR collapse was occupied by articles, monographs, study books, and training aids in the theory of plasticity that summarized the
results of authors' work for many years of the described developments and results
achieved in the Soviet times. This situation was expected and foreseeable since
irrespective of political events, the 1970s saw a stagnation in the development of
plasticity theory [41, p. 3]. As M. Ya. Leonov believes, the reason for the stagnation
was an extreme complexity of the problem of creating a general mathematical
theory of plasticity and traditional disregard of taking into account the physical
mechanisms of the non-elastic deformation process. In these conditions, a danger
was predicted that the mechanics of a deformable solid body would lose the nature
of a classical natural science and will become [41] "…rather abstract, on the one
hand (elasticity theory), and strictly concrete on the other hand (particular problems
of plasticity theory and destruction)."

These concerns were doomed not to become reality partially because the classical
mechanics of deformations was born along with a new science—mesomechanics
(for example, [45, 61, 72])—that studies the phenomenon of non-elastic deformation balancing between continuum mechanics and microcosm physics, more
inclined, however, to solid body physics. Without diminishing the usefulness
of mesomechanics methods, we should admit that in terms of usefulness for
engineering, the most perspective trend in developing non-elasticity theory is an
approach based on the continuum theory of dislocations [10, 51].

Here we will give only some most prominent Russian publications of this half-
period.

First of all, we must note a two-volume edition of collections by A.A. Ilyushin
[18, 19] that will allow newcomer researchers to get familiar with the original works
of the scientist from a single source for over 30 years of his career.

A monument of the ideal plasticity theory was the publication of a two-volume
book by D. D. Ivlev [23, 24]. Volume 1 includes the author's articles dedicated
to building and studying general ratios of the ideal plasticity theory based on a
statically definable system of equations of a hyperbolic type. The general theory is
described for compressible and anisotropic media, solutions are given for a number
of applied problems (pressing-in of indenters, implementation of rigid bodies, etc.).
Along with studies of problems of an ideal elastic–plastic body, the second volume
also studies models of a hardening plastic body and other complex media.

In terms of developing the theory, let us note the works by D. D. Ivlev together
with A. Yu. Ishlinskiy [25], and works with M. A. Artemov [4]. In all the above
works, the theory of ideal plasticity is developed. The publications by A. V. Kuptsov
[37] and N. D. Vereyko [79] are dedicated to the same problem. Unlike these
works, a 3D anisotropic state is considered in the article by Yu. N. Radayev and
S. Murakami [55].

Among experimental studies, the monograph by B. D. Annin and V. M. Zhigalkin
[3] published in 1999 represents special interest. The book presents experimental

results for the elastic–plastic deformation of several construction materials in the case of a complex stressed state. Special attention is paid to experiments where unloading is done on some areas affected by extreme tangential stresses and loading is done on other areas. It was found that in this case, significantly increased resistance of materials to plastic shear was possible, which is manifested by increased strength in specific directions at special loading trajectories. Let us note that such experiments were later duplicated in the PhD thesis by N. S. Adigamov [1] to justify synergetic representations in the theory of non-elastic deformations.

Among training aids, we must note the third edition of the book by Yu. N. Radayev [55] containing a complete and systematic description of methods and results related to a 3D problem of mathematic plasticity theory. From new study literature, we shall name the book by V. T. Sapunov [70, 71]. It introduces primary laws and equations and methods to solve boundary tasks of modern plasticity and creep theory; solutions are given for tasks of elastic plastic deformation and creep for some elements of structures as applicable to structural strength analysis in engineering. Special attention is paid to the simplicity of stating determinant equations of plasticity and creep theory and problems revealing the stress–strain state of structural elements.

According to studies made in the Soviet period, a number of theses in the theory of plasticity were presented in the 1990s. Among them is the paper by A.M. Kovrizhnykh [35] that proposes two models of a plastic body, which correlate in initial representations with the dependencies of A. K. Malmeister [46] and K. N. Rusinko [69], and the thesis by E. N. Erokhina [11] dedicated to the primary development of solution methods to the problems of applied plasticity theory.

In the second post-Soviet half-period, theoretical developments gradually emerge, and in new economic conditions, fundamental science becomes more oriented towards production needs. It must be noted here that a practical trend of Russian academician science was its distinctive feature after its birth. It is sufficient to recall M. V. Lomonosov and his theses in mining, metallurgy, etc. An example of such works in the development of mechanics of non-elastic deformations was the monograph by S. A. Khristianovich [31] published in 2008 that proposed and developed a new model of plastic deformation of materials in the case of complex loading applicable to the problem of sudden bursts of coal, rock, and gas. The proposed model explains the mechanism of a catastrophic phenomenon called a rock burst in geomechanics. In fact, the monograph solves a number of problems of the theory of rock bursts.[2]

New trends in the development of non-elasticity theory in the second half of the post-Soviet period include the so-called theory of processes. The primary ideas of that theory are based on the isotropy postulate of A.A. Ilyushin and the principle

---

[2]The theory of rock bursts was created by S. G. Avershin [5], a founder of the Soviet scientific school studying the process of rock movement and causes of rock bursts under the action of underground excavations.

of delay. The development of the theory of processes is described in the studies by A.A. Ilyushin, V.G. Zubchaninov [82–84], etc.

According to the isotropy postulate, for an isotropic material, the module of stress vector and its orientation angles in the Frenet frame are unambiguously defined by a change in process parameters from its start to the current moment, e.g. they are some functionalities. A complete definition of plasticity functionalities according to experimental data is extremely complicated, and so far building methods are proposed only to some of them. Another property of plasticity of an isotropic material reflects the principle of delay: the values of the orientation angles of the stress vector in the Frenet frame depend on the curvature change not along the entire preceding trajectory of strain, but along its last part, whose length characteristic of this material is called the trace of delay. This property made it possible to distinguish several types of processes (simple strain, low curvature, etc.) for which the ratios between stresses and elastic-strain deformations are established definitively and do not contain functionalities.

Along with strain theory, works are gradually revived, which develop other trends in plasticity theory. Let us note the thesis by I. S. Nikitin that uses a classical model of Batdorf and Budiansky to study deformations of layered and block media. The classical model of sliding is supplemented by an elastic–viscous–plastic model of the process.

Ideas of the theory of elastic–plastic processes served as a platform for a new trend in mechanics of non-elastic deformations that are called endochronic theories. In some theories, the dependencies between stress and strain look as functionalities. The founder of endochronic theory is K. Valanis who published his first works [76–78] dedicated to endochronic theory in the 1980s in collections of the American Society of Mechanical Engineers (ASME). Apart from the new name rooted in the global scientific literature, Valanis's theory contains no significant development of the theory of processes of A. A. Ilyushin, but just one of particular examples of models within this theory. The concept of internal time introduced by the author represents an equivalent of the internal geometry parameter of the strain trajectory, and setting the heritage-type functionality results from Ilyushin's principle of delay.

Among modern non-Russian publications on the endochronic theory of plasticity, we shall indicate the monographs by T. Nakai [57], H.-Y. Fang, J.L. Daniels [12] and the article by R.O. Davis and A.P.S. Selvadurai published at Cambridge University [9]. All three papers are dedicated to the use of endochronic theory in geomechanics.

In Russia, the development of endochronic theory was instigated at North-Western universities (Saint Petersburg, Perm) and then in the center (Moscow). The studies by Yu. I. Kadashevich and S. P. Pomytkin [27–29] formulate the endochronic theory of a tensor-parametric type for large deformations. The variation of the Kadashevich–Pomytkin theory takes into account the effects of time phenomena. This distinguishes the variation from classical phenomenologic theories where modeling of non-elastic deformations was traditionally done under flow theory.

The study of the practicability of the endochronic theory of plasticity is vastly described in the Internet publication by B. E. Melnikov and his colleagues [47] that discusses the possibility of using this theory for studying the deformation of construction materials and soils. The authors see the attractiveness of endochronic theory in the fact that in building determinant ratios, the theory abandons one of the basic principles of classical plasticity theory—the concept of flow surface. A link between stresses and strains in this theory is provided based on implementing new functions of a heritage type, which allows describing loading and unloading using a unified equation system. A conclusion is made that using the theory abstaining from the concept of loading (flow) surface is efficient in describing the deformation of geo-materials and construction materials, since building such a surface with any tolerance is rather complicated for these materials.

12.7   Abstract

To end this short overview, we shall conclude that currently, there are several various approaches to formulating the determinant laws of plasticity theory resulting in the division of plasticity theory into flow type theories, deformation type theories, sliding theories, statistical plasticity theories, endochronic theories of plasticity, etc. Approaches are also suggested to describe the processes of low non-elastic deformations of initially isotropic materials (synergetics).

The existing situation was differently assessed by representatives of various scientific schools: from the most optimistic: "Principal variants of the theory of plastic deformations differ in the completeness and perfection of mathematical formulations" A. A. Ilyshin, [17, p. 5], to the pessimistic one: "Both sliding theory and consideration of the simplest model lead to pessimistic conclusions as to the possible progress of plasticity theory" Yu. N. Rabotnov, [64, p. 104].

In any way, it should be admitted that a desire to describe the entire scope of non-elastic properties of a real body by a unified system of equations has not resulted in sufficiently definitive results. One should pay serious attention to non-competent (but ambitious) assertions about an allegedly created "...expansion of the unified theory of strength in plasticity theory leading to the unified processing of metal plasticity and plasticity of geo-materials in general" [80]. For many decades, the development of mechanics of non-elastic deformations will be represented by various simplified models suitable for describing principal phenomena in the deformation of a specific class of materials.

References

1. N. Adigamov, *Protsessy neobratimogo deformirovaniya i rezervy prochnosti materialov* (Processes of irreversible deformation and reserves of strength of materials). Dr. philos. sci. diss. KRSU Publ., Bishkek (2004), 210 p. PhD thesis, Bishkek, KRSU Publ. fizmathim/com/protsessy-neobratimogo-deformirovanija-i-rezervy-prochnosti-materialov. Accessed 20 Oct 2014

2. A. Alyamovskii, *SolidWorks. Komp'yuternoe modelirovanie v inzhenernoi praktike* (Solid-Works. Computer modeling in engineering practice). (BKHV-Peterburg Publ., Saint-Petersburg, 2006)
3. B. Annin, V. Zhigalkin, *Povedenie materialov v usloviyakh slozhnogo nagruzheniya* (Behavior of materials under complex loading conditions). (Nauka Publ., Novosibirsk, 1999)
4. M. Artemov, D. Ivlev, *O staticheskikh i kinematicheskikh sootnosheniyakh teorii ideal'noi plastichnosti pri kusochno-lineinykh usloviyakh tekuchesti* (On static and kinematic relations of the theory of ideal plasticity under piecewise linear flow conditions). Izv. RAN, MTT, no. 3 (1995), pp. 104–110
5. S. Avershin, *Sdvizhenie gornykh porod pri podzemnykh razrabotkakh* (Movement of rocks in underground mining). (Ugleizdat Publ., Moscow, 1947)
6. S. Batdorf, B. Budiansky, Splitting tests: an alternative to determine the dynamic tensile strength of ceramic materials, in *NACA TC1871* (1949), pp. 1–31
7. G. Chernyi, *slovo o vechno novoi mekhanike. lektsiya v moskovskom politekhnicheskom muzee 20 marta 2008 g.* (Word of forever new mechanics. lecture at the Moscow polytechnic museum March 20, 2008). http://new.math.msu.su/admission/Slovo
8. P. Cicala, On the plastic deformation, in *Atti Accad. naz. Lincei. Rend. Cl. Sci. fis. e. nature.*, V. 8 (1950) pp. 583–586
9. R. Davis, A. Selvadurai, *Plasticity and Geomechanics* (Cambridge University Press, Cambridge, 2002)
10. D. Ehshelbi, *Kontinual'naya teoriya dislokatsii* (Continual theory of dislocations). (IL Publ., Moscow, 1963)
11. E. Erokhina, *Prostranstvennye zadachi statiki sypuchikh sred* (Spatial problems of statics of granular media). Cand. Philos. Sci. Diss. Abstr., PhD thesis, Voronezh, VSU Publ., 2011
12. H.-Y. Fang, J. Daniels, *Introductory Geotechnical Engineering/An Environmental Perspective* (Taylor and Francis Publ., London, 2006)
13. Y. Fridman, *Mekhanicheskie svoistva metallov* (Mechanical properties of metals). (Oborongiz Publ., Moscow, 1952)
14. G. Genki, *Teoriya plastichnosti: sb. perev* (Theory of plasticity: a collection of translations) Moscow, Inostrannaya literatura Publ., chapter *K teorii plasticheskikh deformatsii i vyzyvae-mykh imi v materiale ostatochnykh napryazhenii* (On the theory of plastic deformations and their causes in residual stress material) (1948), pp. 114–135
15. D. Gud'er, F. Khodzh, *Uprugost' i plastichnost'* (Elasticity and plasticity). (Inostrannaya literatura Publ., Moscow, 1960)
16. H. Hencky, Zur theorie plastischer deformationen und der hierdirch im material hervorgerufenen nachspannungen, in *Ztschr. Angew. Math. und Mech. Bd. 4. H. 4* (1924), pp. 323–334
17. A. Il'yushin, *Plastichnost'. (Osnovy obshchei matematicheskoi teorii)* (Plasticity. Fundamentals of general mathematical theory). (Izd-vo AS USSR Publ., Moscow, 1963)
18. A. Il'yushin, *Trudy. T. 1 (1935–1945) / Sostaviteli: E.A. Il'yushina, N.R. Korotkina* (Scientific work. Vol. 1 (1935–1945). Compiled By: E.A. Ilyushina, N.R. Korotkina). (Fizmatlit Publ., Moscow, 2003)
19. A. Il'yushin, *Trudy. T. 2 (1946–1966) / Sostaviteli: E.A. Il'yushina, N.R. Korotkina* (Scientific work. Vol. 2 (1946–1966). Compilers: E.A. Ilyushina, N.R. Korotkina). (Fizmatlit Publ., Moscow, 2004)
20. I. Iosimura, *Mekhanika. Periodich. sb. perev. inostr. statei* (Mechanics. Periodicity. collection of pens. foreign. articles), no. 2 (60), Moscow, Mir Publ., chapter Zamechaniya k teorii skol'zheniya Batdorfa i Budyanskogo (Remarks on the theory of sliding by Batdorf and Budiansky) (1960), pp. 109–116
21. A. Ishlinskii, *Osesimmetrichnaya zadacha teorii plastichnosti i proba Brinellya* (Axisymmetric problem of plasticity theory and the Brinell test). Prikl. matem. i mekh-ka **8**(3), 201–224 (1944)
22. A. Ishlinskii, *Obshchaya teoriya plastichnosti s lineinym uprochneniem* (General theory of plasticity with linear hardening). Ukr. mat. zhurn. **6**(3), 314–325 (1954)
23. D. Ivlev, *Mekhanika plasticheskikh sred: v 2 t. T.1.: Teoriya ideal'noi plastichnosti* (Mechanics of plastic media: in 2 vols. Vol. 1.: Theory of ideal plasticity). (Fizmatlit Publ., Moscow, 2001)

24. D. Ivlev, *Mekhanika plasticheskikh sred: v 2 t. T.2.: Obshchie voprosy. Zhestkoplasticheskoe i uprugo-plasticheskoe sostoyanie tela. Uprochnenie. Deformatsionnye teorii. Slozhnye sredy* (Mechanics of plastic media: in 2 vols. 2.: General questions. Rigid-plastic and elastic-plastic state of the body. Strengthening. Deformation theories. Complex environment). (Fizmatlit Publ., Moscow, 2001)

25. D. Ivlev, A. Ishlinskii, *Polnaya plastichnost' v teorii ideal'no-plasticheskogo tela* (Complete plasticity in the theory of an ideal plastic body). Dokl. RAN (RAS Reports) **368**(3), 333–334 (1999)

26. L. Kachanov, *Issledovaniya po uprugosti i plastichnosti: sb. nauchn. tr. Vyp. 8.* (Studies elasticity and plasticity: collection of scientific. Tr. Vol. 8). Leningrad, Izd-vo LGU Publ., chapter *Ob ehksperimental'nom opredelenii posleduyushchikh poverkhnostei nagruzheniya i ehffekta Baushingera* (About experimental determination of the following loading surfaces and the Bauschinger effect) (1971), pp. 108–112

27. Y. Kadashevich, S. Pomytkin, *Opisanie ehffektov vtorogo poryadka v ramkakh ehndokhronnoi teorii neuprugosti dlya bol'shikh deformatsii* (Description of second order effects within the framework of endochronic theory of inelasticity for large deformations). Izv. RAN. MTT [Izv. Russian Academy of Sciences. MTT] **6**, 123–136 (2010)

28. Y. Kadashevich, S. Pomytkin, *Uprugost' i neuprugost': sb. nauchn. tr.* (Elasticity and inelasticity: a collection of scientific papers). Moscow, Izd-vo MGU Publ., chapter *Ehtapy razvitiya ehndokhronnoi teorii neuprugosti* (Stages of development of the endochronous theory of inelasticity). *Uprugost' i neuprugost': sb. nauchn. tr. [Elasticity and inelasticity: a collection of scientific papers]* (2011), pp. 232–235

29. Y. Kadashevich, S. Pomytkin, *Ehndokhronnaya teoriya plastichnosti, obobshchayushchaya teoriyu Sandersa-Klyushnikova* (Endochronic theory of plasticity generalizing the theory of Sanders-klyuchnikova). Inzh.-stroit. zhurnal [Eng. - Builds. Journal] **1**, 82–86 (2013)

30. R. Khill, *Matematicheskaya teoriya plastichnosti* (Mathematical theory of plasticity). (IL Publ., Moscow, 1956)

31. S. Khristianovich, *Problemy teorii plastichnosti v geomekhanike* (Problems of plasticity theory in geomechanics). (Nauka Publ., Moscow, 2008)

32. S. Khristianovich, S. Mikhlin, B. Devison, *Nekotorye novye voprosy mekhaniki sploshnoi sredy* (Some new questions mechanics of a continuous environment). (Izd. AN SSSR Publ., Moscow, 1938)

33. S. Khristianovich, E. Shemyakin, *O ploskoi deformatsii plasticheskogo materiala pri slozhnom nagruzhenii* (About flat deformation of plastic material under complex loading). Mekh-ka tv. tela (Mechanics of a solid body) **5**, 138–149 (1969)

34. V. Klushnikow, Stability of materials and structures, in *Plasticity and Failure Behaviour of Solids* (1990), pp. 201–211

35. A. Kovrizhnykh, *Dilatansionno-sdvigovaya model' v teorii plastichnosti metallov i geomateri-alov* (Dilatation-shift model in the theory of plasticity of metals and geomaterials). PhD thesis, Novosibirsk, In-t theory. and Appl. mekh-Ki SO an RAS Publ., 1993

36. D. Kuhlmann-Wilsdorf, Stability of materials and structures, in *Plasticity and Failure Behaviour of Solids* (1990), pp. 201–211

37. A. Kuptsov, *Postroenie ustoichivykh konechno-raznostnykh skhem prostranstvennykh zadach ideal'noi plastichnosti pri uslovii Mizesa* (Construction of stable finite-difference schemes for spatial problems of ideal plasticity under the Mises condition), in *Vestnik PMM (Bulletin of the AMM). izz. 6.* (Voronezh, VSU Publ., 2007), pp. 83–89

38. V. Lenskii, *Vvedenie v teoriyu plastichnosti: vyp. 1* (Introduction to the theory of plasticity: vol 1). (MGU Publ., Moscow, 1968)

39. V. Lenskii, *Vvedenie v teoriyu plastichnosti vyp.* (Introduction to the theory of plasticity vol. 2). (MGU Publ., Moscow, 1969)

40. M. Leonov, *Osnovnye postulaty teorii plastichnosti* (The basic postulates of the theory of plasticity). DAN SSSR [Reports of the USSR Academy of Sciences] **199**(1), 51–54.

41. M. Leonov, *Mekhanika deformatsii i razrusheniya* (Deformation and fracture mechanics). (Izd-vo Ilim Publ., Frunze, 1981)

42. M. Leonov, V. Molotnikov, *K teorii deformatsii metallov s yarko vyrazhennym predelom tekuchesti* (On the theory of deformations of metals with bright expressed yield strength), in *Izv. AN Kirg. SSR (Izv. Academy of Sciences of Kyrghyz. SSR)*, no. 6 (1974), pp. 3–10

43. M. Leonov, V. Molotnikov, B. Rychkov, *Polzuchest' tverdogo tela: sb. nauchn. tr.* (Creep solid body: collection of scientific. work). Frunze, Izd-vo Ilim Publ., chapter *Nekotorye obobshcheniya kontseptsii skol'zheniya v teorii plastichnosti* (Some generalizations the concept of slip theory of plasticity) (1974), pp. 12–35

44. M. Leonov, N. Shvaiko, *Slozhnaya ploskaya deformatsiya* (Complex plane deformation), dan sssr (Reports of the USSR academy of sciences). DAN SSSR (Reports of the USSR Academy of Sciences) **159**(5), 1007–1010 (1964)

45. V. Malinin, N. Malinina, *Strukturno-analiticheskaya mezomekhanika deformiruemogo tv"erdogo tela* (Structural and analytical mesomechanics of a deformable solid). Fizicheskaya mezomekhanika [Physical mesomechanics] **8**(5), 31–45 (2005)

46. A. Malmeister, V. Tamuzh, G. Teters, *Soprotivlenie zhestkikh polimernykh materialov* (Resistance of rigid polymeric materials). (Izd-vo Zinatne Publ., Riga, 1972)

47. B. Mel'nikov, N. Chernysheva, N. Podgornaya, I. Chigareva, *Osobennosti primeneniya ehndokhronnoi teorii plastichnosti dlya izucheniya deformirovaniya geomaterialov i stroitel'nykh materialov* (Features of application of endochronous plasticity theory for studying the deformation of geomaterials and building materials). Internet-zhurnal Stroitel'stvo unikal'nykh zdanii i sooruzhenii **2**(7), 71–77 (2013). (Online journal Construction of unique buildings and structures) http://unistroy.spb.ru/. Accessed 18 Oct 2014

48. S. Mikhlin, *Osnovnye uravneniya matematicheskoi teorii plastichnosti* (Basic equations of the mathematical theory of plasticity). (Izd-vo AN SSSR Publ., Leningrad, 1934)

49. S. Mikhlin, *Sovremennoe sostoyanie matematicheskoi teorii plastichnosti* (Current state of the mathematical theory of plasticity), in *Uspekhi matematicheskikh nauk (UMN) [Advances in mathematical Sciences (UMN)]*, no. 3 (1937), pp. 175–193

50. R. Mises, Mechanik der festen körper im plastischdeformablen zustand, in *Göttinger: Königlichen Gesellschaft* (1913), pp. 582–592

51. V. Molotnikov, *Materialy konferentsii molodykh uchenykh (Frunze, Izd-vo MSKh Kirg. SSR)* (Materials of the conference of young people scientists). Frunze, publishing house of the Ministry of agriculture of the Kirg. SSR, chapter *Klinovye dislokatsii v ploskoi zadache mekhaniki neuprugogo tverdogo tela* (Wedge dislocations in a plane problem of mechanics an inelastic solid) (1981), pp. 18–23

52. V. Molotnikov, A. Molotnikova, *Neuprugaya deformatsiya tverdogo tela* (Inelastic deformation of a solid body), in *Sovremennye innovatsionnye tekhnologii v s/kh mashinostroenii: materialy mezhdunar. nauchn.-pr. konf. Interagromash. (Rostov n/D.)* (Modern innovative technologies in agricultural engineering: proceedings of the international. scientific-Ave. Conf. Interagromash. (Rostov-on-Don)]m (2007), pp. 144–145

53. V. Molotnikov, A. Molotnikova, *Plosko-plasticheskaya deformatsiya* (Flat-plastic deformation), in *Sostoyanie i perspektivy razvitiya sel'skokhozyaistvennogo mashinostroeniya: materialy mezhdunar. nauchn.-pr. konf. "Interagromash"* [State and prospects of agricultural engineering development: materials of the international conference. nauchn. - Ave. Conf. "Interagromash"] (2009), pp. 186–189

54. V. Moskvitin, *Plastichnost' pri peremennykh napryazheniyakh* (Plasticity under variable loads). (Izd-vo MGU Publ., Moscow, 1965)

55. S. Murakami, Y. Radaev, *Matematicheskaya model' tr"ekhmernogo anizotropnogo sostoyaniya povrezhd"* (Mathematical model of a three-dimensional anisotropic state of damage) *Izv. RAN, MTT. (Izvestiya RAS, solid state Mechanics)*, no. 4 (1996), pp. 94–110

56. A. Nadai, The forces required for rolling steel strip under tension. J. Appl. Mech **6**(2), 54–62 (1939)

57. T. Nakai, *Constitutive Modeling of Geomaterials. Principles and Applications* (CRC Press, 2010)

58. V. Novozhilov, *I eshche o postulate izotropii* (And more about the postulate of isotropy). *Izv. AN SSSR. OTN. Mekh-ka i mashinostr. [Izvestia of the USSR Academy of Sciences. Department of technical Sciences. Mechanics and mechanical engineering]*, no. 1 (1962), pp. 205–208

59. I. Oding, *Prochnost' metallov* (Strength of metals). (ONTI Publ., Moscow, Leningrad, 1935)
60. F. Odquist, Verfestigung flusseisenahnlichen korpern. Zeitschr. for Angewandte Math. und Mech. **13**(5), 360–363 (1933)
61. V. Panin, *Osnovy fizicheskoi mezomekhaniki* (Fundamentals of physical mesomechanics). Fizicheskaya mezomekhanika (Physical mesomechanics) **1**(1), 5–22 (1998)
62. L. Prandtl, Ueber die hart plastischer korper. *Gottingen Nachrichten* (1920), pp. 74–85
63. L. Prandtl, Spannungsverteilung in plastischen korper, in *Intern. Congr. for Appl. Mech., Delft.* (1924), pp. 43–54
64. Y. Rabotnov, *Polzuchest' ehlementov konstruktsii* (Creep of structural elements). (Nauka Publ., Moscow, 1966)
65. Y. Rabotnov (ed.), *Teoriya plastichnosti: sb. perev. inostr. statei* (Plasticity theory: foreign translation collection. Articles). (IL Publ., Moscow, 1948)
66. A. Reuss, Vereintachte berechnung der plastischen foamänderung in der plastizitetstheorie, in *Zeitsch. angew. Math. und Mech. Bd. 10. H. 2* (1930), pp. 266–274
67. A. Reuss, Anisotropy caused by strain, in *Proc. IV Intern. Congr. for Appl. Mech.* (1935)
68. O. Richmond, Plane strain necking of v-notched and un-notched tensile bars. J. Mech. And Phys. Solids, 17(2), 83–90 (1969)
69. K. Rusinko, E. Blinov, *Analiticheskoe issledovanie sootnosheniya napryazhenie–deformatsiya pri proizvol'noi traektorii nagruzheniya* (Analytical study of the ratio stress-strain at an arbitrary loading path). Mekhanika polimerov (Mechanics of polymers) **6**, 981–986 (1971)
70. V. Sapunov, *Osnovy teorii plastichnosti i polzuchesti: uchebn. posobie* (Fundamentals of the theory of plasticity and creep: textbook. stipend). (Mosk. inzh.-fiz. in-t Publ., Moscow, 2008)
71. V. Sapunov, *Teoriya plastichnosti: ucheb. posobie* (Theory of plasticity: textbook. stipend). (NIYAU MIFI Publ., Moscow, 2010)
72. D. Si, *Mezomekhanika, ponyatie segmentatsii i mul'tiskeilingovyi podkhod: nano-mikro-makro* (Mesomechanics, the concept of segmentation and a multi-skaling approach: nano-micro-macro). Fizicheskaya mezomekhanika (Physical mesomechanics) **11**(3), 5–18 (2008)
73. S. Sobolev, The problem of propagation of plastical state, in *Works Seismological Institute of the USSR Academy of Sciences*, no. 49 (1935), pp. 28–39
74. S. Timoshenko, *Istoriya nauki o soprotivlenii materialov s kratkimi svedeniyami iz istorii teorii uprugosti i teorii sooruzhenii. 2-e izd.* (History of resistance science materials with brief information from the history of the theory of elasticity and the theory of structures. The 2nd prod.). (Moscow, GITTL Publ., 1935)
75. Tomas, The general theory of compatibility conditions. Jnt. I. Eng. Sc. **4**(3), 207–233 (1966)
76. K. Valanis, A theory of viscoplasticity without a yield surfaces. Arch. Mech. stosow. **23**(4), 517–551 (1971)
77. K. Valanis, *Obosnovanie ehndokhronnoi teorii plastichnosti metodami mekhaniki sploshnoi sredy* (Substantiation of the endochronous theory of plasticity by methods of continuum mechanics). *Tr. ASME. Teoreticheskie osnovy inzhenernykh raschetov* (Work ASME. Theoretical foundations of engineering calculations), no.4, vol. 106, (1984), pp. 72–81
78. K. Valanis, J. Fan, *Plastycity today: modelling. meth. And apl. London–New York*, Cambridge University Press, chapter Experimental verification of endochronic plasticy in Spatially varying strain fields (1985), pp. 153–174
79. N. Verveiko, A. Kuptsov, *Pole skorostei prostranstvennoi staticheski opredelimoi zadachi ideal'noi plastichnosti* (The velocity field of a statically determinate problem of ideal plasticity). Vestnik VGU. Seriya fiz.-matem. (VSU Bulletin. Series Phys.-mod.) **2**, 174–179 (2006)
80. M.H. Yu, *The Generalized Plasticity* (Springer, Berlin, 2006)
81. V. Zubchaninov, *Osnovy teorii uprugosti i plastichnosti* (Fundamentals of the theory of elasticity and plasticity). (Vyssh. shk. Publ., Moscow, 1990)
82. V. Zubchaninov, *Problemy mekhaniki deformiruemogo tverdogo tela: sb. nauchn. tr.* (Problems of mechanics of a deformable solid: Sat. scientific tr). SPb, SPBGU Publ., chapter *Gipoteza ortogonal'nosti v teorii plastichnosti* (Orthogonality hypothesis in the theory of plasticity) (2002), pp. 137–140

83. V. Zubchaninov, *Matematicheskaya teoriya plastichnosti* (Mathematical Theory of Plasticity) (TGTU Publ., Tver, 2003)
84. V. Zubchaninov, *Ustoichivost' i plastichnost': v 2 T.: T. 2 Plastichnost'* (Stability and plasticity: in 2 t. T. 2: Plasticity). (Tver', TGTU Publ., 2006)

Chapter 13
Initial Concepts of Plasticity Theory

13.1 Second-Rank Tensor in Euclidean Space

Let us select two arbitrary coordinate systems (x_1, x_2, x_3) and (x_1', x_2', x_3') in the Euclidean space. Let us designate unit vectors as $(\vec{e}_1, \vec{e}_2, \vec{e}_3)$ and $(\vec{e}_1', \vec{e}_2', \vec{e}_3')$.

The vector $\vec{a}$ in the first coordinate system can be written as

$$\vec{a} = \sum_{i=1}^{3} a_i \vec{e}_i = a_i \vec{e}_i, \quad (i = 1, 2, 3),$$

where a_i are the components of the vector $\vec{a}$ in the first coordinate system; the summing symbol in the second form of writing is omitted since it is suggested on default in vector and tensor designations that summing is done over repeated indexes.

Similarly to the previous formula, the vector $\vec{a}$ in the second coordinate system is represented as follows:

$$\vec{a} = a_i' \vec{e}_i',$$

the components (a_i') of the vector $\vec{a}$ in the second system of coordinates are expressed through the components (a_i) of this vector in the first system under the formulas

$$a_i' = \beta_{ik} a_k, \quad \beta_{ik} = \vec{e}_i' \vec{e}_k = \cos(\widehat{\vec{e}_i', \vec{e}_k}). \tag{13.1}$$

In the same manner, the vector components in the first coordinate system are expressed through the components in the second system using the formulas

$$a_i = \beta_{ki} a_k', \quad \beta_{ki} = \vec{e}_k' \vec{e}_i = \cos(\widehat{\vec{e}_k', \vec{e}_i}). \tag{13.2}$$

© The Author(s), under exclusive license to Springer Nature Switzerland AG 2021
V. Molotnikov, A. Molotnikova, *Theory of Elasticity and Plasticity*,
https://doi.org/10.1007/978-3-030-66622-4_13

Dependencies (13.1) and (13.2) are used as a basis to adopt as follows.

Definition 1 A vector is an aggregate of three values a_1, a_2, a_3 related to this coordinate system and transformed when switching to the other system using formulas *(13.1)* and *(13.2)*.

Similarly to the above, the following is adopted.

Definition 2 A second-rank tensor in the Euclidean space is an aggregate of nine values A_{ij}, $(i, j \sim 1, 2, 3)$ related to this coordinate system and transformed when switching to another coordinate system under the law

$$A'_{ij} = \beta_{ik}\beta_{js}A_{ks},\tag{13.3}$$

$$A_{ij} = \beta_{ki}\beta_{sj}A'_{sk}.\tag{13.4}$$

Here A_{ij} are tensor components in the previous coordinate system, A'_{ij} are tensor components in the new coordinate system, β_{ki} as defined above, and summing is done over repeated indexes.

If $A_{ij} = A_{ji}$, the second-rank tensor is called *symmetric*. As known [2], the symmetric matrix A_{ij} can be represented in a diagonal view. To do it, such coordinate axes must be found where all components not located on the diagonal turn to zero. These coordinate axes are called principal. In the principal axes,

$$A_{ij} = \begin{vmatrix} \neq 0 \text{ for } i = j, \\ 0 \ \ \text{ for } i \neq j \end{vmatrix}.\tag{13.5}$$

Tensor components other than zero are referred to as its principal values. They are defined as the own values of the matrix A_{ij} corresponding to the own vectors of this matrix by solving the standard problems for own values using the Jacobi rotation method.

Principal tensor values do not depend on selecting the coordinate system and so they are invariants. Any scalar function from second-rank tensor principal values is also invariant.

13.2   Tensors in Plasticity Theory

Assume that a body is exposed to the action of a system of forces $\{P\}$. The stress and strain state in each point of a body is characterized by stress (T_σ) and strain tensors (T_ε)

$$T_\sigma = \begin{pmatrix} \sigma_x & \tau_{xy} & \tau_{xz} \\ \tau_{yx} & \sigma_y & \tau_{yz} \\ \tau_{zx} & \tau_{zy} & \sigma_z \end{pmatrix}, \quad T_\varepsilon = \begin{pmatrix} \varepsilon_x & \frac{1}{2}\gamma_{xy} & \frac{1}{2}\gamma_{xz} \\ \frac{1}{2}\gamma_{yx} & \varepsilon_y & \frac{1}{2}\gamma_{yz} \\ \frac{1}{2}\gamma_{zx} & \frac{1}{2}\gamma_{zy} & \varepsilon_z \end{pmatrix}. \tag{13.6}$$

To designate stress and strain components, we will use the following designations:

$$\sigma_{ij}, \quad \varepsilon_{ij}, \quad (i, j \sim x, y, z),$$

whereas

$$\sigma_{ij} = \begin{vmatrix} \sigma_i & \text{at } i = j, \\ \tau_{ij} & \text{at } i \neq j; \end{vmatrix} \quad \varepsilon_{ij} = \begin{vmatrix} \varepsilon_i & \text{at } i = j, \\ \frac{1}{2}\gamma_{ij} & \text{at } i \neq j. \end{vmatrix}$$

For reduction purposes, we will write tensors (13.6) as follows:

$$T_\sigma = \{\sigma_{ij}\}, \quad T_\varepsilon = \{\varepsilon_{ij}\}.$$

Below, deformations mean low deformations unless agreed otherwise. The components of low deformations are expressed through the components u_i, ($i \sim x, y, z$) of the displacement vector $\vec{u}$ using formulas (2.2), (2.3) that can be represented as follows:

$$\varepsilon_{ij} = \frac{1}{2}\left(u_{i,j} + u_{j,i}\right), \tag{13.7}$$

where the comma after the first index means differentiation by variable corresponding to the second index; for example,

$$u_{x,y} = \frac{\partial u_x}{\partial y}; \quad u_{z,x} = \frac{\partial u_z}{\partial x};$$

$$\varepsilon_{xx} = \frac{1}{2}\left(\frac{\partial u_x}{\partial x} + \frac{\partial u_x}{\partial x}\right) = \frac{\partial u_x}{\partial x};$$

$$\varepsilon_{zx} = \frac{1}{2}\gamma_{zx} = \frac{1}{2}\left(\frac{\partial u_z}{\partial x} + \frac{\partial u_x}{\partial z}\right).$$

Apart from stress and strain tensors (13.6), plasticity theory also uses *stress and strain rate tensors:*

$$T_{\dot{\varepsilon}} = \{\dot{\varepsilon}_{ij}\}, \quad T_{\dot{\sigma}} = \{\dot{\sigma}_{ij}\}, \tag{13.8}$$

where $\dot{\varepsilon}_{ij}$ are the components of the strain rate tensor

$$\dot{\varepsilon}_{i,j} = \frac{1}{2}\left(\dot{u}_{i,j} + \dot{u}_{j,i}\right), \quad \left(\dot{u} = \frac{du}{dt}\right),$$

and then $\dot{\sigma}_{ij} = d\sigma_{ij}/dt$ is the stress rate, whereas t is the time or any other monotonously increasing parameter. Below we will adopt that a point over the value designation means a time derivative unless indicated otherwise.

Due to the law of twoness of tangential stresses (p. 12) $\sigma_{ij} = \sigma_{ji}$, e.g. the stress tensor T_σ is symmetric. As applicable to the stressed state, formula (13.3) yields

$$\sigma'_{ij} = \beta_{ik}\beta_{js}\sigma_{ks},$$

or in an expanded view

$$\sigma'_{ij} = \beta_{i1}\beta_{js}\sigma_{1s} + \beta_{i2}\beta_{js}\sigma_{2s} + \beta_{i3}\beta_{js}\sigma_{3s}$$

$$= \beta_{i1}\beta_{j1}\sigma_{11} + \beta_{i1}\beta_{j2}\sigma_{12} + \beta_{i1}\beta_{j3}\sigma_{13} + \ldots + \beta_{i3}\beta_{j3}\sigma_{33}$$

$$= \left| \text{we give similar terms, considering that } \sigma_{ij} = \sigma_{ji} \text{ and } \beta_{ij} = \beta_{ji} \right|$$

$$= \beta_{i1}\beta_{j1}\sigma_{11} + \beta_{i2}\beta_{j2}\sigma_{22}$$

$$+ \beta_{i3}\beta_{j3}\sigma_{33} + 2(\beta_{i1}\beta_{j2}\sigma_{12} + \beta_{i2}\beta_{j3}\sigma_{23} + \beta_{i3}\beta_{j1}\sigma_{31}). \qquad (13.9)$$

In the principal axes, the stress tensor will be

$$T_\sigma = \begin{pmatrix} \sigma_1 & 0 & 0 \\ 0 & \sigma_2 & 0 \\ 0 & 0 & \sigma_3 \end{pmatrix}.$$

The principal values (σ_1, σ_2, σ_3) of the stress tensor are referred to as principal stresses. They are defined as roots of a characteristic equation [2]

$$\begin{vmatrix} \sigma_x - \sigma & \tau_{xy} & \tau_{xz} \\ \tau_{yx} & \sigma_y - \sigma & \tau_{yz} \\ \tau_{zx} & \tau_{zy} & \sigma_z - \sigma \end{vmatrix} = 0. \qquad (13.10)$$

Equation (13.10) can also be written as follows:

$$[\sigma_{ij} - \sigma\delta_{ij}] = 0, \qquad (13.11)$$

where δ_{ij} is defined by the formula

$$\delta_{ij} = \begin{vmatrix} 1 \text{ for } i = j, \\ 0 \text{ for } i \neq j, \end{vmatrix}$$

and is referred to as the Kronecker symbol ([5, p. 115]). In the principal axes, the characteristic equation (13.11) looks as follows:

$$(\sigma - \sigma_1)(\sigma - \sigma_2)(\sigma - \sigma_3) = 0, \tag{13.12}$$

or

$$\sigma^3 - I_\sigma \sigma^2 + II_\sigma - III_\sigma = 0, \tag{13.13}$$

where

$$I_\sigma = \sigma_1 + \sigma_2 + \sigma_3,$$
$$II_\sigma = \sigma_1\sigma_2 + \sigma_2\sigma_3 + \sigma_3\sigma_1,$$
$$III_\sigma = \sigma_1\sigma_2\sigma_3 = det[\sigma_{ij}], \tag{13.14}$$

whereas σ_1, σ_2, σ_3 are principal stresses. The values $I_\sigma, \ldots$ defined by formulas (13.14) are referred to as *principal invariants of the stress tensor.*

The characteristic equation and principal invariants of the strain tensor are written in a similar way.

$$(\varepsilon - \varepsilon_1)(\varepsilon - \varepsilon_2)(\varepsilon - \varepsilon + 3) = 0,$$

or

$$\varepsilon^3 - I_\varepsilon \varepsilon^2 + II_\varepsilon \varepsilon - III_\varepsilon = 0,$$

where

$$I_\varepsilon = \varepsilon_1 + \varepsilon_2 + \varepsilon_3,$$
$$II_\varepsilon = \varepsilon_1\varepsilon_2 + \varepsilon_2\varepsilon_3 + \varepsilon_3\varepsilon_1,$$
$$III_\varepsilon = \varepsilon_1\varepsilon_2\varepsilon_3 = det[\varepsilon_{ij}]. \tag{13.15}$$

13.3 Decomposition of Stress and Strain Tensors

A fundamental hypothesis in the theory of elastic–plastic deformations of solid bodies is a suggestion that the strain tensor beyond the elastic limit (T_ε) can be decomposed into the tensor of elastic deformations (T_{ε^y}) and tensor of plastic deformations (T_{ε^e}):

$$T_\varepsilon = T_{\varepsilon^y} + T_{\varepsilon^e} = \{\varepsilon_{ij}^y\} + \{\varepsilon_{ij}^e\}. \tag{13.16}$$

This hypothesis propagates phenomena of disappearance of elastic deformation and preservation of non-elastic deformations observed in uniaxial elongation experiments in the case of full unloading after pre-elongation beyond the elastic limit.

Let us designate the components of plastic deformation as e_{ij}, assuming for example:

$$\text{it } i = j \text{ then } e_{xx} = \varepsilon_x^e,$$
$$\text{if } i \neq j \text{ then } e_{xy} = \frac{1}{2}\gamma_{xy}^e.$$

Since the components of elastic deformation are unambiguously defined through stresses under Hooke's law, the fundamental problem of plasticity theory is defining the link between the components of plastic deformation (the rate of plastic deformation, to be more precise), the stressed state at this point of time, and *loading history*.

Let us represent the stress tensor T_σ as the sum of two tensors:

$$T_\sigma = S_\sigma + D_\sigma, \tag{13.17}$$

where

$$S_\sigma = \begin{pmatrix} \sigma_0 & 0 & 0 \\ 0 & \sigma_0 & 0 \\ 0 & 0 & \sigma_0 \end{pmatrix}, \quad \left(\sigma_0 = \frac{1}{3}(\sigma_x + \sigma_y + \sigma_z)\right) \tag{13.18}$$

is referred to as *spherical stress tensor*, and D_σ is the *stress deviator* earlier found from the first formula (1.18), (p. 16):

$$D_\sigma = \begin{pmatrix} \sigma_x - \sigma_0 & \tau_{xy} & \tau_{xz} \\ \tau_{yx} & \sigma_y - \sigma_0 & \tau_{yz} \\ \tau_{zx} & \tau_{zy} & \sigma_z - \sigma_0 \end{pmatrix}.$$

To designate them, the following writing will be also used:

$$S_\sigma = \{\sigma_0\delta_{ij}\}, \quad D_\sigma = \{\sigma'_{ij}\} = \{\sigma_{ij} - \sigma_0\delta_{ij}\}. \tag{13.19}$$

The experiments of Bridgman [1], Nadai [3] et al. proved that for plastic materials, hydrostatic stress (spherical tensor) did not affect the conditions of occurrence and development of plastic deformation, and plastic deformation did not affect the volume change. This means that first of all, the volumetric deformation of plastic materials within a wide range of pressures is elastic and related to the stress of the first dependency (1.16):

$$\varepsilon_0 = \frac{\sigma_0}{K},$$

and second, that the deviator of plastic deformation coincides with the plastic
deformation tensor.

$$D_{\varepsilon^p} = T_{\varepsilon^p} = \begin{pmatrix} e_{xx} & e_{xy} & e_{xz} \\ e_{yx} & e_{yy} & e_{yz} \\ e_{zx} & e_{zy} & e_{zz} \end{pmatrix}, \quad \left(e_{xx} = \varepsilon_x^p, \ e_{xy} = \frac{1}{2}\gamma_{xy}^p, \ \ldots \right), \qquad (13.20)$$

since

$$e_0 = \frac{1}{3}(e_{xx} + e_{yy} + e_{zz}) = 0;$$

$$\varepsilon_0 = \frac{1}{3}(\varepsilon_x + \varepsilon_y + \varepsilon_z) = \frac{1}{3}(\varepsilon_x^y + \varepsilon_y^y + \varepsilon_z^y). \qquad (13.21)$$

13.4 Other Invariants in Plasticity Theory

We have already talked (p. 166) that other invariants can be formed from the
principal tensor values. In the mechanics of elastic deformations, the following three
invariants of the stress tensor are frequently used, which are expressed through the
previously defined I_σ, II_σ, III_σ, (13.14),

$$J_1 = \sigma_{ii} = I_\sigma,$$

$$J_2 = \sigma_{ij}\sigma_{ij} = I_\sigma^2 - 2II_\sigma,$$

$$J_3 = I_\sigma^3 - 3I_\sigma II_\sigma + 3III_\sigma = \sigma_{ij}\sigma_{jk}\sigma_{ki}. \qquad (13.22)$$

Similar invariants can be written for the strain tensor.

Since $I_{\sigma'} = 0$, invariants of type (13.22) for the stress deviator will be

$$J_1' = I_{\sigma'} = 0,$$

$$J_2' = \sigma_{ij}'\sigma_{ij}' = -2II_{\sigma'},$$

$$J_3' = \sigma_{ij}'\sigma_{jk}'\sigma_{ki}'. \qquad (13.23)$$

In an expanded form, formula (13.23) can be written as follows:

$$J_2' = \sigma_x'^2 + \sigma_y'^2 + \sigma_z'^2 + 2(\tau_{xy}^2 + \tau_{yz}^2 + \tau_{zx}^2)$$

$$= \sigma_1'^2 + \sigma_2'^2 + \sigma_3'^2 = \frac{1}{3}\Big[(\sigma_x - \sigma_y)^2 + (\sigma_y - \sigma_z)^2$$

$$+ (\sigma_z - \sigma_x)^2 + 6(\tau_{xy}^2) + \tau_{yz}^2 + \tau_{zx}^2\Big]$$

$$= \frac{1}{3}\Big[(\sigma_1 - \sigma_2) + (\sigma_2 - \sigma_3) + (\sigma_3 - \sigma_1)\Big];$$

$$J_3' = (\sigma_1'^3 + \sigma_2'^3 + \sigma_3'^3)$$

$$= \frac{1}{9}(2\sigma_1 - \sigma_2 - \sigma_3)(2\sigma_2 - \sigma_3 - \sigma_1)(2\sigma_3 - \sigma_2 - \sigma_1). \tag{13.24}$$

Similar ratios between invariants can be written for the strain tensor and its deviator, as well as their expanded expression through the components ε_{ij}. These expressions are obtained from the respective expressions of the tensor invariants and stress deviator by substituting $(\sigma_x, \sigma_y, \sigma_z)$ with $(\varepsilon_x, \varepsilon_y, \varepsilon_z)$ and $(\tau_{xy}, \tau_{yz}, \tau_{zx})$ with $\left(\frac{1}{2}\gamma_{xy}, \frac{1}{2}\gamma_{yz}, \frac{1}{2}\gamma_{zx}\right)$.

Together with these, plasticity theory uses other invariants. Stresses on areas equally inclined to the principal axes are very important. Eight such areas can be made. For clarity, they are routed not through the reference point, but as shown in Fig. 13.1 so that they form an octahedron. Therefore, the areas are called *octahedral*. Guiding cosines of the normal line to the front octahedral area

$$n_1 = n_2 = n_3 = \frac{1}{\sqrt{3}}. \tag{13.25}$$

Let us designate the stress vector on this face as **S**:

$$\mathbf{S} = \frac{1}{\sqrt{3}}\left[\sigma_1\mathbf{e}_1 + \sigma_2\mathbf{e}_2 + \sigma_3\mathbf{e}_3\right], \tag{13.26}$$

where $\mathbf{e}_i$, $(i \sim 1, 2, 3)$ are single vectors of the principal axes of the stress tensor T_σ.

Fig. 13.1 Octahedral planes

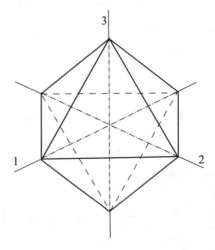

Normal stress (σ) on the octahedral area will be

$$\sigma = \mathbf{S}\mathbf{n} = \frac{1}{3}(\sigma_1 + \sigma_2 + \sigma_3) = \sigma_0. \tag{13.27}$$

The result (13.27) expresses that normal stress on octahedral areas will be equal to the mean arithmetic from principal stresses. As said before (p. 15), this stress is also called hydrostatic.

Let us calculate the tangential stress (τ_0) on the octahedral area. To do it, let us use the formula

$$\tau_0 = \sqrt{S^2 - \sigma_0^2}, \tag{13.28}$$

where we assume that

$$S^2 = \frac{1}{3}[\sigma_1^2 + \sigma_2^2 + \sigma_3^2],$$

$$\sigma_0^2 = \frac{1}{9}[\sigma + 1^2 + \sigma_2^2 + \sigma_3^2 + 2\sigma_1\sigma_2 + 2\sigma_2\sigma_3 + 2\sigma_3\sigma_1].$$

By substituting the last expression into formula (13.28), we obtain

$$\tau_0^2 = \frac{2}{9}[\sigma_1^2 + \sigma_2^2 + \sigma_3^2 - \sigma_1\sigma_2 - \sigma_2\sigma_3 - \sigma_3\sigma_1]$$

$$= \frac{1}{9}\left[(\sigma_1 - \sigma_2)^2 + (\sigma_2 - \sigma_3)^2 + (\sigma_3 - \sigma_1)^2\right], \tag{13.29}$$

or otherwise

$$\tau_0^2 = \frac{1}{3}J_2'. \tag{13.30}$$

The tangential stress on the octahedral area is referred to as *octahedral tangential stress*. Based on the results (13.29) and (13.30) it is an invariant of the stress tensor. The octahedral tangential stress modulus is defined by the formula

$$\tau_0 = \frac{1}{3}\sqrt{(\sigma_1 - \sigma_2)^2 + (\sigma_2 - \sigma_3)^2 + (\sigma_3 - \sigma_1)^2}, \tag{13.31}$$

and its tangential cosines can be calculated as ratios of the vector components $\mathbf{S} \times \mathbf{n}$ to its module. In this manner, let us designate the unit vector of octahedral tangential stress as $\vec{\beta}(\beta_1, \beta_2, \beta_3)$:

$$\beta_1 = \frac{\sigma_2 - \sigma_3}{\sqrt{(\sigma_1 - \sigma_2)^2 + (\sigma_2 - \sigma_3)^2 + (\sigma_3 - \sigma_1)^2}},$$

$$\beta_2 = \frac{\sigma_3 - \sigma_1}{\sqrt{(\sigma_1 - \sigma_2)^2 + (\sigma_2 - \sigma_3)^2 + (\sigma_3 - \sigma_1)^2}}, \qquad (13.32)$$

$$\beta_3 = \frac{\sigma_1 - \sigma_2}{\sqrt{(\sigma_1 - \sigma_2)^2 + (\sigma_2 - \sigma_3)^2 + (\sigma_3 - \sigma_1)^2}}.$$

Similarly to the strain tensor, we introduce the concept of *octahedral shift* (γ_0) that is defined through principal deformations ε_1, ε_2, ε_3 using the formula

$$\gamma_0 = \frac{2}{3}\sqrt{(\varepsilon_1 - \varepsilon_2)^2 + (\varepsilon_2 - \varepsilon_3)^2 + (\varepsilon_3 - \varepsilon_1)^2}. \qquad (13.33)$$

Along with the octahedral tangential stress and octahedral shift, the theory of non-elastic deformations also uses values proportional to them: *stress intensity* $\sigma_и$ and *strain intensity* $\varepsilon_и$:

$$\sigma_и = \frac{3}{\sqrt{2}}\tau_0 = \frac{1}{\sqrt{2}}\sqrt{(\sigma_1 - \sigma_2)^2 + (\sigma_2 - \sigma_3)^2 + (\sigma_3 - \sigma_1)^2},$$

$$\varepsilon_и = \frac{1}{\sqrt{2}}\gamma_0 = \frac{\sqrt{2}}{3}\sqrt{(\varepsilon_1 - \varepsilon_2)^2 + (\varepsilon_2 - \varepsilon_3)^2 + (\varepsilon_3 - \varepsilon_1)^2}. \qquad (13.34)$$

The multiplier $3/\sqrt{2}$ in front of τ_0 in formula (13.30) is selected so that in the case of uniaxial elongation of a rod with the stress σ, the stress intensity $\sigma_?$ coincides with σ. In a similar way, the multiplier $1/\sqrt{2}$ in front of γ_0 is selected such that in the case of uniaxial elongation ($\nu = 0, 5$), the strain intensity $\varepsilon_?$ coincides with the relative linear deformation in the direction of the rod axis.

13.5   On the Criterion of Similarity of Stress and Strain Deviators

Let us consider the equilibrium of body elements restricted by the coordinate planes and octahedral plane (Fig. 13.2).

Fig. 13.2 To the similarity of stress and strain deviators

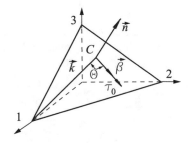

Let us designate the area of the triangle 123 using ω and areas of triangles lying in the coordinate planes with normal lines 1, 2, 3 using ω_1, ω_2, ω_3. We have

$$\omega_1 = \omega_2 = \omega_3 = \omega \cdot \frac{1}{\sqrt{3}}. \qquad (13.35)$$

Assume that $\vec{n}$ is a normal line to the octahedral plane. Its guiding cosines are defined by formula (13.25). As above (p. 173), $\vec{\beta}$ in Fig. 13.2 designates a unit vector of octahedral tangential stress. Let us write equilibrium equations of the considered triangle pyramid in projections on the principal axes of the stress tensor as follows:

$$\sigma_i' = \sqrt{3}\tau_0\beta_i, \quad (i \sim 1, 2, 3), \qquad (13.36)$$

where $\sigma_i' = \sigma_i - \sigma_0$, and σ_0 is the mean (hydrostatic) stress. Let us connect the point C, Fig. 13.2 with a point of crossing between the axis 1 and octahedral plane 123 using a straight line and designate the unit vector of that line as $\vec{k}$. We have

$$k_1 = \frac{2}{\sqrt{6}}, \quad k_2 = k_3 = \frac{1}{\sqrt{6}}, \quad \vec{k}\vec{\beta} = \cos\Theta, \qquad (13.37)$$

where Θ is the angle between the vectors $\vec{k}$ and $\vec{\beta}$.

Let us express the guiding cosines of the unit vector $\vec{\beta}$ through the angle Θ. We can write

$$\beta_1^2 + \beta_2^2 + \beta_3^2 = 1,$$

$$\beta_1 + \beta_2 + \beta_3 = 0, \qquad (13.38)$$

$$2\beta_1 - \beta_2 - \beta_3 = \sqrt{6}\cos\Theta.$$

The second equation of the system (13.38) is a consequence of orthogonality of the vectors $\vec{n}$ and $\vec{\beta}$, and the third one represents a result of the scalar product of the unit vectors $\vec{k}$ and $\vec{\beta}$.

The system (13.38) is solved by the formulas

$$\beta_1 = \sqrt{\frac{2}{3}}\cos\Theta,$$

$$\beta_2 = \sqrt{\frac{2}{3}}\cos\left(\Theta - \frac{2\pi}{3}\right), \qquad (13.39)$$

$$\beta_3 = \sqrt{\frac{2}{3}}\cos\left(\Theta - \frac{4\pi}{3}\right).$$

By substituting Eqs. (13.39) into equilibrium equations (13.36), we obtain

$$\sigma_1' = \frac{2}{3}\sigma_? \cos \Theta,$$

$$\sigma_2' = \frac{2}{3}\sigma_? \cos \left(\Theta - \frac{2\pi}{3} \right), \qquad (13.40)$$

$$\sigma_3' = \frac{2}{3}\sigma_? \cos \left(\Theta - \frac{4\pi}{3} \right).$$

The octahedral tangential stress is substituted here with stress intensity from the first of the formulas (13.34).

From formula (13.40) it follows that *the stress deviator is fully defined by five values: the direction of three principal areas, scalar $\sigma_?$ and angle Θ*, selected as shown in Fig. 13.2.

Definition 1 A director tensor of the stress deviator is the tensor whose principal components are defined by the formula

$$\bar{\sigma}_i' = \frac{\sigma_i'}{\sigma_?}, \quad (i \sim 1, 2, 3). \qquad (13.41)$$

Definition 2 Stress deviator tensors are called congruent if their principal axes coincide and angles Θ are equal.

The definition shows that congruent tensor deviators can differ from each other by a scalar multiplier only.

Noting that stress deviator components in the principal axes are represented as $\sigma_i' = \sigma_i - \sigma_0$ and using the expression (13.24) of the third deviator invariant, we can write as follows:

$$J_3' = \frac{2}{9}\sigma_?^3 \cos \Theta.$$

Taking into account that stress intensity $\sigma_?$ is expressed though the second variant of the stress deviator using the formula (13.34), the previous dependency can be written as

$$\cos \Theta = \frac{J_3'}{3J_2'^{3/2}}. \qquad (13.42)$$

After all the previous calculations as applicable to the stress tensor deviator, similar to formula (13.42), we can get

$$\cos \eta = \frac{I_3'}{3I_2'^{3/2}}, \qquad (13.43)$$

where η is similar to the angle Θ defined for the strain deviator, and I_2', I_3' are invariants of the strain deviator.

According to Definition 2, if the angles Θ and η are equal, and the principal axes coincide, the stress deviator and strain deviator are similar. According to Definition 1, the director tensors of these deviators are equal. For similar deviators, Lode and Nadai introduced [4] invariants

$$\mu_\sigma = 2\frac{\sigma_2 - \sigma_3}{\sigma_1 - \sigma_3} - 1, \quad \mu_\varepsilon = 2\frac{\varepsilon_2 - \varepsilon_3}{\varepsilon_1 - \varepsilon_3} - 1. \tag{13.44}$$

Invariants defined by formulas (13.44) are referred to as Lode and Nadai parameters. It can be shown that they can be expressed through the second and third invariants of the respective deviators and the areas of changes in their values are defined by the formulas

$$-1 \leqslant \mu_\sigma \leqslant 1, \quad \mu_\varepsilon \leqslant 1. \tag{13.45}$$

13.6 Stress Diagrams and Their Idealization

In the mechanics of non-elastic deformations, many positions are obtained as a result of summarizing observations in the case of uniaxial elongation of material samples. Therefore, it is reasonable to return to the analysis of stress diagrams $\sigma \sim \varepsilon$.

Previously, we talked of the stress diagram (Fig. 1.3) for a material with the yield area. However, not all plastic materials find the yield area on the stress diagram. In many cases, this diagram looks as shown in Fig. 13.3.

Fig. 13.3 Diagram "elongation—unloading—compression"

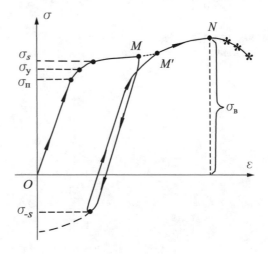

The diagram may have all points corresponding to the proportionality limit $\sigma_?$, elastic limit σ_y, and yield point σ_s. All the three characteristics are not absolute and have a relative sense because their values substantially depend on measurement accuracy, micro-damages of the material structure, accuracy tolerance for the geometric size of the sample, and other factors. The ultimate strength $\sigma_?$ is more stable in elongation experiments due to a higher certainty of the position of the point N on the stress diagram.

In the case of loading beyond the yield strength, unloading, and repeated loading, a hysteresis loop is formed, which describes energy dissipation in the case of non-elastic deformation. In the case of sample compression after preliminary elongation beyond the yield strength, the proportionality limit, elastic limit, and yield strength are decreased. This phenomenon is referred to as the *Bauschinger effect*. For all metals and most other materials $\sigma_{-s} \leqslant \sigma_s$.

Currently, it is impossible in plasticity theory to use a real diagram of stresses (Fig. 13.4), and sometimes the approximation accuracy of real diagrams is not reasonable. Therefore, the Prandtl diagram is used, Fig. 13.5, which is an idealization of the real diagram. Effects in the vicinity of the elastic limit are ignored, and it is assumed that $\sigma_? = \sigma_y = \sigma_s$. Moreover, it is believed that unloading after previous loading beyond the yield strength occurs under the elastic law, there is no hysteresis phenomenon, so further loading along the trajectory also follows Hooke's law.

This idealization is permitted by low-carbon steels having a yield area in the case of moderate deformations.

For some materials, the diagram $\sigma \sim \varepsilon$ permits sound approximation of a two-link polyline. In this case, they say that the medium has linear strengthening (Fig. 13.6). Two-link approximation is applicable to aluminum and its alloys, some plastics and other structural materials.

Speaking of elongation diagrams, we suggested that the loading process is isothermal at moderate strain rates and normal external conditions, since the diagram $\sigma \sim \varepsilon$ substantially depends on test conditions. It should also be noted that a characteristic feature of elastic–plastic deformation is no unique dependence

Fig. 13.4 Real stress diagram

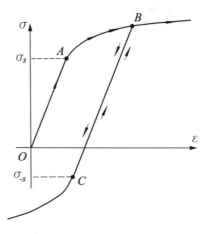

Fig. 13.5 Ideal
elastic–plastic body

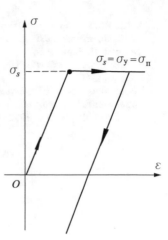

Fig. 13.6 Material with
linear strengthening

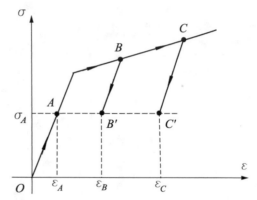

between stresses and strains. Figure 13.6 shows that the same stress σ_A can correspond to different strains ε_A, ε_B, or ε_C. Everything depends on how we reached the stress σ_A. The strain ε_A is obtained during the elastic work of the material and is unambiguously defined by the stress σ_A. The strains ε_B and ε_C are obtained by the elongation of the material beyond the yield strength and further unloading along the trajectories BB' and CC' to the stress σ_A. As we see, each of these cases corresponds to their own strain values; however, the stress is the same. In this connection, the concept of *"loading history" is introduced in the mechanics of non-elastic deformations*. They also say that materials have memory.

References

1. P. Bridzhmen, *Issledovaniya bol'shikh plasticheskikh deformatsii i razryva : Vliyanie vysokogo gidrostaticheskogo davleniya na mekhanicheskie svoistva materialov* [Studies of large plastic deformations and impact of high hydrostatic pressure on the mechanical properties of materials] (Izdatel'skaya gruppa URSS Publ., Moskow, 2010)

2. A. Kurosh, *Kurs vysshei algebry* [Course of higher algebra] (Moscow, Nauka Publ., 1968)
3. A. Nadai, *Plastichnost' i razrushenie tverdykh tel* [Plasticity and destruction of solids] (Moscow, IL Publ., 1954)
4. Y. Rabotnov, *Polzuchest' ehlementov konstruktsii* [Creep of structural elements] (Nauka Publ., Moscow, 1966)
5. I.E.A. Vinogradov, *Matematicheskaya ehntsiklopediya, t.3, red. I. M. Vinogradov [i dr.] [Mathematical Encyclopedia, vol. 3, editor I.M. Vinogradov (and others)]* (Sovetskaya ehntsiklopediya Publ., Moscow, 1982)

Chapter 14
On the Plasticity Conditions of an Isotropic Body

14.1 General Considerations

Conditions that must be satisfied by stresses corresponding to the appearance of plastic strains are usually called *plasticity conditions.*

In the case of uniaxial elongation–compression, the plasticity condition looks as follows:

$$|\sigma| = \sigma_s, \tag{14.1}$$

where σ_s is the plasticity limit for elongation–compression.

In the case of pure shift, the plasticity condition is as follows:

$$\tau_{max} = \tau_s, \tag{14.2}$$

where τ_s is the shear yield stress (torsion of a thin-wall tube).

Setting plasticity conditions for a complex stressed state is still a complicated task. It is solved by various hypotheses, assumptions, and idealizations. Before describing plasticity conditions, let us note as follows.

14.2 General Notes

Physical representations result in a number of requirements that the plasticity condition must satisfy. Let us consider these requirements.

The plasticity conditions of an initially isotropic body must not depend on selecting the coordinate system. Consequently, it can be represented in a general form by the following equation:

© The Author(s), under exclusive license to Springer Nature Switzerland AG 2021
V. Molotnikov, A. Molotnikova, *Theory of Elasticity and Plasticity*,
https://doi.org/10.1007/978-3-030-66622-4_14

$$f(\sigma_1, \sigma_2, \sigma_3) = 0, \tag{14.3}$$

whereas f is a symmetric function of stress invariants. Condition (14.3) can also be written as follows:

$$f(J_1, J_2, J_3) = 0, \tag{14.4}$$

where J_i are stress tensor invariants.

　　If we neglect the effect of hydrostatic stress on the plasticity condition, arguments of the function f in the ratio (14.4) will be deemed the second and third invariants of the stress deviator, e.g. assume that

$$f(J_2', J_3') = 0. \tag{14.5}$$

Equations (14.3) and (14.5) describe some surface in stress space. Let us study the characteristics of this surface. Let us select the coordinate system $O123$ (Fig. 14.1) coinciding with the main stress tensor axes, and let us represent the vectors

$$\mathbf{S} = \mathbf{e}_i\sigma_i, \quad \mathbf{P} = \mathbf{e}_i\sigma_i', \quad \mathbf{Q} = \mathbf{e}_i\sigma_0, \quad (i \sim 1, 2, 3), \tag{14.6}$$

where $\sigma_i = \sigma_i' - \sigma_0$, and summing is done upon repeated indexes. From formulas (14.6), we see that the vector $\mathbf{P}$ is an equivalent of the stress deviator, and the vector $\mathbf{Q}$ is an equivalent of the spherical tensor. In this manner,

$$\mathbf{S} = \mathbf{P} + \mathbf{Q}. \tag{14.7}$$

　　Let us draw a plane π equally inclined to the coordinate axes through the point O (Fig. 14.1). The equation of this plane will be

$$\sigma_1 + \sigma_2 + \sigma_3 = 0. \tag{14.8}$$

The vector $\mathbf{P}$ lies in the deviator plane since

Fig. 14.1 To the general characteristics of yield conditions

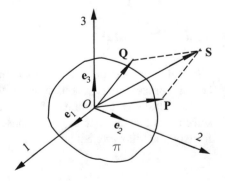

Fig. 14.2 Trace of the yield
surface on the octahedral
plane

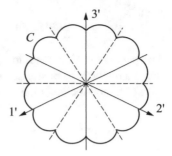

$$\sigma_1' + \sigma_2' + \sigma_3' = 0. \tag{14.9}$$

Equations (14.9) result in that the components of the vector **P** satisfy Eq. (14.8) of the plane π, e.g. **P** $\in \pi$.

Based on Eq. (14.5), the plasticity condition does not depend on **Q** and is fully defined by the vector **P**. Consequently, the plasticity surface (14.5) is a cylinder whose generatrix is perpendicular to the plane π. To build this surface, it is sufficient to build its trace on the plane π.

This building is done in Fig. 14.2. Here the plane π is aligned with the figure plane. The axes $1'$, $2'$, $3'$ are orthogonal projections of the coordinate axes 1, 2, and 3 on the plane π. C is the trace of the plasticity surface on the plane π.

Let us show that to build the line C, it is sufficient to have experimental data for $1/12$ of its part.

Indeed, first of all, the C trace must be symmetric relative to the axes $1'$, $2'$, $3'$ since the plasticity surface view cannot depend on axes numbering.

Second, the C line must not go through the origin of the coordinate system since zero stresses in real solid bodies do not cause plastic deformations.

Third, any beam originating in the origin crosses C only once. Otherwise, there would be 2 and more stress states corresponding to the occurrence of plasticity, which is impossible.

Fourth, since initially isotropic materials are considered, which equally resist elongation and compression, the C trace must be symmetric relative to the axes perpendicular to the axes $1'$, $2'$, $3'$. The latter are shown in Fig. 14.2 by dash lines. It can be easily found that the Lode-Nadai parameter μ_σ (p. 177) on the indicated part $1/12$ of the trace C changes from 0 to 1.

14.3 Tresca Plasticity Condition

We have already mentioned (p. 147) that historically, the first plasticity condition was formulated by Tresca in 1864. Observing the extrusion of materials through dies, Tresca found that the plasticity phenomenon occurred when the maximum tangential stress (τ_{max}) reached a specific value:

$$\tau_{max} = \frac{1}{2} max\left[(\sigma_1 - \sigma_2), (\sigma_2 - \sigma_3), (\sigma_3 - \sigma_1)\right] = K. \tag{14.10}$$

The parameter K in condition (14.10) is a constant value of the material. It does not depend on the type of stressed state and is found experimentally. In particular, in the case of uniaxial elongation, we have

$$\sigma_1 = \sigma_s; \quad \sigma_2 = \sigma_3 = 0; \quad \tau_{max} = \frac{\sigma_1 - \sigma_3}{2} = \frac{\sigma_s}{2} = K. \tag{14.11}$$

In the second partial case, pure shear, we obtain as follows:

$$\sigma_1 = -\sigma_3 = \tau_{max} = \tau_s. \tag{14.12}$$

In this manner, for the plasticity condition of Tresca[1], formulas (14.11) and (14.12) result in

$$\sigma_s = 2\tau_s. \tag{14.13}$$

Later, we will talk of compliance between dependency (14.13) and experimental data (p. 186).

Let us consider a geometric interpretation of the Tresca plasticity condition. In the space of principal stresses, condition (14.10) corresponds to a hexagonal prism equally inclined to the coordinate axes (Fig. 14.3). Indeed, condition (14.10) can be written as follows:

Fig. 14.3 Coulomb–Tresca prism

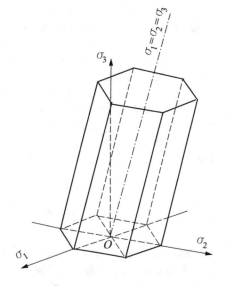

---

[1]Condition (14.10) is also called the Coulomb–Tresca condition; it can be expressed also through stress deviator invariants.

Fig. 14.4 Prism trace in planes 1–2

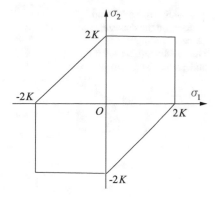

$$\sigma_1 - \sigma_2 = \pm 2K,$$
$$\sigma_2 - \sigma_3 = \pm 2K, \qquad\qquad (14.14)$$
$$\sigma_3 - \sigma_1 = \pm 2K.$$

Each of these equations depicts two planes parallel to each other. On the octahedral plane π, the Tresca prism trace represents an equilateral hexagon whose apexes are located on the projections of the coordinate axes in the plane π. Figure 14.4 depicts a prism trace in the plane of principal stresses $\sigma_1 \sim \sigma_2$.

14.4   Huber–Mises Plasticity Condition

The specific abstractedness of the initial pre-requisites of the Tresca plasticity condition induced researchers to find another (in a known sense) physically justified condition. As such a condition, Huber [8] and then Mises [7] proposed a so-called energetic plasticity condition. Apart from the noted physicality, the Huber–Mises yield condition was free from other disadvantages of the Tresca condition. In particular, the Tresca yield condition expressed by formulas (14.10) or (14.14) when solving spatial problems of plasticity theory leads [5, p. 43] to significant mathematical complications.

In the Huber–Mises plasticity condition, the Tresca hexagonal prism is replaced by the described circular cylinder. The sectioning of this cylinder by a deviator plane is a circumference circumscribed around the Tresca prism trace (Fig. 14.5). Mathematically, it represents any of the following writings:

$$\tau_0 = const;$$
$$\sigma_i = const;$$
$$J_2' = const; \qquad\qquad (14.15)$$
$$u_f = const,$$

Fig. 14.5 Traces of yield surfaces on the octahedral plane according to Huber–Mises (circumference) and Tresca (hexagon)

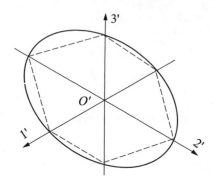

where u_f is the specific potential energy of shape changing [8]

$$u_f = \frac{1+\mu}{6E}\left[(\sigma_1 - \sigma_2)^2 + (\sigma_2 - \sigma_3)^2 + (\sigma_3 - \sigma_1)^2\right].$$ (14.16)

In the case of uniaxial elongation,

$$\sigma_1 = \sigma_s; \quad \sigma_2 = \sigma_3 = 0;$$

then using formulas (13.34) and (13.31), we obtain

$$\sigma_i = \sigma_s,$$

and the second condition (14.15) can be written as

$$\sigma_i^2 = \frac{1}{2}\left[(\sigma_1 - \sigma_2)^2 + (\sigma_2 - \sigma_3)^2 + (\sigma_3 - \sigma_1)^2\right] = \sigma_s^2.$$ (14.17)

In the case of pure shear,

$$\sigma_1 = -\sigma_3 = \tau_s.$$

By substituting this result to formula (14.17), we obtain

$$\tau_s = \frac{1}{\sqrt{3}}\sigma_s \approx 0,577\sigma_s,$$ (14.18)

whereas, from the Tresca plasticity condition, formula (14.13) gives $\tau_s = 0,5\sigma_s$.

By comparing formulas (14.16) and (14.17), we notice that expressions in their square brackets coincide. Therefore, the Huber–Tresca condition is sometimes called energetic and is formulated as follows: *the yield state of the material is achieved at some constant specific potential energy of shape changing.*

Multiple experiments to check plasticity conditions have shown that the Huber–Mises condition is fulfilled for poly-crystalline materials (in particular, for steels) somewhat better than the condition of the constancy of the maximum tangential stress. Together with that, for some alloys, the Tresca condition better conforms with experimental data. Due to the closeness of results defined by these criteria and the unavoidable error of the experiment, the Huber–Mises and Tresca conditions can be viewed as equal formulations of the yield condition. Kachanov suggests [5] decreasing the difference between the Tresca and Mises conditions by assuming a cylinder as the yield surface, whose trace on the deviator plane lies in the middle between the Huber circumference and a circumference described within the Tresca hexagon.

However, there are other proposals. One of them suggests that plastic strain occurs when the reduced stress (S_{max}) reaches a specific value constant for the material:

$$S_{max} = max\,\{|\sigma_1 - \sigma_0|,\ |\sigma_2 - \sigma_0|,\ |\sigma_3 - \sigma_0|\} = const. \tag{14.19}$$

Condition (14.19) also slightly differs from the Tresca and Huber–Mises conditions. On the deviator plane, it circumscribes an equilateral hexagon, for which the Huber circumference is described.

14.5 Experimental Study of the Elastic–Plastic Properties of Materials

For an experimental study of strain laws in general and for the establishment and check of plasticity conditions in particular, power units, measurement equipment, and specially prepared material specimens are used. The latter are the object of the research since the two first belong to test equipment. To understand modern test equipment, Fig. 14.6 gives a general view of test systems supplied by Schenck, [4]. Position a depicts a unit with a horizontal arrangement of the test specimen, and position b shows a four-column machine with a vertical arrangement of the specimen.

These devices belong to servohydraulic machines. Depending on the model, they can develop axial forces of up to 2000 kN in the specimen. There are principally different unique testing machines of special purposes. Among them is a Russian hydraulic machine of proportional loading designed by VNIMI (Saint Petersburg) scientists and engineers. The device is intended to test specimens of hard rocks in axial compression imposed over confining pressure (up to 3000 MPa!).

In the experiment intended to study the mechanical properties of materials, first of all, we must select a geometric shape of the specimen and then set its loading method. Frequently, one tries to achieve a homogeneous stress–strain state in the working area of the specimen. Currently, researchers avail of a program

a   b

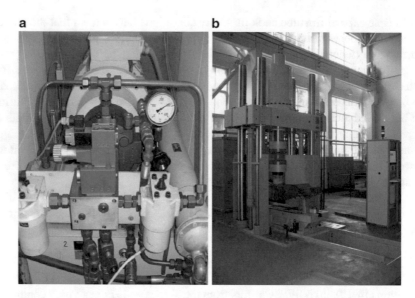

Fig. 14.6 Servohydraulic testing machines

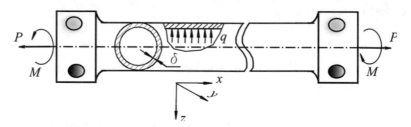

Fig. 14.7 Tubular specimen for tests

of an arbitrary plane stressed state with an independent change of stress tensor components. The implementation of an arbitrary plane stressed state is still an unsolved problem. Therefore, strain laws are checked when testing thin-wall tubular specimens (Fig. 14.7).

During tests, the axial force P, twisting moment M, and inner pressure q can be set as arbitrary functions of time from the range of values permitted by the loading device characteristics. For the directions of the axes x, y, and z shown in Fig. 14.7, the stress components will be as follows:

$$\sigma_x = \frac{P}{F} = \frac{P}{2\pi R\delta}, \quad \sigma_y = \frac{qR}{\delta}, \quad \tau_{xy} = \frac{M}{2\pi R^2\delta}, \tag{14.20}$$

where R is the average tube radius and δ is the thickness of its wall (see the superimposed cross-section in Fig. 14.7). Apart from the three stress components given in the formulas (14.20), for $q \neq 0$ there is also stress σ_z that equals $-q$ on the inner surface of the tube and turns zero on the outer surface. Due to a small size of the component σ_z as compared to three other components of stress, $\sigma_z = 0$ is adopted everywhere in the specimen body.

In this manner, in the case of complicated loading of a tubular specimen, an arbitrary plane stressed state can be achieved. The axis z is the main axis of the tensor state. It is substantial that each stress component depends on one force only, and therefore, each component can be changed irrespective of each other.

The specimen strain will be represented by the components ε_x, ε_y abd γ_{xy}. To determine them experimentally, we must measure the elongation Δl of the working part (l) of the specimen, the change of the radius ΔR, as well as the angle (φ) of turning of one fixed section of the tube relative to the other one distanced from the first one by l_1. Then,

$$\varepsilon_x = \frac{\Delta l}{l}, \quad \varepsilon_y = \frac{\Delta(2\pi R)}{2\pi R} = \frac{\Delta R}{R}, \quad \gamma_{xy} = \frac{R\varphi}{l_1}. \tag{14.21}$$

The strain ε_z can be determined by measuring the change in the tube wall thickness or calculated using the formula

$$\varepsilon_z = -\varepsilon_x - \varepsilon_y + \frac{\sigma_0}{3K}, \tag{14.22}$$

where K is the volumetric elasticity modulus defined by formula (1.17). Equation (14.22) is a consequence of the first of formulas (1.16).

Tubular specimens are tested upon various programs. Most frequent are $(P - q)$-experiments, $(M = 0)$ and $(P - M)$-experiments, $(q = 0)$. In the stress space, arbitrary loading may correspond to the line called *the loading trajectory*. In $(P-q)$-experiments, the loading trajectory is located in the plane $\sigma_x \sim \sigma_y$ (Fig. 14.8a), and in $(P - M)$-experiments, it is in the plane $\sigma_x \sim \tau_{xy}$ (Fig. 14.8b).

Proportional (or ordinary) *loading* is such loading when the stress deviator components change proportionally to the same parameter. This loading in the stress deviator space corresponds to a linear trajectory. Otherwise, the loading is called *complex*.

Experiments with thin-wall tubes allow for an experimental study of strain laws at ordinary and complex loading. As said above, these experiments are also used

Fig. 14.8 Loading trajectories: (**a**) in $(P - q)$-experiments ; (**b**) in $(P - M)$-experiments

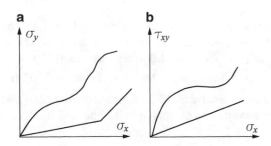

for the experimental check of plasticity conditions. In particular, in $(P - M)$-experiments, the plasticity conditions look as follows:

Tresca condition

$$\sqrt{\sigma_x^2 + 4\tau_{xy}^2} = \sigma_s;$$

Huber condition

$$\sqrt{\sigma_x^2 + 3\tau_{xy}^2} = \sigma_s.$$

In $(P - q)$-experiments, similar conditions will be

Tresca condition

$$\sigma_x - \sigma_y = \pm\sigma_s;$$

Mises condition

$$\sqrt{\sigma_x^2 + \sigma_y^2 - \sigma_x\sigma_y} = \sigma_s.$$

14.6   Volumetric Elasticity of Materials

Most homogeneous and initially isotropic materials comply with the *volumetric elasticity law:* a relative change in the material volume in the case of the isometric strain process is a specific function of only hydrostatic stress σ_0, and the strain process is reversible. The linkage between σ_0 and ε_0 is set through hydrostatic compression experiments. We have already said (p. 16) that such experiments were done in a sufficient number by Bridgeman [1]. The experiments found that the absolute majority of materials and chemical elements in all-around compression behaved as elastic bodies: the volume decreased with a pressure rise and restored with pressure relief, e.g.

$$3\varepsilon_0 = \Theta = \frac{\sigma_0}{K}\left(1 + \frac{\sigma_0}{K_1}\right), \tag{14.23}$$

where K and K_1 are the material constant values. It has been found that most materials have the constant value K of about 10^5 MPa and the constant value K_1 of about 10^4 MPa. By assessing all the parameters of formula (14.23), we can approximately find that for $\sigma_0 \simeq 300$ MPa $\frac{\sigma_0}{K_1} \simeq 0,01 \div 0,02$. Therefore, the second addend in brackets of formula (14.23) is usually neglected due to its smallness as compared to the unit, and it is assumed as follows:

$$\Theta \approx \frac{\sigma_0}{K} \quad or \quad \sigma_0 = 3K\varepsilon_0. \tag{14.24}$$

When the principal stresses are not equal between each other, as the experiments of Davidenkov [2] et al. showed, the volumetric elasticity law (14.23) (14.24) also remains true in the general case of stressed state.

By noting that $K \gg \sigma_0$ for the stress σ_0 of about σ_s, the ratio σ_0/K equals several fractions of a percent. Therefore, in plasticity theory (especially in the case of developed plastic strains), volumetric compression of materials is usually neglected and it is assumed as follows:

$$\varepsilon_x - \varepsilon_0 \approx \varepsilon_x, \ldots.$$

Moreover, let us also recall (p. 170) that all-around compression slightly affects the plasticity condition and ratios between stresses and strains beyond the yield limit. However, the effect of hydrostatic pressure on material plasticity is rather high. For example, brittle bodies such as hard rocks (marble, granite, etc.) exposed to high all-around pressure by additional forces can acquire high residual deformations.

14.7 Invariant Form of Hooke's Law

Previously, we gave (p. 14) two forms of the mathematical representation of Hooke's law. The given formulas are

$$\begin{aligned}
\sigma_x &= \lambda\Theta + 2G\varepsilon_x, \\
\sigma_y &= \lambda\Theta + 2G\varepsilon_y, \\
\sigma_z &= \lambda\Theta + 2G\varepsilon_z; \\
\tau_{xy} &= G\gamma_{xy}, \\
\tau_{yz} &= G\gamma_{yz}, \\
\tau_{zx} &= G\tau_{zx},
\end{aligned} \tag{14.25}$$

which designates

$$\lambda = \frac{2G\nu}{1 - 2\nu}.$$

The shift modulus G included in the last formulas is sometimes denoted by the letter μ, and [6, 9] used the parameters λ, μ name "Lame constants."

By summing the three first ratios (14.25), we obtain

$$\sigma_0 = K\varepsilon_0, \quad \left(K = \frac{E}{1 - 2\nu}\right),$$

which coincides with formulas (1.16) and (1.17). By subtracting the mean stress from the three first Eqs. (14.25), we obtain

$$
\begin{aligned}
\sigma_x - \sigma_0 &= 2G(\varepsilon_x - \varepsilon_0), \\
\sigma_y - \sigma_0 &= 2G(\varepsilon_y - \varepsilon_0), \\
\sigma_z - \sigma_0 &= 2G(\varepsilon_z - \varepsilon_0); \\
\tau_{xy} &= 2G\varepsilon_{xy}, \\
\tau_{yz} &= 2G\varepsilon_{yz}, \\
\tau_{zx} &= 2G\varepsilon_{zx},
\end{aligned}
\tag{14.26}
$$

where

$$
\varepsilon_{ij} = \frac{1}{2}\gamma_{ij}, \quad (i, j \sim x, y, z).
$$

In the tensor form, dependencies (14.26) are represented by formulas (1.16)

$$
D_\sigma = 2G D_\varepsilon, \quad \sigma_0 = K\varepsilon_0.
$$

Let us designate deviator guides of the stress and strain tensor using $\overline{D}_\sigma$ and $\overline{D}_\varepsilon$, respectively.

$$
\overline{D}_\sigma = \frac{1}{\tau_0}D_\sigma, \quad \overline{D}_\varepsilon = \frac{2}{\gamma_0}D_\varepsilon.
\tag{14.27}
$$

Taking into account that in the elastic stage of strain

$$
\tau_0 = G\gamma_0,
\tag{14.28}
$$

the law (1.16), taking into account formulas (14.27) and (14.28), can finally be represented as follows:

$$
\overline{D}_\sigma = \overline{D}_\varepsilon.
\tag{14.29}
$$

In this manner, Hooke's law formulation is represented by the following three dispositions.

1. Linear invariants of the stress (T_σ) and strain (T_ε) tensors are proportional; or otherwise, changes in the body element volume are proportional to the mean nominal stress:

$$
\sigma_0 = K\varepsilon_0.
\tag{14.30}
$$

2. Guides of the stress and strain tensors coincide

$$
\overline{D}_\sigma = \overline{D}_\varepsilon.
\tag{14.31}
$$

3. Quadratic invariants of the stress (D_σ) and strain (D_ε) deviators are proportional

$$\tau_0 = G\gamma_0,$$

or

$$\sigma_? = 3G\varepsilon_?. \tag{14.32}$$

The formulation of Hooke's law in the form of theses 1–3 does not contain coordinates and depends only on invariants of the stress and strain tensors. The second advantage of such formulation is that it is possible to naturally make a bridge between elasticity theory that is physically based on Hooke's law and the simplest so-called deformational theory of plasticity named as the theory of low elastic–plastic strains by Ilyushin [3]. The following chapter describes the primary ratios of this theory. Its applicability limits are also described.

References

1. P. Bridzhmen, *Issledovaniya bol'shikh plasticheskikh deformatsii i razryva: Vliyanie vysokogo gidrostaticheskogo davleniya na mekhanicheskie svoistva materialov* (Studies of large plastic deformations and impact of high hydrostatic pressure on the mechanical properties of materials). (Izdatel'skaya gruppa URSS Publ., Moskow, 2010)
2. N. Davidenkov, *Mekhanicheskie svoistva i ispytanie metallov* (Mechanical properties and testing of metals). (Oniks Publ., Moscow, 2012)
3. A. Il'yushin, *Mekhanicheskie svoistva i ispytanie metallov* (Mechanical properties and testing of metals). (OGIZ Publ., Leningrad, Moscow, 1948)
4. *Ispytatel'nye mashiny Schenck* (Schenck test machines) (2014). https://schenck-rotec.de/
5. L. Kachanov, *Fundamentals of Plasticity Theory* (Dover, New York, 2004)
6. A. Lyav, *Matematicheskaya teoriya uprugosti* (Mathematical theory of elasticity). (ONTI NKTP SSSR Publ., Moscow, Leningrad, 1935)
7. R. Mises, Mechanik der festen körper im plastischdeformablen zustand, in *Göttinger: Königlichen Gesellschaft* (1913), pp. 582–592
8. V. Molotnikov, *Mekhanika konstruktsii* (Mechanics of structures). (SPb., Moscow, Krasnodar, Lan' Publ., 2012)
9. N. Muskhelishvili, *Nekotorye osnovnye zadachi matematicheskoi teorii uprugosti* (Some main problems of mathematical theory elasticity). (Nauka Publ., Moscow, 1966)

Chapter 15
Plasticity Theory of Henky–Nadai–Ilyushin

15.1 Laws of Active Elastic–Plastic Deformation

Assume that a body element is subject to simple strain and the intensity of stresses $\sigma_i(t)$ changes with time t. Let us fix two moments in time t_1 and t_2, so that $t_2 > t_1$. If $\sigma_i(t_2) > \sigma_i(t_1)$, the element strain is called *active*. Active strain is called *the loading process*. If $\sigma_i(t_2) < \sigma_i(t^*)$ at $t^* < t_2$, the strain is called *passive*. In this case, they say that there is a *process of unloading*.

One of the primary provisions[1] of the strain theory of plasticity is an assumption that for active strain, plastic strain rises beyond the elastic limit. Based on this hypothesis and using equivalents of theses $1 - 3$ from the previous paragraph, the following primary *laws of the theory of low elastic-elastic deformations are formulated.*

1. *Hardening law.* The intensity of stresses is universal and is a function (Φ) of strain intensity that does not depend on the type of stressed state:

$$\sigma_i = \Phi(\varepsilon_i). \tag{15.1}$$

A possible form of this universal function for hardening material is shown in Fig. 15.1. It shows the diagram for active loading and a trajectory section in the case of unloading.

2. *Law of volumetric strain elasticity.* The volumetric strain of a body complies with Hooke's law:

$$\sigma_0 = K\varepsilon_0. \tag{15.2}$$

---

[1] This assertion is true for simple loading of a hardening material, but in general it is incorrect for complex loading.

V. Molotnikov, A. Molotnikova, *Theory of Elasticity and Plasticity*,
https://doi.org/10.1007/978-3-030-66622-4_15

Fig. 15.1 Exemplary form of
the function $\Phi(\varepsilon_?)$

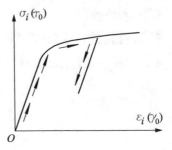

3. *Director stress and strain tensors coincide:*

$$\overline{D}_\sigma = \overline{D}_\varepsilon, \tag{15.3}$$

$$\frac{1}{\tau_0} D_\sigma = \frac{2}{\gamma_0} D_\varepsilon, \tag{15.4}$$

or else

$$D_\sigma = \frac{2\sigma_i}{3\varepsilon_i} D_\varepsilon. \tag{15.5}$$

When switching from the tensor form to a regular one, formula (15.5) can be
written as follows:

$$
\begin{aligned}
\sigma_x - \sigma_0 &= \frac{2\sigma_i}{3\varepsilon_i}(\varepsilon_x - \varepsilon_0), \\
&\cdots\cdots\cdots\cdots\cdots\cdots\cdots, \\
&\cdots\cdots\cdots\cdots\cdots\cdots\cdots; \\
\tau_{xy} &= \frac{2\sigma_i}{3\varepsilon_i}\varepsilon_{xy} = \frac{\sigma_i}{3\varepsilon_i}\gamma_{xy}, \\
&\cdots\cdots\cdots\cdots\cdots\cdots\cdots, \\
&\cdots\cdots\cdots\cdots\cdots\cdots\cdots.
\end{aligned}
\tag{15.6}
$$

Formulas (15.3)–(15.5) result in that all the main axes of stress and strain tensors,
as well as the ratios of the main tangential stresses to the main shears, coincide,
e.g.

$$\frac{\tau_{12}}{\gamma_{12}} = \frac{\tau_{23}}{\gamma_{23}} = \frac{\tau_{31}}{\gamma_{31}} = \frac{\sigma_i}{3\varepsilon_i}.$$

This formula is a consequence of dependencies (15.6). Ratios (15.6) contain five
independent equations expressing a link between six independent components of
the stress tensor T_σ and six independent components of the strain tensor T_ε. The
sixth equation gives the hardening law (15.1).

Multiple experiments conducted for decades showed that the primary provisions of the strain theory of plasticity in some cases gave rather satisfactory

similarity with the experiment, but in others they could not be deemed even rough approximation to reality.

15.2 Defining the Universal Hardening Function

For the experimental definition of the universal hardening function $\sigma_i = \Phi(\varepsilon_i)$, we can use an experiment for thin-wall tube twisting (p. 188). We use this experiment to define the diagram $\tau \sim \gamma$, (Fig. 15.2, lower curve) $\tau = \Phi_1(\gamma)$.

In this case, all other tensor components T_σ and T_ε, except for τ and γ, equal zero.

Then let us calculate

$$\sigma_i = \sqrt{3}\tau, \quad \varepsilon_i = \frac{1}{\sqrt{3}}\gamma.$$

In this manner, we find as follows from the last two formulas:

$$\sigma_i = \sqrt{3}\Phi_1(\sqrt{3}) \equiv \Phi(\varepsilon_i).$$

The upper curve in Fig. 15.2 corresponds to the function found.

Dependency $\sigma_i \sim \varepsilon_i$ can be also obtained from uniaxial elongation experiments. Apart from the elongation diagram $\sigma_1 = \Phi_2(\varepsilon_1)$, where σ_1, ε_1 is axial stress and strain, we must also measure lateral strain ε_2 to obtain the Poisson ratio $\nu_p = -\varepsilon_2/\varepsilon_1$. An exemplary form of the curve $\nu_p \sim \varepsilon_1$ is shown in Fig. 15.3. ε_s designates the axial strain at the yield point.

In the uniaxial compression experiment, principal stresses and strains will be

$$\sigma_1 \neq 0; \quad \sigma_2 = \sigma_3 = 0;$$
$$\varepsilon_1 \neq 0; \quad \varepsilon_2 = \varepsilon_3 = -\nu_p\varepsilon_1.$$

Expressions for stress and strain intensity look as follows:

Fig. 15.2 Experimental definition of the universal hardening function

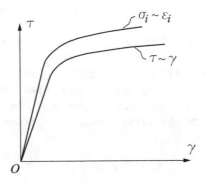

Fig. 15.3 Change of the
lateral strain coefficient
(Poisson ratio) of steel

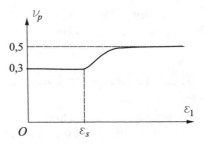

$$\sigma_i = \sigma_1, \quad \varepsilon_i = \frac{2}{3}(1 + \nu_p)\varepsilon_1.$$

For the hardening function, we obtain as follows:

$$\sigma_i = \Phi_2\left[\frac{3}{2}\frac{\varepsilon_i}{(1 + \nu_p)}\right] \equiv \Phi(\varepsilon_i).$$

For the developed plastic strains, we can assume $\nu_p = \dfrac{1}{2}$, (see Fig. 15.3). For this condition

$$\sigma_i = \Phi_2(\varepsilon_1).$$

Let us note that the dependency $\sigma_i \sim \varepsilon_i$ can be built without measuring lateral strains and we can use the dependency $\sigma_1 \sim \varepsilon_1$ involving Hooke's law for volumetric strain (see [4, p. 171]).

15.3 Some Properties of the Hardening Function

In the elastic stage of the material work, Hooke's law is correct (14.32)

$$\sigma_i = 3G\varepsilon_i.$$

This means that the tangent of the angle $M'OM''$ (Fig. 15.4) equals the triple shear modulus, e.g. tg $\angle M'OM'' = 3G$.

Let us take an arbitrary point M (Fig. 15.4) on the non-elastic section of the hardening diagram. Let us draw a beam OM from the reference point O to this point, and draw a tangent line to the curve $\sigma_i \sim \varepsilon_i$ in the point M. The incline angle tangent of the beam OM to the coordinate axis is called *a secant modulus E_s*

$$E_s = \text{tg} \angle MOM'' = \frac{\sigma_i}{\varepsilon_i},$$

Fig. 15.4 Hardening
function

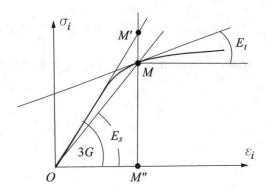

and the tangent of the incline angle of the tangent line in the point M *is the tangent
modulus E_t*

$$E_t = \frac{d\sigma_i}{d\varepsilon_i}.$$

For real hardening materials (at least in the case of proportional loading), there are
ratios

$$3G \geqslant \frac{\sigma_i}{\varepsilon_i} \geqslant \frac{d\sigma_i}{d\varepsilon_i},$$

or

$$3G \geqslant E_s \geqslant E_t. \tag{15.7}$$

Geometrically, the ratios (15.7) express the camber of the curve $\sigma_i \sim \varepsilon_i$.

Let us represent the dependency $\sigma_i = \Phi(\varepsilon_i)$ as follows

$$\sigma_i = 3G\varepsilon_i[1 - \omega(\varepsilon_i)], \tag{15.8}$$

where the function $\omega(\varepsilon_i)$ is expressed by the formula

$$\omega(\varepsilon_i) = \frac{3G\varepsilon_i - \Phi(\varepsilon_i)}{3G\varepsilon_i}. \tag{15.9}$$

From formula (15.9), it follows that for an elastic section of the dependency, $\sigma_i \sim \varepsilon_i$
$\omega(\varepsilon_i) = 0$. A geometric interpretation of Eq. (15.9) is the ratio (Fig. 15.4)

$$\omega(\varepsilon_i) = \frac{MM'}{M'M''}.$$

From ratios (15.7) for the function $\omega(\varepsilon_?)$, we can obtain

$$1 \geqslant \omega + \varepsilon_i \frac{d\omega}{d\varepsilon_i} \geqslant \omega \geqslant 0.$$

When the diagram $\sigma_i \sim \varepsilon_i$ is expressed by a two-link polyline, we obtain as follows for the function ω:

$$\omega = \begin{cases} 0 & \text{for the } \varepsilon_i \leqslant \varepsilon_s, \\ \lambda \left(1 - \dfrac{\varepsilon_s}{\varepsilon_i} \right) & \text{for the } \varepsilon_i > \varepsilon_s, \end{cases} \tag{15.10}$$

where

$$\lambda = 1 - \frac{1}{3G} \frac{d\sigma_i}{d\varepsilon_i} = 1 - \frac{E_t}{3G}.$$

N o t e . The above properties are true at least at proportional loading.

15.4   Another Form of Strain Ratios

The equation system (15.6) can be represented as follows:

$$\sigma_x = \left(K - \frac{2\sigma_i}{3\varepsilon_i} \right) \Theta + \frac{2\sigma_i}{3\varepsilon_i} \varepsilon_x,$$
$$\dotfill,$$
$$\dotfill, \tag{15.11}$$
$$\tau_{xy} = \frac{2\sigma_i}{3\varepsilon_i} \varepsilon_{xy}.$$

If the stress intensity σ_i is understood as its explicit expression through the intensity of strain ε_i, only strain components are included in the right parts of the system equations (15.11). By solving this system relative to the strain components, we obtain as follows:

$$\varepsilon_x = \frac{3\varepsilon_i}{2\sigma_i} \sigma_x - \left(\frac{3\varepsilon_i}{2\sigma_i} - \frac{1}{3K} \right) \sigma_0,$$
$$\dotfill,$$
$$\dotfill, \tag{15.12}$$
$$\varepsilon_{xy} = \frac{1}{2} \gamma_{xy} = \frac{3\varepsilon_i}{2\sigma_i} \tau_{xy}.$$

Here, the ratio ε_i / σ_i is deemed to be the function of the stress intensity only. Let us find the explicit expression of this ratio. The properties (15.7) allow solving the ratio $\sigma_i = \Phi(\varepsilon_i)$ relative to the strain intensity and writing is as follows:

$$\varepsilon_i = \Phi^-(\sigma_i).$$

In this manner,

$$\frac{\varepsilon_i}{\sigma_i} = \frac{\Phi^-(\sigma_i)}{\sigma_i}. \tag{15.13}$$

For an ideally plastic material, formula (15.13) loses any sense.

As we said before (p. 169), full strain can be represented as the sum of elastic and plastic components:

$$\varepsilon_x = \varepsilon_x^y + \varepsilon_x^e = \varepsilon_x^y + e_x,$$
$$\dots\dots\dots\dots\dots\dots\dots\dots\dots\dots,$$
$$\dots\dots\dots\dots\dots\dots\dots\dots\dots\dots,$$
$$\gamma_{xy} = \gamma_{xy}^y + \gamma_{xy}^e = \gamma_{xy}^y + e_{xy}.$$

Elastic strain components are defined through the stresses under Hooke's law both within elasticity and beyond it. Based on the previous dependencies, plastic strain components can be defined as a difference of full and respective elastic components:

$$e_x = \varepsilon_x - \varepsilon_x^y,$$
$$\dots\dots\dots\dots\dots,$$
$$\dots\dots\dots\dots\dots, \tag{15.14}$$
$$e_{xy} = \gamma_{xy} - \gamma_{xy}^y.$$

Instead of full strains, let us substitute their expressions according to (15.12) into formulas (15.14), and their expressions (1.11)–(1.12) under Hooke's law instead of elastic strain components. After simple conversions, we obtain as follows:

$$e_x = \frac{\varphi(\sigma_i)}{3G}\left[\sigma_x - \frac{1}{2}(\sigma_y + \sigma_z)\right],$$
$$\dots\dots\dots\dots\dots\dots\dots\dots\dots\dots,$$
$$\dots\dots\dots\dots\dots\dots\dots\dots\dots\dots, \tag{15.15}$$
$$e_{xy} = \frac{\varphi(\sigma_i)}{G}\,\tau_{xy},$$

where

$$\varphi(\sigma_i) = \frac{3G\varepsilon_i - \sigma_i}{\sigma_i} = \frac{\omega}{1-\omega}.$$

From formulas (15.15), it follows that plastic strain components form a deviator, e.g.

$$e_x + e_y + e_z = 0,$$

and full strain intensity can be represented as a sum of intensities of its elastic and plastic components

$$\varepsilon_i = \varepsilon_i^y + \varepsilon_i^p.$$

15.5   Unloading Laws

We have already said (p. 195) that the unloading condition in the strain theory of plasticity is defined by the inequation

$$\frac{d\sigma_i}{dt} < 0, \tag{15.16}$$

where t is the time or any other monotonously growing parameter [1, 2]. Assume that active loading was done to some point $M(\varepsilon_i; \sigma_i)$ (Fig. 15.5), then the condition (15.16) is fulfilled, and unloading is done via the trajectory $MM' \parallel OA$.

For an arbitrary $\tilde{\sigma}_i < \sigma_i$ the following first *unloading law is observed:*

$$\tilde{\sigma}_i = 3G(\tilde{\varepsilon}_i - \varepsilon_i). \tag{15.17}$$

Apart from the law (15.17), we should also write the *second unloading law*—the law of proportionality of the stress deviator and the elastic strain deviator

$$D_{\tilde{\sigma}} = 2G D_{\tilde{\varepsilon}}, \tag{15.18}$$

as well as the *third law*—the law of elasticity of volumetric strain

$$\tilde{\sigma}_0 = 3K\tilde{\varepsilon}_0 = 3K(\tilde{\varepsilon}_0 - \varepsilon_0) = 3K\tilde{\varepsilon}_0. \tag{15.19}$$

Fig. 15.5 To the formulation of unloading laws

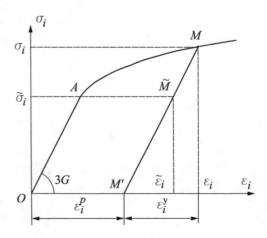

Ratios (15.17)–(15.19) fully define the link between stresses and strains in the case of unloading. In the coordinate form, these ratios can be written as follows:

$$\sigma_x - \tilde{\sigma}_x = \lambda(\Theta - \tilde{\Theta}) + 2G(\varepsilon_x - \tilde{\varepsilon}_x),$$
$$\dots\dots\dots\dots\dots\dots\dots\dots\dots\dots\dots\dots,$$
$$\dots\dots\dots\dots\dots\dots\dots\dots\dots\dots\dots\dots,$$ \hfill (15.20)
$$\tau_{xy} - \tilde{\tau}_{xy} = G(\gamma_{xy} - \tilde{\gamma}_{xy}).$$

If we assume $\tilde{\sigma}_x = \tilde{\sigma}_y = \dots = \tilde{\tau}_{zx} = 0$ in formulas (15.20), the components $\tilde{\varepsilon}_x = \varepsilon_x^p, \dots$ will define residual plastic strains in the case of full unloading.

15.6 Work of Stresses, Potential Energy, and Potentials

The work of stresses when the body element goes from a non-strained state O into the strained state M is defined by the integral

$$W = \int_O^M (\sigma_x \delta\varepsilon_x + \dots + \tau_{zx}\delta\gamma_{zx}) = \int_O^M \sigma_{ij}\delta\varepsilon_{ij}. \tag{15.21}$$

Let us show that the sub-integral expression is a perfect differential, e.g. the work W does not depend on the form of the integration path and is defined by the position of the points O and M.

Let us designate

$$\delta' W = \sigma_x \delta\varepsilon_x + \dots + \tau_{zx}\delta\gamma_{zx}, \tag{15.22}$$

and prove that this expression represents a perfect differential. In formula (15.22), let us go to the components of the strain and stress deviator by assuming that

$$\sigma_x = \sigma_x' + \sigma_0, \quad \varepsilon_x = \varepsilon_x' + \varepsilon_0, \dots \tag{15.23}$$

By substituting expressions (15.23) into formula (15.22), we obtain

$$\delta' W = \sigma_x' \delta\varepsilon_x' + \dots + \tau_{zx}\delta\gamma_{zx} + $$
$$+ \sigma_0(\delta\varepsilon_x' + \delta\varepsilon_y' + \delta\varepsilon_x') + (\sigma_x' + $$
$$+ \sigma_y' + \sigma_z')\delta\varepsilon_0 + 3\sigma_0\delta\varepsilon_0.$$

Here the equations in the brackets equal zero since the first bracket is a variation of volumetric plastic strain and the second one is a linear invariant of the stress deviator. Consequently,

$$\delta' W = \sigma'_x \delta \varepsilon'_x + \ldots + \tau_{zx} \delta \gamma_{zx} + 3\sigma_0 \delta \varepsilon_0. \tag{15.24}$$

Let us now express the stresses in formula (15.24) through strains using the strain theory ratios

$$\sigma_0 = K\Theta; \quad \sigma'_x = \sigma_x - \sigma_0 = \frac{2\sigma_i}{3\varepsilon_i}(\varepsilon_x - \varepsilon_0) = \frac{2\sigma_i}{3\varepsilon_i}\varepsilon'_x. \tag{15.25}$$

We obtain

$$\delta' W = \frac{2\sigma_i}{3\varepsilon_i}(\varepsilon'_x \delta \varepsilon'_x + \ldots + 2\varepsilon'_{xy} \delta \varepsilon'_{xy} + \ldots) + K\Theta\delta\Theta. \tag{15.26}$$

Let us show that the expression in the brackets of formula (15.26) is the perfect differential. Indeed

$$(\varepsilon'_x \delta \varepsilon'_x + \ldots + 2\varepsilon'_{xy} \delta \varepsilon'_{xy} + \ldots) = \frac{1}{2}\delta(\varepsilon'^2_x + \ldots + 2\varepsilon'^2_{zx}) = \frac{3}{4}\delta\varepsilon^2_? = \frac{3}{2}\varepsilon_?\delta\varepsilon_?.$$

Taking into account the last result, formula (15.26) gives

$$\delta' W = \sigma_i \delta \varepsilon_i + \sigma_0 \delta \Theta. \tag{15.27}$$

Having in mind that $\sigma_i = \Phi(\varepsilon_i)$ and $\sigma_0 = K\Theta$, formula (15.27) can be written as follows

$$\delta W = \Phi(\varepsilon_i)\delta \varepsilon_i + K\Theta\delta\Theta. \tag{15.28}$$

The result shows that a perfect differential is written in the right part so the dash in variation is omitted. By integrating formula (15.28) and taking into account the work at a hydrostatic change in the element volume, we will finally obtain

$$W = \int_0^{\varepsilon_i} \sigma_i \delta \varepsilon_i + \frac{1}{2}\sigma_0\Theta. \tag{15.29}$$

Let us note that the first addend in formula (15.29) represents the work of changing the element shape, and the second is the work of changing its volume. Second, as follows from (15.29), the work of the stresses W is a function

$$W = W(\varepsilon_i, \Theta) = W(\varepsilon_x, \ldots, \gamma_{zx}). \tag{15.30}$$

15.6.1   Stress Potential

Let us vary the function (15.30):

$$\delta W = \frac{\partial W}{\partial \varepsilon_i}\delta\varepsilon_i + \frac{\partial W}{\partial \Theta}\delta\Theta = \frac{\partial W}{\partial \varepsilon_x}\delta\varepsilon_x + \ldots + \frac{\partial W}{\partial \gamma_{zx}}\delta\gamma_{zx}. \tag{15.31}$$

By comparing formulas (15.31), (15.22), and (15.27), we come to the formulas:

$$\sigma_i = \frac{\partial W}{\partial \varepsilon_i}, \quad \sigma_0 = \frac{\partial W}{\partial \Theta}, \quad \sigma_x = \frac{\partial W}{\partial \varepsilon_x}, \quad \ldots, \quad \tau_{zx} = \frac{\partial W}{\partial \gamma_{zx}}. \tag{15.32}$$

The obtained result means (15.32) that the function W is *the stress potential.*

Definition   The potential energy of a single body element is that part of the work W that will be returned by the element at full loading.

It follows from this expression that the potential energy (W_e) equals the work of stresses with elastic deformations, e.g.

$$W_e = \frac{1}{2}(\sigma_x\varepsilon_x + \ldots + \tau_{zx}\gamma_{zx}) = \frac{1}{2}\sigma_i\varepsilon_i + \frac{1}{2}\sigma_0\Theta. \tag{15.33}$$

In the right part of formula (15.33), the first addend gives the potential energy of the shape change, and the second addend gives the potential energy of the volume change. Taking into account that

$$\sigma_i = 3G\varepsilon_i, \quad \sigma_0 = K\Theta,$$

the expression for the potential energy W_e can be written as

$$W_e = \frac{1}{6G}\sigma_i^2 + \frac{1}{2K}\sigma_0^2. \tag{15.34}$$

For incomplete unloading, the potential energy reserve $(\tilde{W}_e)$ in the body element can be defined using the formula

$$\tilde{W}_e = \frac{1}{6G}\tilde{\sigma}_i^2 + \frac{1}{2K}\tilde{\sigma}_0^2. \tag{15.35}$$

The irreversible (dispersed) part of work will be

$$W_p = W - W_e = \int_0^{\varepsilon_i} \sigma_i d\varepsilon_i - \frac{\sigma_i^2}{6G}.$$

Fig. 15.6 To the conclusion
of strain potential

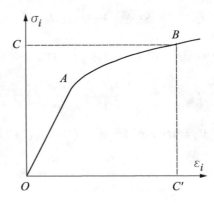

15.6.2 Potential of Strains

Let us find the hardening curve OAB Fig. 15.6. The area of the figure $OABC$ is represented as

$$S_{OABC} = W' - \frac{1}{2}\sigma_0\Theta, \tag{15.36}$$

where W' is some function. The same area can be written otherwise:

$$S_{OABC} = \sigma_i\varepsilon_i - \int_0^{\varepsilon_i} \sigma_i d\varepsilon_i = \int_0^{\sigma_i} \varepsilon_i d\sigma_i. \tag{15.37}$$

By equating the right parts of formulas (15.36) and (15.37), we obtain

$$W' = \int_0^{\sigma_i} \varepsilon_i d\sigma_i + \frac{1}{2}\sigma_0\Theta. \tag{15.38}$$

By comparing formulas (15.38) and (15.29), we note that the function W' is obtained from W, if we swap stresses σ_i and strains ε_i in the latter formula. However, W is a potential for stresses. Hence it follows that the function W' is a potential for strains, e.g.

$$\varepsilon_i = \frac{\partial W'}{\partial \sigma_i}, \quad \Theta = \frac{\partial W'}{\partial \sigma_0},$$
$$\varepsilon_x = \frac{\partial W'}{\partial \sigma_x}, \dots, \gamma_{zx} = \frac{\partial W'}{\partial \tau_{zx}}. \tag{15.39}$$

Having the formulas for calculating the work spent for deforming the body element, now we can calculate the full work (A) of strain of any homogeneous body. By designating the element volume as dV and the entire body volume as V, we can write[2]

$$A = \iiint_V W \, dV = \iiint_V \left[\int_0^{\varepsilon_i} \sigma_i \, d\varepsilon_i + \frac{1}{2} K \Theta^2 \right] dV. \qquad (15.40)$$

The potential energy that can be separated from the deformed body will be

$$A_e = \iiint_V W_e \, dV = \frac{1}{6G} \iiint_V \sigma_i^2 \, dV + \frac{1}{2K} \iiint_V \sigma^2 \, dV.$$

Using (15.35), we obtain the remaining potential energy (A_0) for partial loading.

$$A_0 = \frac{1}{6G} \iiint_V \tilde{\sigma}_i \, dV + \frac{1}{2K} \iiint_V \tilde{\sigma}_0^2 \, dV.$$

15.7   Theorem of the Minimal Work of Inner Forces

Assume that some body is subject to the action of the system of surface forces. Let us designate the projections onto the axes of the Cartesian system of the coordinates $Oxyz$ of the external load acting in the point with the normal line ν as X_ν, Y_ν, Z_ν. Assume that an arbitrary point of the body M receives displacement under the action of forces

$$\mathbf{r} = \mathbf{i}u + \mathbf{j}v + \mathbf{k}w. \qquad (15.41)$$

For the known $\mathbf{r}$, we can find strains using Cauchy formulas, and then find stresses using strains.

Kinematically, let the possible body state be called a state defined by the vector

$$\mathbf{r}' = \mathbf{r} + \delta \mathbf{r}, \qquad (15.42)$$

where $\delta \mathbf{r}$ is the virtual displacement defined by the formula

---

[2]The triple integral is taken upon the initial body volume. It is suggested that changes in the shape and size of the body are small. This assumption is justified for low strains. For high strains, it may lead to significant errors.

$$\delta \mathbf{r} = \mathbf{i}\delta u + \mathbf{j}\delta v + \mathbf{k}\delta w.$$

Apart from continuity, the virtual displacement property is that it turns zero at the body boundary. Possible strains and stresses are defined through the vector $\mathbf{r}'$ in the same manner as actual stresses and strains are defined through the vector $\mathbf{r}$.

Theorem *The true state of body equilibrium differs from any kinematically possible body in that the work of inner forces*

$$A(\varepsilon_i, \Theta) = \iiint\limits_V \left[\int\limits_0^{\varepsilon_i} \sigma_i d\varepsilon_i + \frac{1}{2} K \Theta^2 \right] dV$$

has a minimum.

Not to overload the book with mathematical calculations, we do not provide proofs of this theorem here. If the reader is interested, it can be found in the monograph by A. A. Ilyushin [3, pp. 112–115] (the proof is given for the condition of active loading $d\sigma_i/d\varepsilon_i > 0$).

15.8 Lagrange Equilibrium Variation Equation

Assume that the body is under the action of external surface and mass forces. The mechanical condition of the body is defined by the elastic characteristics E, G, the hardening curve $\sigma_i \sim \varepsilon_i$, and the laws of linkage between stresses and strains. It is required to find stresses and strains in the body.

Let us write the Lagrange equilibrium variation equation that is true for both elastic and plastic bodies. Let us designate the displacements of points of a deformed body as $u = u(x, y, z, t)$, $v = v(x, y, z, t)$ and $w = w(x, y, z, t)$. *Virtual displacements* are any variation δu, δv, δw compatible with links superimposed on the body and its parts. Assume that the function

$$A = A(\varepsilon_i, \Theta) = A(\varepsilon_x, \ldots, \gamma_{zx})$$

designates the work of body strain.

The Lagrange equilibrium variation principle asserts as follows.

The variation of the work of inner forces during virtual displacements of body particles equals the work of outer surface and mass forces on displacement variations:

$$\delta A = \iiint_V \delta W dV = \iiint_V \left[\frac{\partial W}{\partial \varepsilon_x} \delta \varepsilon_x + \ldots \right] dV =$$

$$= \iiint_V \rho(X\delta u + Y\delta v + Z\delta w)dV + \iint_S (X_v\delta u + Y_v\delta v + Z_v\delta w)dS, \tag{15.43}$$

where X_v, Y_v, Z_v are still projections onto the axes of the Cartesian coordinate system $Oxyz$ of the external surface loading acting in the point with the normal line v; ρX, ρY, ρZ are the projections of mass forces on the same axes, and S is the surface confining the area V.

Let us substitute strains in formula (15.43) with their expressions through displacements under the formulas

$$\varepsilon_x = \frac{\partial u}{\partial x}, \quad \ldots, \quad \gamma_{zx} = \frac{\partial u}{\partial z} + \frac{\partial w}{\partial x}.$$

Furthermore, we will substitute derivatives and variations in the second integral of these formulas with the following expressions:

$$\frac{\partial W}{\partial \varepsilon_x} = \sigma_x, \ldots, ; \quad \delta \varepsilon_x = \delta \frac{\partial u}{\partial x} = \frac{\partial}{\partial x} \delta u, \ldots.$$

By substituting the last expressions into formula (15.43) and grouping addends for the variations δu, δv, and δw, we obtain as follows

$$\delta A = \iiint_V \left[\left(\sigma_x \frac{\partial}{\partial x} + \tau_{xy} \frac{\partial}{\partial y} + \tau_{xz} \frac{\partial}{\partial z} \right) \delta u \right.$$

$$+ \left(\tau_{yx} \frac{\partial}{\partial x} + \sigma_y \frac{\partial}{\partial y} + \tau_{yz} \frac{\partial}{\partial z} \right) \delta v$$

$$\left. + \left(\tau_{zx} \frac{\partial}{\partial x} + \tau_{zy} \frac{\partial}{\partial y} + \sigma_z \frac{\partial}{\partial z} \right) \delta w \right] dV. \tag{15.44}$$

For further transformation of expression (15.44), let us use the Ostrogradsky–Green formula

$$\iiint_V \frac{\partial P(x, y, z)}{\partial x} dV = \iint_S P(x, y, z)lds; \quad [l = \cos(v,\hat{}x)], \tag{15.45}$$

as well as equations

$$\frac{\partial}{\partial x}(\sigma_x \delta u) = \frac{\partial \sigma_x}{\partial x} \delta u + \sigma_x \frac{\partial}{\partial x} \delta u;$$

$$\sigma_x \frac{\partial}{\partial x} \delta u = \frac{\partial}{\partial x}(\sigma_x \delta u) - \frac{\sigma_x}{\partial x} \delta u. \tag{15.46}$$

Assume in formula (15.45) $P(x, y, z) = \sigma_x \delta u$; and we will obtain

$$\iiint\limits_V \sigma_x \frac{\partial}{\partial x} \delta u \, dV = \iint\limits_S \sigma_x l \delta u \, dS - \iiint\limits_V \frac{\partial \sigma_x}{\partial x} \delta u \, dV.$$

Such formulas can be obtained for each of the addends of the sub-integral expression of formula (15.44). By substituting these formulas into the variation of work (15.44), the latter is transformed into

$$\delta A = \iiint\limits_V \left[\left(\frac{\partial \sigma_x}{\partial x} + \frac{\partial \tau_{xy}}{\partial y} + \frac{\partial \tau_{xz}}{\partial z} \right) \delta u + (\ldots\ldots) \delta v + (\ldots\ldots) \delta w \right] dV$$

$$- \iint\limits_S \left[(\sigma_x l + \tau_{xy} m + \tau_{yz} n) \delta u + (\ldots\ldots) \delta v + (\ldots\ldots) \delta w \right] dS;$$

$$[m = \cos(\hat{v,} y); \quad n = \cos(\hat{v,} z); \quad l = \cos(\hat{v,} x)]. \tag{15.47}$$

If the volumetric integral in the left part of formula (15.43) is substituted according to formula (15.47) and addends are grouped with equal variations, we obtain

$$\delta A = \iiint\limits_V \left[\left(\frac{\partial \sigma_x}{\partial x} + \frac{\partial \tau_{xy}}{\partial y} + \frac{\partial \tau_{xz}}{\partial z} + \rho X \right) \delta u \right.$$

$$+ (\ldots + \rho Y) \delta v + (\ldots + \rho Z) \delta w \bigg] dV$$

$$- \iint\limits_S \left[(\sigma_x l + \tau_{xy} m + \tau_{xz} n - X_v) \delta u \right.$$

$$+ (\ldots - Y_v) \delta v + (\ldots - Z_v) \delta w \bigg] dS = 0. \tag{15.48}$$

Due to the independence of displacement variations in any point of the body, the equation of (15.48) is possible provided all the co-factors of variations equal zero. By zeroing the brackets for the variations δu, δv, and δw, we obtain equilibrium equations (2.1) (taking into account mass forces)

$$\frac{\partial \sigma_x}{\partial x} + \frac{\partial \tau_{xy}}{\partial y} + \frac{\partial \tau_{xz}}{\partial z} + \rho X = 0,$$

$$\frac{\partial \tau_{yx}}{\partial x} + \frac{\partial \sigma_y}{\partial y} + \frac{\partial \tau_{yz}}{\partial z} + \rho Y = 0, \tag{15.49}$$

$$\frac{\partial \tau_{zx}}{\partial x} + \frac{\partial \tau_{zy}}{\partial y} + \frac{\partial \sigma_z}{\partial z} + \rho Z = 0,$$

and boundary conditions

$$\sigma_x l + \tau_{xy} m + \tau_{xz} n = X_\nu,$$
$$\tau_{yx} l + \sigma_y m + \tau_{yz} n = Y_\nu, \tag{15.50}$$
$$\tau_{zx} l + \tau_{zy} m + \sigma_z n = Z_\nu.$$

In the short form, conditions (15.49)–(15.50) look as follows:

$$\sigma_{ij,j} + \rho X_i = 0;$$
$$\sigma_{ij} n_j = X_{i\nu}. \tag{15.51}$$

Here $i, j \sim x, y, z$; $n_j \sim l, m, n$; $X_i \sim X, Y, Z$ and $X_{i\nu} \sim X_\nu, Y_\nu, Z_\nu$. Summing is done upon repeated indexes, and the comma between indexes means differentiation upon the coordinate corresponding to the index after the comma.

In this manner, the variation equation of Lagrange equilibrium includes equilibrium equations and boundary conditions. If we assume that all the values in formula (15.48) or in formulas (15.49)–(15.50) are expressed in displacements, these equations are sufficient to solve the problem set in the beginning of this paragraph (p. 208). If the system (15.49)–(15.50) is considered as equations in stresses, they must be added to equations of joint strain. To do it, let us solve the conformity conditions (2.10)–(2.11) relative to strains; we obtain as follows:

$$\frac{\partial^2 \gamma_{xy}}{\partial x \partial y} = \frac{\partial^2 \varepsilon_x}{\partial y^2} + \frac{\partial^2 \varepsilon_z}{\partial x^2},$$
$$\dots\dots\dots\dots\dots\dots\dots,$$
$$\dots\dots\dots\dots\dots\dots\dots; \tag{15.52}$$
$$2\frac{\partial^2 \varepsilon_x}{\partial y \partial z} = \frac{\partial}{\partial z}\left(-\frac{\partial \gamma_{yz}}{\partial x} + \frac{\partial \gamma_{zx}}{\partial y} + \frac{\partial \gamma_{xy}}{\partial z}\right),$$
$$\dots\dots\dots\dots\dots\dots\dots\dots\dots,$$
$$\dots\dots\dots\dots\dots\dots\dots\dots\dots\dots$$

Since the problem is solved in stresses, strains in conditions (15.52) must be expressed through stresses. For example, for deformational theory

$$\varepsilon_x = \frac{3\varepsilon_i}{3\sigma_i}\sigma_x - \left(\frac{3\varepsilon_i}{2\sigma_i} - \frac{1}{3K}\right)\sigma_0,$$
$$\dots\dots\dots\dots\dots\dots\dots\dots\dots$$

In this manner, the conformity conditions are rather large and we do not give them.

15.9 Setting Boundary Problems of Plasticity Theory

Based on the previous equations, three primary problems of plasticity theory at active loading are formulated.

1. *Find three functions u, v, w so that for arbitrary continuums with continuous derivatives of variations δu, δv, δw, there is a variation equation of equilibrium* (15.48).

In this setting, one of the variation methods can be used to solve the problem, for example, the Ritz method. Let us explain the essence of the method.

Let us select a full orthogonal system of functions $f_n(x, y, z)$ in the area D occupied by the body and represent the sought displacements with rows

$$u = \sum a_n f_n, \quad v = \sum b_n f_n, \quad w = \sum c_n f_n, \tag{15.53}$$

where a_n, b_n, c_n are yet unknown coefficients.

Let us find variations of functions (15.53)

$$\delta u = \sum f_n \delta a_n, \quad \delta v = \sum f_n \delta b_n, \quad \delta w = \sum f_n \delta c_n \tag{15.54}$$

and calculate the work of inner forces on displacement variations (15.54):

$$A = \iiint\limits_V W dV = A(a_n, b_n, c_n); \tag{15.55}$$

$$\delta A = \frac{\partial A}{\partial a_n} \delta a_n + \frac{\partial A}{\partial b_n} \delta b_n + \frac{\partial A}{\partial c_n} \delta c_n. \tag{15.56}$$

By substituting variations (15.54) into formula (15.43) and equating the right part to the right part of expression (15.56), we obtain a system of three equations

$$\frac{\partial A}{\partial a_n} = \iiint\limits_V \rho X f_n dV + \iint\limits_S X_v f_n dS,$$

$$\dots\dots\dots\dots\dots\dots\dots\dots\dots\dots\dots\dots\dots, \tag{15.57}$$

$$\dots\dots\dots\dots\dots\dots\dots\dots\dots\dots\dots\dots\dots$$

The system (15.57) can be written for each n. In this manner, there will be as many equations of type (15.57) as indefinite coefficients a_n, b_n, c_n. Consequently, the obtained system of linear algebraic equations allows defining these specific coefficients in principle.

2. *Find three functions u, v, w satisfying differential equations of equilibrium expressed in displacements and boundary conditions.*

This problem with arbitrary outer loads can have a solution in those cases only when equilibrium equations in the Lame form [5] will have elliptical type. An efficient method to solve problems in setting 2 is a so-called method of elastic solutions that we will consider later.

3. *Let us find six functions σ_x, σ_y, $\dots$, τ_{zx} satisfying equilibrium equations in stresses, conformity conditions, and boundary conditions.*

The existence and singularity of solving this problem have been proved.

15.10   Theorem of Simple Loading

Multiple experimental studies have shown that theory laws of small elastic–plastic strains take place at least when the loading of a body element is simple or close to simple.

There is a question: are there such loads applied to the body so that in each point there is a process of simple loading, e.g. the director stress tensor is constant for each point during the entire loading process (director tensor can change from point to point)?

In the case of homogeneous strain, the answer to this question is trivial: the entire body will endure simple loading if external forces change in proportion to the same parameter, though all-around volumetric compression can change under an arbitrary law.

In a general case, this question is rather complicated. So far, we have the following *theorem of simple loading proved by A.A. Ilyushin.*

Theorem *Assume that the external forces $X(x, y, z)$, X_ν change in proportion to the same parameter*

$$X(x, y, z) = \lambda X^*(x, y, z),$$

$$X_\nu = \lambda X^*(x, y, z), \tag{15.58}$$

where λ is the parameter defining the consecutive values of forces applied to the body (such as time).

Simple loading in each body point will take place if the dependency $\sigma_i \sim \varepsilon_i$ can be represented by the power law

$$\sigma_i = A\varepsilon_i^\chi, \quad (A, \chi - const), \tag{15.59}$$

and the condition of non-compressibility is met

$$\Theta = 3\varepsilon_0 = 0. \tag{15.60}$$

Conditions (15.58)–(15.60) are sufficient but not necessary.

The proof of this theorem in these strict conditions is elementary. Assume that the plasticity theory problem is solved for a body at the fixed value of the parameter λ. For certainty, let us adopt $\lambda = 1$. This means that in each body point, stresses $\sigma_x^*, \ldots, \tau_{zx}^*$, strains $\varepsilon_x^*, \ldots, \gamma_{zx}^*$, intensities $\sigma_?^*$, $\varepsilon_?^*$, and displacements u^*, v^*, w^* are defined. The obtained solutions turn the equilibrium equations, boundary conditions, and condition (15.60) into identical equations:

$$\frac{\partial \sigma_x^*}{\partial x} + \frac{\tau_{xy}^*}{\partial y} + \frac{\partial \tau_{xz}^*}{\partial z} = 0,$$

$$\dots\dots\dots\dots\dots\dots\dots,$$

$$\sigma_x^* l + \tau_{xy}^* m + \tau_{xz}^* = X_v,$$

$$\dots\dots\dots\dots\dots\dots\dots,$$ (15.61)

$$\frac{\sigma_x^* - \sigma_0}{\sigma_i} = \frac{2}{3} \frac{\varepsilon_x^*}{\varepsilon_i},$$

$$3\varepsilon_0 = \frac{\partial u^*}{\partial x} + \frac{\partial v^*}{\partial y} + \frac{\partial w^*}{\partial z} = 0.$$

Let us try to find the solution for some $\lambda \neq 1$. Let us represent the solution as follows:

$$\sigma_x = \lambda \sigma_x^*, \dots; \ \varepsilon_x = \mu \varepsilon_x^*;$$
$$\sigma_i = \lambda \sigma_i^*; \dots; \ \varepsilon_i = \mu \varepsilon_x^*,$$ (15.62)

where μ is a so-far not defined function of a single λ.

In the presence of identical equations (15.61), the solution (15.62) will also turn the equilibrium equations, boundary conditions, and the incompressibility condition into identical equations: The last equation that must be satisfied is condition (15.59). We have

$$\lambda \sigma_i^* = \mu^\chi A(\varepsilon_i^*)^\chi.$$ (15.63)

However, according to condition (15.59) $\sigma_i^* = A(\varepsilon_i^*)^\chi$. To fulfill Eq. (15.63), we must assume $\lambda = \mu^\chi$. The components of the director stress tensor

$$\frac{\sigma_x - \sigma_0}{\sigma_i}, \dots, \frac{\varepsilon_x}{\varepsilon_i}$$

will not depend on the parameter λ.

Remarks

1. The theorem is proved for any volumetric stressed state. In partial cases of volumetric stressed state, this theorem is proved for the arbitrary dependency $\sigma_i = \Phi(\varepsilon_i)$.
2. The ratio (15.59) allows describing the experimental dependency $\varepsilon_i \sim \sigma_i$ for a wide class of materials (Fig. 15.7).

Fig. 15.7 Family of
hardening curves

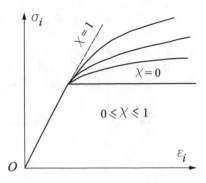

15.11   Theorem of Unloading

Let us consider an arbitrary body exposed to the action of a defined system of forces X, Y, Z; X_ν, Y_ν, Z_ν described by the parameter η. Let the plasticity theory problem be solved for the body, e.g. σ_x, $\ldots$; ε_x, $\ldots$; σ_i; ε_i; u, v, w are known. Let us study the unloading process

$$\frac{d\sigma_i}{dt} < 0.$$

Let us call the unloading simple if all external forces decrease in proportion to the same parameter. All the values belonging to the unloading process will be denoted with a twiddle on top. At the beginning of unloading ($\tilde{\eta} = \eta$) we have:

$$\tilde{\sigma}_x = \sigma_x, \ldots, \tilde{w} = w.$$

Both stresses σ_x, $\ldots$ and stresses $\tilde{\sigma}_x$, $\ldots$ satisfy the equilibrium equations and boundary conditions. So for any $\tilde{\eta} \leqslant \eta$, their difference also satisfies these equations and we can write as follows:

$$\frac{\partial(\sigma_x - \tilde{\sigma}_x)}{\partial x} + \frac{\partial(\tau_{xy} - \tilde{\tau}_{xy})}{\partial y} + \frac{\partial(\tau_{xz} - \tilde{\tau}_{xz})}{\partial z} + \rho(X - \tilde{X}) = 0,$$

$$\ldots\ldots\ldots\ldots\ldots\ldots\ldots\ldots\ldots\ldots\ldots\ldots\ldots\ldots\ldots\ldots ,$$

$$\ldots\ldots\ldots\ldots\ldots\ldots\ldots\ldots\ldots\ldots\ldots\ldots\ldots\ldots\ldots\ldots ;$$

$$\tag{15.64}$$

$$(\sigma_x - \tilde{\sigma}_x)l + (\tau_{xy} - \tilde{\tau}_{xy})m + (\tau_{xz} - \tilde{\tau}_{xz})m = (X_\nu - \tilde{X}_\nu),$$

$$\ldots\ldots\ldots\ldots\ldots\ldots\ldots\ldots\ldots\ldots\ldots\ldots\ldots\ldots\ldots\ldots ,$$

$$\ldots\ldots\ldots\ldots\ldots\ldots\ldots\ldots\ldots\ldots\ldots\ldots\ldots\ldots\ldots\ldots$$

$$\tag{15.65}$$

Apart from ratios (15.64) and 15.65), during unloading, there can be the following law of linkage between stresses and strains:

$$\sigma_x - \tilde{\sigma}_x = \lambda(\Theta - \tilde{\Theta}) + 2G(\varepsilon_x - \tilde{\varepsilon}_x), \tag{15.66}$$

where λ is the Lame constant (p. 191).

If dependencies (15.66) are substituted to Eqs. (15.64), we obtain the equilibrium equations in the Lame form [5]:

$$(\lambda + G)\frac{\partial}{\partial x}(\Theta - \tilde{\Theta}) + 2G\Delta(u - \tilde{u}) + \rho(X - \tilde{X}) = 0,$$
$$\dots\dots\dots\dots\dots\dots\dots\dots\dots\dots\dots\dots\dots\dots\dots\dots\dots\dots\dots\dots\dots, \tag{15.67}$$
$$\dots\dots\dots\dots\dots\dots\dots\dots\dots\dots\dots\dots\dots\dots\dots\dots\dots\dots\dots\dots$$

Equations (15.67) together with boundary conditions (15.65) have a single solution that can be found by elasticity theory methods. This solution can be represented:

$$u_1 = u - \tilde{u}, \quad v_1 = v - \tilde{v}, \quad w_1 = w - \tilde{w}. \tag{15.68}$$

The solution (15.68) allows finding strains and displacements.

In this manner, the following is proved.

Theorem *Displacements of a body point* $(\tilde{u}, \tilde{v}, \tilde{w})$ *at some point of the unloading stage differ from their values* (u, v, w) *at the beginning of unloading by the values of elastic displacements* (u_1, v_1, w_1) *that would occur in the body if the forces* $(X - \tilde{X})$, $(X_\nu - \tilde{X}_\nu)$ are applied to it in a naturally non-stressed state [6].

In particular, if we assume $\tilde{X} = \tilde{X}_\nu = 0$, there is a case of full unloading and residual displacements of body points will be

$$\tilde{u} = u - u_1, \quad \tilde{v} = v - v_1, \quad \tilde{w} = w - w_1.$$

The below follows from the proved theorem.

If the plasticity theory problem is solved for the body and the defined values of the system of forces (X, X_ν) *correspond to the true state* (S), *and if the fictitious problem of elasticity theory is solved for the body, e.g. the same system of forces* (X, X_ν) *is assigned with the fictitious state* (S_1), *as a result of the full unloading of the body, it will still have displacements, strains, and stresses equal to the differences of their values in the states* (S) *and* (S_1).

It is suggested that residual stresses do not go beyond the elasticity limits. The formulated conclusion is illustrated by the example of an elastic–plastic pure bending of the beam depicted in Fig. 15.8. The epure S represents a distribution of stresses at the elastic–plastic bending of a real beam. The epure S_1 represents a fictitious solution to the elastic problem at the assumption there are no plastic strains and the elastic properties of the material are preserved at any loads. The epure $(S) - (S_1)$ represents residual stresses after full unloading. The full solution to the problem is given in the next chapter of the book.

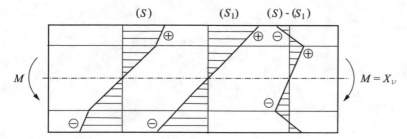

Fig. 15.8 To determination of residual stresses

References

1. H. Hencky, The general theory of compatibility conditions. Zeitschr. für angew. Math. und Mechanik. **3**(4), 241–251 (1923)
2. H. Hencky, Zur theorie plastischer deformationen und der hierdirch im material hervorgerufenen nachspannungen. Ztschr. angew. Math. und Mech. Bd. **4**(4), 323–334 (1924)
3. A. Il'yushin, *Mekhanicheskie svoistva i ispytanie metallov* [Mechanical properties and testing of metals] (OGIZ Publ., Moscow, Leningrad, 1948)
4. A. Il'yushin, V. Lenskii, *Soprotivlenie materialov* [Strength of materials] (Fizmatlit Publ., Moscow, 1959)
5. A. Lyav, *Matematicheskaya teoriya uprugosti* [Mathematical theory of elasticity] (ONTI NKTP SSSR Publ., Moscow, Leningrad, 1935)
6. A. Nadai, *Plastichnost' i razrushenie tverdykh tel* [Plasticity and destruction of solids] (IL Publ., Moscow, 1954)

Chapter 16
Solution of the Simplest Problems for the Strain Theory of Plasticity

16.1 Pure Bending of a Straight Beam

Let a straight beam be subject to bending by two pairs of M as shown in Fig. 16.1a. For simplicity, we will assume that the beam cross-section has two symmetry axes (Fig. 16.1b), and there is an elongation diagram for the beam material (Fig. 16.1c).

Let us designate the main central axes of the beam cross-section as x and y and align the axis z with the lateral axis of the beam so that zy is the bending plane. Let us designate the cross-section height and width as $2h$ and b, respectively, and the curvature of the deformed beam axis as χ.

The problem is solved in stresses. We have equilibrium equations, strain conformity equations, and boundary conditions. By assuming the hypotheses of plane cross-sections [3], we have

$$\varepsilon_z = \varepsilon_1 = \chi y. \tag{16.1}$$

The condition of material non-compressibility gives

$$\varepsilon_x + \varepsilon_y + \varepsilon_z = 0. \tag{16.2}$$

For the entire cross-section, let us assume the Poisson ratio as 0.5. In the following, we will frequently use this simplification since its effect on the results is insignificant [2]. Taking into account assumptions for strains in the transverse direction, we have

$$\varepsilon_x = \varepsilon_y = -\frac{1}{2}\chi y. \tag{16.3}$$

Since strains are linear functions y, conformity equations are identically satisfied. We also note that formulas (16.1)–(16.3) represent a law of uniaxial strain. So only

© The Author(s), under exclusive license to Springer Nature Switzerland AG 2021
V. Molotnikov, A. Molotnikova, *Theory of Elasticity and Plasticity*,
https://doi.org/10.1007/978-3-030-66622-4_16

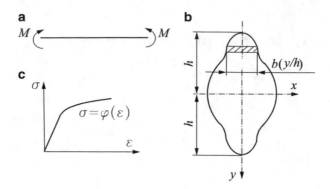

Fig. 16.1 Pure elastic–plastic bending of a straight beam

the component of stress σ_z differs from zero, and it does not depend on the axial coordinate z.

In this manner, the problem is solved by the single ratio between the bending moment and curvature that is obtained from the equilibrium of moments of inner and outer forces:

$$M = \int_{-h}^{h} \sigma(\varepsilon) \cdot y \cdot b\left(\frac{y}{h}\right) dy. \tag{16.4}$$

The substitution of the dependency $\sigma = \varphi(\varepsilon)$ to formula (16.4) gives

$$M = \int_{-h}^{h} \varphi(\varepsilon) b\left(\frac{y}{h}\right) y \, dy = \int_{-h}^{h} \varphi(\chi y) b\left(\frac{b}{h}\right) y \, dy. \tag{16.5}$$

Let us make the following transformations in the integral (16.5):

$$\varepsilon = \chi y; \quad y = \frac{\varepsilon}{\chi}; \quad b\left(\frac{y}{h}\right) = b\left(\frac{\varepsilon}{\varepsilon_h}\right),$$

where $\varepsilon_h = \chi h$ is the strain of the marginal (lower) fiber of the beam. By substituting these values into formula (16.5), we can write as follows:

$$M = \frac{2}{\chi^2} \int_{0}^{\varepsilon_h} \varphi(\varepsilon) b\left(\frac{\varepsilon}{\varepsilon_h}\right) \varepsilon \, d\varepsilon = \frac{2h^2}{\varepsilon_h^2} \int_{0}^{\varepsilon_h} \varphi(\varepsilon) b\left(\frac{\varepsilon}{\varepsilon_h}\right) \varepsilon \, d\varepsilon. \tag{16.6}$$

Let us designate

$$\frac{1}{\varepsilon_h^2} \int_{0}^{\varepsilon_h} \varphi(\varepsilon) b\left(\frac{\varepsilon}{\varepsilon_h}\right) \varepsilon \, d\varepsilon = \Phi(\varepsilon_h). \tag{16.7}$$

Then the dependency (16.6) will look like

$$\frac{M}{2h^2} = \Phi(\varepsilon_h).$$ (16.8)

For the known function $\varphi(\varepsilon)$, formula (16.8) allows finding the strain ε_h and therefore the curvature χ corresponding to the set value of the moment M. Formula (16.8) is correct in the elastic and elastic–plastic zone of the beam material.

Let us consider partial cases:

1. A rectangular-section beam $2h$ high and b wide is made of an ideally plastic material:

$$\sigma = \begin{vmatrix} E\varepsilon & \text{when } \varepsilon < \varepsilon_s; \\ \sigma_s & \text{when } \varepsilon \geqslant \varepsilon_s, \end{vmatrix}$$

where ε_s and σ_s are the strain and stress at the time of yield occurrence, respectively. In the considered case, we find

$$\Phi(\varepsilon_h) = \frac{1}{\varepsilon_h^2}\left[\int_0^{\varepsilon_s} Eb\varepsilon^2 d\varepsilon + \int_{\varepsilon_s}^{\varepsilon_h} \sigma_s b\varepsilon d\varepsilon\right].$$ (16.9)

The following results from formula (16.9):

(a) For the elastic work of the beam,

$$\Phi(\varepsilon_h) = \frac{bE\varepsilon_h}{3}, \quad (\varepsilon_h < \varepsilon_s).$$ (16.10)

In this case, formula (16.8) gives a result known from the strength of materials:

$$\chi = \frac{\varepsilon_h}{h} = \frac{M}{EJ_x}, \quad \left(J_x = \frac{b(2h)^3}{12}\right).$$

(b) After yield occurrence $\varepsilon_h \gg \varepsilon_s$. Then, by neglecting the square of the ratio $\varepsilon_s/\varepsilon_h$, we obtain as follows from formula (16.9):

$$\Phi(\varepsilon_h) = \frac{b\sigma_s}{2}.$$

By substituting this result to formula (16.8), we obtain the limit bending moment M_{max} when yield occurs in all lateral fibers of the beam

$$M_{max} = bh^2\sigma_s.$$ (16.11)

Fig. 16.2 Dependency
between the curvature and
bending moment

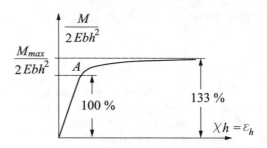

The moment (16.11) is referred to as the bearing capacity of the beam in pure bending.

Figure 16.2 depicts the dependency between the curvature and bending moment built based on formulas (16.9) and (16.11). The point A corresponds to the time of yield occurrence in marginal fibers of the beam ($y = \pm h$). The asymptotic value of the bending moment M_{max} (bearing capacity) is one-third higher than the moment in the point A.

2. The same rectangular-section beam with the length $2h$ and width b but made of a material with linear hardening:

$$\sigma = \varphi(\varepsilon) = \begin{vmatrix} E\varepsilon & \text{when } \varepsilon < \varepsilon_s; \\ \sigma_s + E_t(\varepsilon - \varepsilon_s) & \text{when } \varepsilon \geqslant \varepsilon_s, \end{vmatrix} \qquad (16.12)$$

where E_t is the tangential modulus on the elongation diagram. By substituting expression (16.12) into formula (16.7), we obtain as follows after calculation of quadratures:

$$\Phi(\varepsilon_h) = \frac{bE_t\varepsilon_h}{3} + \frac{b\varepsilon_s^3}{3\varepsilon_h^2}(E - E_t) + \frac{b(\varepsilon_h^2 - \varepsilon_s^2)}{2\varepsilon_h^2}(\sigma_s - E_t\varepsilon_s). \qquad (16.13)$$

In the case of elastic work of the beam ($\varepsilon_h < \varepsilon_s$), the solution accurately coincides with the formulas of case (a) studied above.

If $\varepsilon_h \geqslant \varepsilon_s$, formulas (16.8) and (16.13) give the following value of the bearing capacity of the beam:

$$M_{max} = bh^2\left(\sigma_s + \frac{2}{3}E_t\varepsilon_h\right). \qquad (16.14)$$

Here the second addend in the brackets reflects the effect of linear hardening.

16.2 Torsion of a Round-Section Beam

The solution to the problem of elastic–plastic torsion of a straight beam of a round cross-section is obtained fundamentally in the suggestion that the hypothesis of plane sections and no radius curvature is true.

Let us separate an elemental tube from the beam using two coaxial surfaces with radii r and $r+dr$ and two cross-sections distanced from each other by dz (Fig. 16.3).

Due to the assumptions, the tube element $abcd$ limited by two sufficiently closely distanced axial sections will be subject to shear so that the radii Oa and Od will turn by the same angle $d\vartheta$. The generatrixes ba and cd will turn by the angle γ. By expressing the arch length aa' from the triangles Oaa' and baa' and equaling these expressions, we obtain

$$r\,d\vartheta = \gamma\,dz.$$

Hence we find as follows:

$$\gamma = r\Theta, \tag{16.15}$$

which designates

$$\Theta = \frac{d\vartheta}{dz}. \tag{16.16}$$

The value Θ defined by formula (16.16) is referred to as [3] the linear torsion angle or simply *a twist*.

If the diagram $\tau = \varphi(\gamma)$ is obtained from the experiment, formula (16.15) gives the solution of the problem. It is required to find the dependency $(\Theta \sim M)$ between the twist and the torque. Equilibrium equations and conformance conditions are satisfied since γ is the linear function and the stress τ does not depend on the section position along the beam axis. Let us satisfy the boundary condition by assuming that

Fig. 16.3 Torsion of
elementary tube

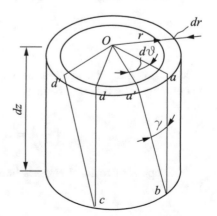

$$M = \int_0^R \tau r \cdot 2\pi r dr = 2\pi \int_0^R \varphi(\gamma) r^2 dr$$

$$= \frac{2\pi R^3}{\gamma_R^3} \int_0^{\gamma_R} \varphi(\gamma)\gamma^2 d\gamma = \frac{2\pi}{\phi^3} \int_0^{\gamma_R} \varphi(\gamma)\gamma^2 d\gamma, \qquad (16.17)$$

where R is the cross-section radius and γ_R is the shear on the outline and is designated as

$$\phi = \frac{\gamma_R}{R}.$$

Similarly to formula (16.8), let us write the dependency (16.17) as

$$\frac{M}{2\pi R^3} = \Phi(\gamma_R); \quad \Phi(\gamma_R) = \frac{1}{\gamma_R^3} \int_0^{\gamma_R} \varphi(\gamma)\gamma^2 d\gamma. \qquad (16.18)$$

Let, for example, the function $\varphi(\gamma)$ satisfy the conditions of the simple loading theorem (p. 213). Assume that

$$\tau = \varphi(\gamma) = \tau_\mathrm{T} \left(\frac{\gamma}{\gamma_\mathrm{T}} \right)^m, \quad (0 < m < 1), \qquad (16.19)$$

where τ_T is the shear yield stress, and γ_T is its respective shear strain. By substituting the function (16.19) into formulas (16.18) in the considered partial case, we obtain

$$\Phi(\gamma_R) = \frac{\tau}{\gamma_R^3 \gamma^m} \int_0^{\gamma_R} \gamma^{2+m} d\gamma = \frac{\tau}{3+m} \left(\frac{\gamma_R}{\gamma} \right)^m;$$

$$\frac{M}{2\pi R^3} = \frac{\tau}{3+m} \left(\frac{\gamma_R}{\gamma} \right)^m. \qquad (16.20)$$

Having the solution (16.20), we can calculate residual stresses $(\tilde{\tau})$ in the case of full unloading using the formula

$$\tilde{\tau} = \tau - \frac{Mr}{J_p}, \qquad (16.21)$$

where J_p is the polar moment of inertia of the beam cross-section. Assuming for uncertainty $m = 0.5$, we obtain as follows using formulas (16.20)–(16.21):

$$\left(\frac{\tilde{\tau}}{\tau_\mathrm{T}} \right) = \sqrt{\frac{2r}{R}} \left(1 - \frac{8}{7}\sqrt{\frac{r}{R}} \right). \qquad (16.22)$$

The epure of residual stresses built upon the dependency (16.22) is shown in Fig. 16.4.

Fig. 16.4 Epure of residual
stresses

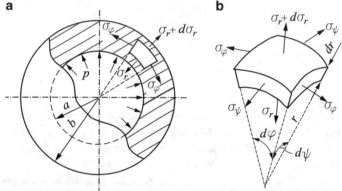

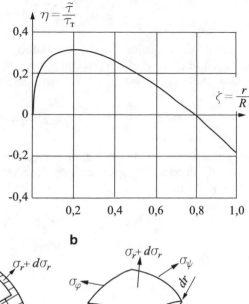

a   **b**

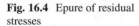

Fig. 16.5 Spherical vessel (**a**) and its element (**b**)

16.3 Elastic–Plastic Inflation of a Spherical Vessel

A spherical vessel with an external radius b and internal radius a is under the action
of even internal pressure p (Fig. 16.5). Due to central symmetry, the principal axes
of stress and strain tensors will be the direction of the central radius r and any two
directions perpendicular to it on the sphere $r = const$. Let us designate two latter
directions using the indexes φ and ψ (Fig. 16.5b). Then radial stresses, strains, and
displacements will be designated as σ_r, ε_r, and $u = u(r)$, respectively, and tangential stresses, strains, and displacements as $\sigma_\varphi = \sigma_\psi$, $\varepsilon_\varphi = \varepsilon_\psi$, and $v = w = 0$.

All components of stresses, strains, and displacements other than zero are
functions of the variable r only. Based on formulas (1.24) and (1.25) in any point of
the sphere wall, we have

$$\varepsilon_r = \frac{du}{dr}; \quad \varepsilon_\varphi = \varepsilon_\psi = \frac{u}{r}.$$

Equilibrium equations in the considered problem result into a single one that is obtained from the condition of equilibrium of forces acting on the element shown in Fig. 16.5b. By projecting all forces on the central radius direction, we obtain

$$\frac{d\sigma_r}{dr} + 2\frac{\sigma_r - \sigma_\varphi}{r} = 0. \tag{16.23}$$

As noted above, all the three strain components other than zero are expressed via a single displacement component. Therefore, there are only two conformity equations in the problem:

$$\varepsilon_\varphi = \varepsilon_\psi; \\ \frac{d}{dr}(r\varepsilon_\varphi) - \varepsilon_r = 0 \tag{16.24}$$

or

$$\frac{d\varepsilon_\varphi}{dr} + \frac{\varepsilon_\varphi - \varepsilon_r}{r} = 0. \tag{16.25}$$

In this manner, we have initial equations of the problem (16.23) and (16.25). To these equations, we must add: Hooke's law in the case of an elastic problem; the law of link between stresses and strains adopted in strain theory in the case of a plastic problem. To do it, we must ensure the appropriateness of using the ratios of this theory, namely: is there simple loading? To do it, let us calculate the Lode–Nadai parameter (13.44); we have

$$\sigma_1 = \sigma_2 = \sigma_\varphi = \sigma_\psi; \quad \sigma_3 = \sigma_r; \tag{16.26}$$

then

$$\mu_\sigma = 2\frac{\sigma_2 - \sigma_3}{\sigma_1 - \sigma_3} - 1 = 2\frac{\sigma_\varphi - \sigma_r}{\sigma_\varphi - \sigma_r} - 1 = 2 = const.$$

Consequently, loading in each point of the vessel is simple, and strain theory can be applied without any exceptions.

1. *Elastic state.* Let us use the same method of solution as in the Lame problem (p. 18). Based on Eqs. (16.23) and (16.25) and Hooke's law, we obtain

$$\sigma_r = p_0\left(1 - \frac{b^3}{r^3}\right), \\ \sigma_\varphi = p_0\left(1 + \frac{1}{2}\frac{b^3}{r^3}\right), \tag{16.27}$$

where

$$p_0 = p \frac{a^3}{b^3 - a^3}.$$

Let us define pressure when the plastic strain occurs. Taking into account formulas (16.26), the solution (16.27) allows making a conclusion that plastic strain will start developing from the inner surface of the sphere $r = a$. Let us use the Huber–Mises plasticity condition:

$$\sigma_i = \frac{\sqrt{2}}{2} \sqrt{(\sigma_1 - \sigma_2)^2 + (\sigma_2 - \sigma_3)^2 + (\sigma_3 - \sigma_1)^2} = \sigma_s,$$

where σ_s is the plasticity limit of the sphere material for elongation. By substituting the values of main stresses according to formulas (16.26) into this condition, we obtain the plasticity condition as follows:

$$\sigma_i = \sigma_\varphi - \sigma_r = \sigma_s. \tag{16.28}$$

Using the solution (16.27) and condition (16.28), we find pressure p_0^s when plastic strain occurs on the inner surface of the sphere:

$$p_0^s = \frac{2a^3}{3b^3} \sigma_s. \tag{16.29}$$

2. *Elastic–plastic solution of the problem.* We will consider the elastic–plastic stage of vessel work assuming the ideal plasticity of the material. In this case, the condition (16.28) is true not only at the start of plastic strain but also in the process of its development everywhere where the intensity of stresses (σ_i) has reached the yield strength (σ_s).

For a plastic zone, the equilibrium equation (16.23) taking into account the plasticity condition (16.28) looks as follows:

$$\frac{d\sigma_r}{dr} - \frac{2\sigma_s}{r} = 0,$$

or

$$\sigma_r = 2\sigma_s \frac{dr}{r}. \tag{16.30}$$

The solution to equation (16.30) is the function

$$\sigma_r = 2\sigma_s \ln r + C_1, \quad (C_1 - const). \tag{16.31}$$

Let us define the constant C_1 in two cases: (a) the plastic zone covers the entire thickness of the spherical vessel wall, and (b) the plastic zone has space only in the case of $a \leqslant r = c < b$.

(a) *The entire sphere material is covered by yield.* There are two boundary conditions:

$$r = a: \quad \sigma_s = -p; \quad r = b: \quad \sigma_r = 0. \tag{16.32}$$

By using the condition (16.32) from the solution (16.31), we have

$$\begin{aligned} -p_{lim} &= 2\sigma_s \ln a + C_1, \\ 0 &= 2\sigma_s \ln b + C_1, \end{aligned} \tag{16.33}$$

where p_{lim} is the limit pressure when the sphere strains are unlimitedly increased.

From the system (16.33), we find

$$\begin{aligned} C_1 &= -[2\sigma_s \ln a + p_{lim}], \\ p_{lim} &= 2\sigma_s \ln \frac{b}{a}. \end{aligned} \tag{16.34}$$

Then we will find as follows for stresses in the limit state:

$$\begin{aligned} \sigma_r^{lim} &= \sigma_\varphi^{lim} - \sigma_s, \\ \sigma_\varphi^{lim} &= 2\sigma_s \ln \frac{r}{a} - p_{lim}. \end{aligned} \tag{16.35}$$

(b) *Only a part of the sphere is covered by yield.* Let us designate the pressure intensity of the elastic zone on the plastic zone at the common boundary as q.

For the elastic zone, we can immediately write the problem solution. To do it, let us use the solution (16.27):

$$\begin{aligned} \sigma_r &= q_0 \left(1 - \frac{b^3}{r^3} \right), \\ \sigma_\varphi &= q_0 \left(1 + \frac{1}{2} \frac{b^3}{r^3} \right), \end{aligned} \tag{16.36}$$

where

$$q_0 = q \, \frac{c^3}{b^3 - c^3}.$$

For the plastic zone, let us use the solution (16.31) in the boundary conditions

$$r = a: \ \sigma_r = -p; \quad r = c: \ \sigma_r = -q. \tag{16.37}$$

We obtain

$$-p = 2\sigma_s \ln a + C_1,$$
$$-q = 2\sigma_s \ln c + C_1. \tag{16.38}$$

Equations (16.38) contain three unknown values: C_1, q, and c. To find them, we must make another equation. The condition of displacement continuity u at the boundary of the elastic and plastic zones can be used. However, the condition $\sigma_\varphi - \sigma_r = \sigma_s$ at $r \to c + 0$ is more convenient. Indeed,

$$(\sigma_\varphi - \sigma_r)\Big|_{r=c+0} = q_0 \left[1 - \frac{b^3}{c^3} - 1 + \frac{b^3}{c^3} \right] = \sigma_s,$$

for example,

$$\frac{3}{2} q_0 \frac{b^3}{c^3} = \sigma_s. \tag{16.39}$$

The system of equations (16.38)–(16.39) is complete to find C_1, c, q. Therefore, the problem can be deemed solved. Figure 16.6a gives an epure of the distribution

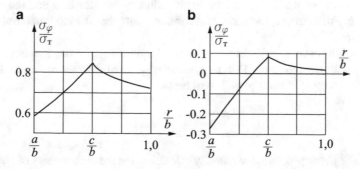

Fig. 16.6 Epure of tangential (**a**) and residual (**b**) stresses

of tangential stresses for the case when there are elastic ($c \leqslant r \leqslant b$) and plastic ($a \leqslant r \leqslant c$) zones in the vessel wall at the same time.

By analyzing the given solution, we can note the following properties:

1. In the solution to the problem set by formulas (16.36) and (16.27), there is no Poisson coefficient v. Consequently, the compressibility (non-compressibility) of a material does not affect the stress distribution in the sphere. When finding strains or displacements, this coefficient would appear in calculation formulas and its effect on the magnitude of displacements is insignificant.

2. Residual stresses after complete unloading can be calculated using the formulas:

$$\begin{aligned} \tilde{\sigma}_\varphi &= \sigma_\varphi - \sigma_\varphi^y, \\ \tilde{\sigma}_r &= \sigma_r - \sigma_r^y, \end{aligned} \tag{16.40}$$

where the elastic stresses σ_φ^y and σ_r^y are defined by formulas (16.27), and instead of σ_φ and σ_r, it is required to substitute the solution (16.36) at $c \leqslant r \leqslant b$ or solution (16.31), (16.28) at $a \leqslant r \leqslant c$. Formulas (16.40) are true for the condition that for complete unloading $\tilde{\sigma}_i < \sigma_s$. The epure of tangential residual stresses is shown in Fig. 16.6b.

3. As seen from the epure of residual stresses (Fig. 16.6b), after complete unloading, there are compression stresses from the inner surface of the sphere. In the case of repeated loading with the same pressure p exerted while loading, the intensity of stresses at $r = a$ will be less than the yield point σ_s and the vessel remains elastic. This effect is frequently used by process engineers in machine engineering where it is known as *autofrettage*.

16.4   Symmetric Strain of a Cylindrical Tube

Let us define stresses and strains of a thick-wall cylindrical tube loaded by internal (p_a) and external (p_b) pressures and by the axial elongating force (P). As in the Lame problem, we will designate the inner radius of the tube as a and the external radius as b. Furthermore, let us suggest for simplicity that the tube material is non-compressible.

Let us introduce a cylindrical system of coordinates r, φ, z whose axis z coincides with the tube axis. Due to symmetry, normal stresses σ_r, σ_φ and σ_z are principal. The equilibrium equation will be

$$\frac{d\sigma_r}{dr} + \frac{\sigma_r - \sigma_\varphi}{r} = 0. \tag{16.41}$$

As in the Lame problem, the Cauchy equations for the components of strain and radial displacement u look as follows :

$$\varepsilon_r = \frac{du}{dr}; \quad \varepsilon_\varphi = \frac{u}{r}; \quad \varepsilon_z = const. \tag{16.42}$$

The conformity equation for the considered problem coincides with Eq. (16.25):

$$\frac{d\varepsilon_\varphi}{dr} + \frac{\varepsilon_\varphi - \varepsilon_r}{r} = 0. \tag{16.43}$$

The principal equations of the strain theory of plasticity as applicable to the solved problem taking into account material's non-compressibility will be written as follows:

$$\sigma_r - \sigma_0 = \frac{2\sigma_i}{3\varepsilon_i}\varepsilon_r; \quad \sigma_\varphi - \sigma_0 = \frac{2\sigma_i}{3\varepsilon_i}\varepsilon_\varphi; \quad \sigma_z - \sigma_0 = \frac{2\sigma_i}{3\varepsilon_i}\varepsilon_z. \tag{16.44}$$

Subsequently, we will suppose that from experiments (for example, in torsion of thin-wall tubes) we have found the dependency between the tangential stress τ and shear γ:

$$\tau = f(\gamma). \tag{16.45}$$

We have the following boundary conditions on the inner and outer outlines of the tube section:

$$\sigma_r(a) = -p_a; \quad \sigma_r(b) = -p_b. \tag{16.46}$$

By equalizing the principal vector of forces acting in any transverse section of the tube to the lateral force P, we obtain as follows:

$$P = 2\pi \int_a^b \sigma_z r\, dr. \tag{16.47}$$

The highest tangential stress (τ) in any point of the tube can be expressed through principal stresses in this point:

$$\tau = \frac{\sigma_\varphi - \sigma_r}{2}. \tag{16.48}$$

By subtracting the first of these equations from the second equation of the system (16.44), we obtain

$$\sigma_\varphi - \sigma_r = \frac{2\sigma_i}{3\varepsilon_i}(\varepsilon_\varphi - \varepsilon_r). \tag{16.49}$$

Or taking into account formula (16.48)

$$2\tau = \frac{2\sigma_i}{3\varepsilon_i}(\varepsilon_\varphi - \varepsilon_r). \tag{16.50}$$

According to the strain theory of plasticity (see p. 197)

$$\tau = \frac{\sigma_i}{3\varepsilon_i}\gamma.$$

Then the previous formula looks as follows:

$$\gamma = \varepsilon_\varphi - \varepsilon_r. \tag{16.51}$$

Let us write the non-compressibility condition:

$$\varepsilon_\varphi + \varepsilon_r + \varepsilon_z = 0. \tag{16.52}$$

Using the Cauchy formulas (16.42), we obtain as follows from the condition (16.52):

$$\frac{du}{dr} + \frac{u}{r} = -\varepsilon_z. \tag{16.53}$$

A common solution of equation (16.53) is the function

$$u = -\frac{1}{2}\varepsilon_z r + \frac{B}{r}, \quad (B - const).$$

Using formulas (16.42), we will find

$$\varepsilon_r = \frac{du}{dr} = -\frac{1}{2}\varepsilon_z - \frac{B}{r^2}; \quad \varepsilon_\varphi = \frac{u}{r} = -\frac{1}{2}\varepsilon_z + \frac{B}{r^2}.$$

By substituting the obtained strain expressions into formula (16.51), we obtain

$$\gamma = \frac{2B}{r^2}. \tag{16.54}$$

Let us integrate the equilibrium equation (16.41)

$$\int_a^r d\sigma_r = \int_a^r (\sigma_r - \sigma_\varphi)\frac{dr}{r},$$

or, using the boundary condition at $r = a$

$$\sigma_r = -p_a + \int_a^r (\sigma_\varphi - \sigma_r)\frac{dr}{r}.$$

Taking into account formulas (16.49) and (16.50), the last dependency can be written as follows:

$$\sigma_r = -p_a + 2\int_a^r f(\gamma)\frac{dr}{r}.$$

Let us substitute the integration variable r with γ in this formula. To do it, let us differentiate formula (16.54); we will obtain

$$\frac{dr}{r} = -\frac{d\gamma}{2\gamma}.$$

By substituting from the last formula, we obtain

$$\sigma_r = -p_a - \int_{\gamma_a}^{\gamma} f(\gamma)\frac{d\gamma}{\gamma}. \tag{16.55}$$

Having in mind that

$$\frac{\sigma_\varphi - \sigma_r}{2} = \tau = f(\gamma),$$

and taking into account formula (16.55), we will obtain

$$\sigma_\varphi = -p_a - \int_{\gamma_a}^{\gamma} f(\gamma)\frac{d\gamma}{\gamma} + 2f(\gamma). \tag{16.56}$$

Due to formula (16.54), the integration limits in formulas (16.55) and (16.56) are related by the dependency

$$\gamma = \gamma_a \frac{a^2}{r^2}.$$

Let us now determine the axial normal stress σ_z. To do it, let us use an intact boundary condition at the outer outline of the tube section. From formula (16.55), we obtain

$$p_b - p_a = \int_{\gamma_a}^{\gamma_b} f(\gamma)\frac{d\gamma}{\gamma}. \tag{16.57}$$

Let us write the ratio (16.49) as

$$\frac{2\sigma_i}{3\varepsilon_i} = \frac{\sigma_\varphi - \sigma_r}{\varepsilon_\varphi - \varepsilon_r}.$$

By substituting this expression into formula (16.44), we obtain

$$\sigma_z - \sigma_0 = \frac{\sigma_\varphi - \sigma_r}{\varepsilon_\varphi - \varepsilon_r} \varepsilon_z, \tag{16.58}$$

where σ_0 is the mean stress

$$\sigma_0 = \frac{1}{3}(\sigma_r + \sigma_\varphi + \sigma_z).$$

From formula (16.58), let us find

$$\sigma_z = -p_a + f(\gamma) - \int_{\gamma_a}^{\gamma} f(\gamma)\frac{d\gamma}{\gamma} + 3\frac{f(\gamma)}{\gamma}\varepsilon_z. \tag{16.59}$$

Formulas (16.55), (16.56), and (16.59) include two undefined values: γ_a and ε_z. To determine γ_a, we can resolve the dependency (16.57) with the known function $f(\gamma)$. Let us use the condition of equilibrium (16.47)

$$P = 2\pi \int_a^b \left[-p_a + f(\gamma) + \int_{\gamma_a}^{\gamma_b} f(\gamma)\frac{d\gamma}{\gamma} + \frac{3f(\gamma)}{\gamma}\varepsilon_z \right] r dr.$$

The substitution of the variable r with γ in the formula and using the dependency (16.57) will give

$$P = \pi \gamma_a a^2 \int_{\gamma_a}^{\gamma_b} \left[-p_b - f(\gamma) - \frac{3f(\gamma)}{\gamma}\varepsilon_z \right] \frac{d\gamma}{\gamma^2}. \tag{16.60}$$

If there is a complex chart $\tau = f(\gamma)$, the integral in formulas (16.57) and (16.60) can be calculated by numerical integration. If it is possible to approximate this chart by power-law relation (15.59), for example,

$$\tau = \tau_s \left(\frac{\gamma}{\gamma_s} \right)^m, \tag{16.61}$$

all integrals of formulas (16.55)–(16.60) are calculated in quadratures and the solving formulas look as follows

$$p_b - p_a = \frac{\tau_s}{m\gamma_s^m}\gamma_a^m \left(\frac{a^{2m}}{b^{2m}} - 1 \right);$$

$$P = \pi a^2 \left[\frac{b^2 p_b}{b^2 - a^2} + \frac{\tau_s \gamma_a^m}{(m-1)\gamma_s^m}\left(1 - \frac{a^{2m-2}}{b^{2m-2}} \right) \right.$$

$$\left. + \frac{3\tau_s \varepsilon_z \gamma_a^{m-1}}{(m-2)\gamma_s^m}\left(1 - \frac{a^{2m-4}}{b^{2m-4}} \right) \right];$$

$$\sigma_r = -p_a + \frac{\tau_s}{m\gamma_s^m}\gamma_a^m\left(\frac{a^{2m}}{r^{2m}} - 1\right);$$

$$\sigma_\varphi = -p_a + \frac{\tau_s}{m\gamma_s^m}\gamma_a^m\left(\frac{a^{2m}}{r^{2m}} - 1\right) + \frac{2\tau_s}{\gamma_s^m}\gamma_a^m\frac{a^{2m}}{r^{2m}};$$

$$\sigma_z = -p_a + \frac{\tau_s}{m\gamma_s^m}\gamma_a^m\left(\frac{a^{2m}}{r^{2m}} - 1\right) + \frac{\tau_s}{\gamma_s^m}\gamma_a^m\frac{a^{2m}}{r^{2m}} + \frac{3\tau_s\varepsilon_z}{\gamma_s^m}\gamma_a^{m-1}\frac{a^{2m-2}}{b^{2m-2}}.$$

By directly checking, one can make sure that for $m = 1$, $(\tau_s/\gamma_s = G)$ the last dependencies coincide with the known formulas of elasticity theory.

Another partial case $m = 0$, $(\tau = f(\gamma) = \tau_s)$ corresponds to the ideally plastic material. We have

$$p_b - p_a = 2\tau_s \ln\frac{b}{a};$$

$$P = \pi a^2\left[\frac{p_b b^2}{b^2 - a^2} - \tau_s\left(1 - \frac{b^2}{a^2}\right) - 3\frac{\tau_s\varepsilon_z}{2\gamma_a}\left(1 - \frac{b^4}{a^4}\right)\right];$$

$$\sigma_r = -p_a + 2\tau_s \ln\frac{r}{a};$$

$$\sigma_\varphi = -p_a + 2\tau_s\left(1 + \ln\frac{r}{a}\right);$$

$$\sigma_z = -p_a + 2\tau_s\left(\frac{1}{2} + \ln\frac{r}{a}\right) + \frac{3\tau_s}{\gamma_a}\varepsilon_z.$$

In the case of no external pressure in plain strain conditions, we obtain

$$\sigma_r = -2\tau_s \ln\frac{b}{r},$$
$$\sigma_\varphi = 2\tau_s\left(1 - \ln\frac{b}{r}\right), \tag{16.62}$$
$$\sigma_z = 2\tau_s\left(\frac{1}{2} - \ln\frac{b}{r}\right).$$

Figure 16.7 shows epures of the distribution of elastic σ_φ^y, plastic σ_φ, and residual $\tilde{\sigma}_\varphi$ (after complete unloading) tangential stresses. The analysis of the second formula (16.62) leads to conclusions that for $b/a < e$, $(\ln e = 1)$, the stress σ_φ remains positive across the entire tube wall thickness. Otherwise, this component is negative on the internal outline of section and positive on the outer one. Second, the distribution nature of stresses in the case of tube plastic strain is directly opposite to the distribution of elastic stresses, and most dangerous conditions occur on the outer tube surface. This fact is experimentally confirmed by Bridgeman [2].

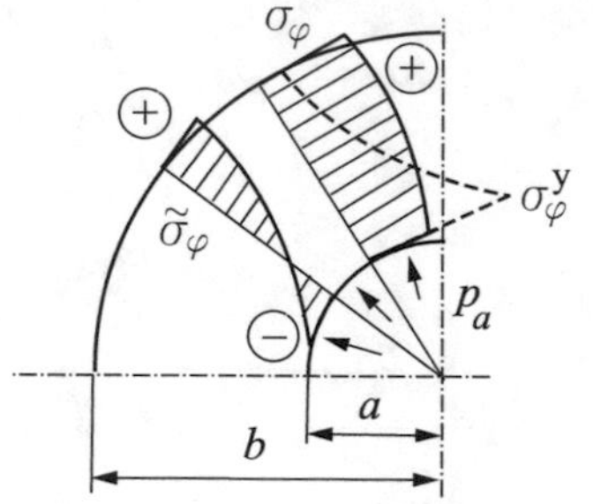

Fig. 16.7 Epures of
tangential stresses

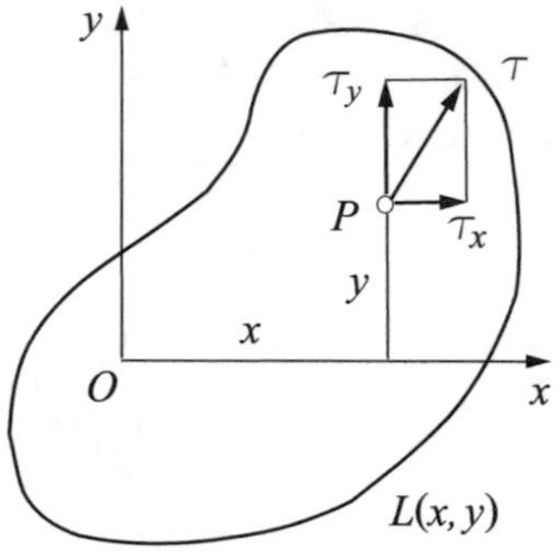

Fig. 16.8 Torsion of a
prismatic beam

16.5   Torsion of a Beam of Ideally Plastic Material

16.5.1   Elastic Torsion: Prandtl Analogy

At first, let us consider the elastic problem of torsion of a prismatic beam of an
arbitrary sine-link cross-section (Fig. 16.8). To find the resulting tangential stress τ
in an arbitrary point p, let us use a rectangular system of coordinates x, y, z by
taking its origin in the point O of the axis relative to which the beam is twisted, and
let us align it with the last axis z.

Let us decompose the tangential stress τ in the point P into the components τ_x
and τ_y in the directions of the axes x and y. Based on formulas (2.1), the equilibrium
equation for the beam element dx, dy, dz looks as follows:

$$\frac{\partial \tau_x}{\partial x} + \frac{\partial \tau_y}{\partial y} = 0. \tag{16.63}$$

Let us designate the projections of displacement of the points x, y, z to the
directions of the selected coordinate axes as ξ, η, ζ. When twisting, the cross-
section $z = const$ runs around the axis z and warps, turning into some surface.
By designating this surface as $\varphi(x, y)$, and the relative angle of twisting as θ, we
have for the given displacements as follows:

$$\xi = -\theta yz, \quad \eta = \theta xz, \quad \zeta = \theta \varphi(x, y). \tag{16.64}$$

By using Cauchy formulas, let us define the shift corresponding to displacements (16.64)

$$
\left.
\begin{aligned}
\gamma_{xz} &= \frac{\partial \xi}{\partial z} + \frac{\partial \zeta}{\partial x} = \theta \left(\frac{\partial \varphi}{\partial x} - y \right), \\
\gamma_{yz} &= \frac{\partial \eta}{\partial z} + \frac{\partial \zeta}{\partial y} = \theta \left(\frac{\partial \varphi}{\partial y} + x \right).
\end{aligned}
\right\}
\tag{16.65}
$$

The tangential stresses τ_x and τ_y are defined under Hooke's law:

$$
\left.
\begin{aligned}
\tau_x &= G\gamma_{xz} = G\theta \left(\frac{\partial \varphi}{\partial x} - y \right), \\
\tau_y &= G\gamma_{yz} = G\theta \left(\frac{\partial \varphi}{\partial y} + x \right).
\end{aligned}
\right\}
\tag{16.66}
$$

Formulas (16.66) show that in each point x, y of the beam cross-section, the following condition is fulfilled:

$$
\frac{\partial \tau_x}{\partial y} - \frac{\partial \tau_y}{\partial x} = -2G\theta = const.
\tag{16.67}
$$

To satisfy Eq. (16.67), let us assume

$$
\tau_x = \frac{\partial \Pi}{\partial y}, \quad \tau_y = -\frac{\partial \Pi}{\partial x},
\tag{16.68}
$$

where the function $\Pi(x, y)$ is referred to *as the Prandtl stress function*. To find it, we can obtain as follows from formulas (16.67) and (16.68) :

$$
\frac{\partial^2 \Pi}{\partial x^2} + \frac{\partial^2 \Pi}{\partial y^2} = -2G\theta.
\tag{16.69}
$$

Since the side surface of the beam is free from stresses, full tangential stress τ in neither point of the outline L of the section shall have a component normal to the outline in this point. Let us write this condition. Assume that the outline is defined by the equation $y = f(x)$. Then along the outline

$$
\frac{\tau_y}{\tau_x} = \frac{dy}{dx}.
$$

The substitution of the expressions (16.68) to the last equation gives

$$
-\tau_y dx + \tau_x dy = \frac{\partial \Pi}{\partial x} dx + \frac{\partial \Pi}{\partial y} dy = 0.
\tag{16.70}
$$

Equation (16.70) means that the function Π on the outline L must have a constant value. In particular, let us assume

$$\Pi\Big|_L = 0. \tag{16.71}$$

The problem of finding the function Π from Eq. (16.69) and condition (16.71) permits the following analogy proposed by Prandtl. Assume that a flexible membrane (soap film) is pulled over the outline L, to which an equal pressure p is applied in perpendicular to its plane, which causes stress T in the membrane. Buckling of such membrane must satisfy [4] the equation $\Delta u = const$, whereas bucklings have a positive value on the outline (for a single-link outline $u = 0$).

In this manner, the soap film surface generally satisfies the same conditions as the function of stresses Π in the case of elastic torsion of the beam of the same cross-section. On the outline of this surface, we have $u = const$ and $\Pi = const$. In each point of the outline, according to Eq. (16.70), the following equation must also be fulfilled:

$$\frac{dy}{dx} = -\frac{\partial \Pi}{\partial x} : \frac{\partial \Pi}{\partial y} = \frac{\tau_y}{\tau_x},$$

which expresses that the tangential stresses τ are directed to the horizontals $\Pi(x,\, y) = const.$ along the tangential lines. Moreover, we have

$$\tau^2 = \tau_x^2 + \tau_y^2 = \left(\frac{\partial \Pi}{\partial x}\right)^2 + \left(\frac{\partial \Pi}{\partial y}\right)^2. \tag{16.72}$$

Formula (16.72) gives a square of the highest surface incline $\Pi(x,\, y)$. Consequently, the full tangential stress τ in any point inside L equals the maximum incline of the surface of stresses in any point. Horizontal lines (level lines) of the stress surface $\Pi(x,\, y)$ depict the stress trajectories in a cross-section of the twisted beam. As we said, in each point of the section, the tangential line to the horizontal defines the direction of the resulting tangential stress τ, whereas the magnitude of the stress is proportional to the highest surface incline $\Pi(x,\, y)$ above this point. It can be shown that the torsion moment M equals the double volume limited the stress surface $\Pi(x,\, y)$.

16.5.2   Elastic–Plastic Beam Torsion

Let us remind that the beam material is deemed ideally plastic. Therefore, the plasticity condition can be written as[1]

---

[1] Plasticity conditions of Tresca and Huber–Mises coincide for the problem under consideration.

$$\tau^2 = \tau_x^2 + \tau_y^2 = \tau_s^2, \tag{16.73}$$

where τ_s is the shear yield stress.

In the case of elastic–plastic torsion, the stresses τ_x and τ_y must satisfy the equilibrium equation (16.63), and this requires fulfillment of Eqs. (16.68). The latter means that the newly implemented function $\Pi(x, y)$ can be interpreted as a surface supported by the cross-section outline, and it can be called the stress function in plastic torsion of the beam of this section. Let us use the stress function $\Pi(x, y)$ for the elastic and elastic–plastic strain of the beam. Based on formulas (16.68) and (16.73), the plasticity condition looks as follows:

$$\left(\frac{\partial \Pi}{\partial x}\right)^2 + \left(\frac{\partial \Pi}{\partial y}\right)^2 = \tau_s^2. \tag{16.74}$$

It means that the problem of elastic–plastic torsion of the beam comes to finding the function $\Pi(x, y)$ that satisfies Eq. (16.69) in the plastic area, Eq. (16.74) in the plastic area, and to the boundary condition (16.71).

Furthermore, the stress τ must change continuously at the boundary between elastic and plastic areas. If we designate belonging of the values of the elastic area using the indexes "y" and those in the plastic area as "p," the continuity condition of tangential stresses at this boundary can be written as follows:

$$\left(\frac{\partial \Pi}{\partial x}\right)^p = \left(\frac{\partial \Pi}{\partial x}\right)^y, \quad \left(\frac{\partial \Pi}{\partial y}\right)^p = \left(\frac{\partial \Pi}{\partial y}\right)^y. \tag{16.75}$$

The left part of Eq. (16.74) represents a square of the modulus of vector **grad** Π that expresses the highest incline of the surface Π. In this manner, the following equation must be fulfilled in all points of the section with plastic strain:

$$|\mathbf{grad}\ \Pi| = \tau_s = const.$$

Finally, according to formula (16.70), the tangential stress τ at the boundary of the plastic zone of cross-section is directed along the tangential line to the outline $y = f(x)$, which equals the condition $\Pi = const$ at this boundary.

Form the above properties of the function Π, it follows that the stress function in plastic torsion represents a surface of the highest incline that can be built on the outline of the beam cross-section. As per formula (16.74), in the plastic area, the tangent of the highest incline angle equals $\pm\tau_s$, whereas this tangent in the elastic area is less than in τ_s. This requires using the method of experimental determination of tangential stresses in elastic and plastic areas of the section and determination of these areas.

The method consists in the following. A surface (roof) of equal slope must be built above the cross-section. The basis of this surface (outline L, Fig. 16.8) is pulled over by a membrane to which even pressure p is applied. If the membrane does not touch the equal slope surface, there is elastic torsion of the beam.

Fig. 16.9 Elastic–plastic
torsion of the prismatic beam

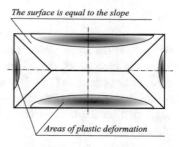

In the case of contact zones between the membrane and the equal slope surface, projections of these zones onto the section plane define the plastic yield area.

Such structures for a rectangular-section beam are made in Fig. 16.9. Zones of plastic condition of the material are highlighted where the condition (16.74) is fulfilled.

When the torque M rises to the limit, the membrane can touch all points of the equal slope roof. In this idealized case for a rectangular section, the lines designating the apex and braces of the equal slope will be break lines (from τ_s to $-\tau_s$) of tangential stresses.

The torque transmitted by the beam can be calculated using the formula

$$M = \iint_S (\tau_y x - \tau_x y)dxdy = - \iint_S \left(\frac{\partial \Pi}{\partial x} + \frac{\partial \Pi}{\partial y} \right) dxdy,$$

where S is the cross-section area. Taking into account the outline condition (16.71), after integration we will obtain

$$M = 2 \iint_S \Pi(x, y)dxdy. \tag{16.76}$$

From the result (16.76), it follows that the limit torque equals the double volume confined between the section and the equal slope roof.

16.6 Rod of a Variable Section: Method of Elastic Solutions

16.6.1 Preparation of Initial Ratios

Let us consider the rod whose axis is aligned with the coordinate axis x (Fig. 16.10). Let us place the reference point in the left end section. The rod cross-section area is designated as $F(x)$. Assume the rod is elongated by the axial forces P_0, P_1 and by the mass force $Q(x)$ per unit acting in the direction of the axis x.

The problem of the elastic–plastic equilibrium of the rod will be solved in displacements using the above-mentioned (p. 212) method of elastic solutions. In

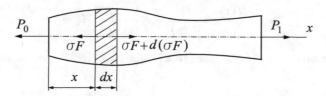

Fig. 16.10 Elastic–plastic equilibrium of a variable-section rod

this manner, we assume that equilibrium equations and boundary conditions are defined and expressed in displacements. It is required to find the displacements u, v, w.

The lateral force N in the arbitrary cross-section of the rod will be

$$N(x) = P_0 - \int_0^x Q(\zeta)d\zeta, \tag{16.77}$$

whereas

$$\frac{d(\sigma F)}{dx} = -Q(x), \tag{16.78}$$

and $\sigma = \sigma_x$ are normal stresses in the rod cross-section. From the condition (16.77), we have as follows:

$$P_1 = P_0 - \int_0^l Q(\zeta)d\zeta, \tag{16.79}$$

where l is the road length.

Assuming that the hypothesis of plane sections is used in rod strain, we can write as follows:

$$u = u(x); \quad \varepsilon_x = \varepsilon = \frac{du}{dx}. \tag{16.80}$$

Normal stresses in cross-sections are represented as

$$\sigma = F\varepsilon[1 - \omega(\varepsilon)], \tag{16.81}$$

where

$$\omega(\varepsilon) = \left|\begin{array}{l} 0 \quad \text{with } \varepsilon < \varepsilon_s, \\ > 0 \text{ with } \varepsilon > \varepsilon_s. \end{array}\right.$$

Stresses in any section taking into account formulas (16.77) will be

$$\sigma = \frac{N(x)}{F(x)} = \frac{P_0}{F(x)} - \frac{1}{F(x)} \int_0^x Q(\zeta)d\zeta. \qquad (16.82)$$

By using expressions (16.80) and (16.81), formula (16.82) can be re-written as follows:

$$\sigma = E\frac{du}{dx}\left[1 - \omega\left(\frac{du}{dx}\right)\right] = \frac{P_0}{F(x)} - \frac{1}{F(x)} \int_0^x Q(\zeta)d\zeta. \qquad (16.83)$$

Let us express as follows:

$$\frac{du}{dx} = \frac{P_0}{EF} - \frac{1}{EF} \int_0^x Q(\zeta)d\zeta + \omega\left(\frac{du}{dx}\right) \cdot \frac{du}{dx}. \qquad (16.84)$$

Let us integrate two parts of Eq. (16.84) upon the variable x from zero to x; now, we have

$$u - u_0 = \frac{P_0}{E}\psi(x) - \frac{1}{E}\int_0^x [\psi(x) - \psi(\eta)]\,Q(\eta)d\eta + \int_0^x \omega\left(\frac{du}{dx}\right)\frac{du}{dx}dx, \qquad (16.85)$$

where $u_0 = u(0)$ and $\psi(x)$ designates the integral

$$\psi(x) = \int_0^x \frac{dx}{F(x)}.$$

The dependencies (16.77), (16.84), and (16.85) form a system of ratios sufficient to solve the problems under consideration.

16.6.2 Specification of Problem Setting

By having Eqs. (16.77), (16.84), and (16.85), we can set the following problems.

Problem 1 P_0, P_1, and $Q(x)$ are set. It is required to find stresses, strains, and displacements.

As for stresses, this task is statically definable.

Problem 2 P_0 and P_1 are not defined in a clear form. Some restrictions that impede displacement are placed on the left and right end of the rod, e.g., the following function is set:

$$u_1 - u_0 = f(P_0, P_1), \quad [u_1 = u(l)]. \qquad (16.86)$$

For example, if both ends of the rod are terminated, $f = 0$. In the case of rigid termination on the left end and elastic attachment of the right end $u_0 = 0$; $u_1 = -kP_1$, where k is the elastic compliance coefficient.

In the case of Problem 2, let us re-write the expression (16.85) as

$$f(P_0, P_1) = \frac{P_0}{E}\psi(l) - \frac{1}{E}\int_0^l [\psi(x) - \psi(\eta)]\, Q(\eta)d\eta + \int_0^l \omega\left(\frac{du}{dx}\right)\frac{du}{dx}dx.$$
(16.87)

16.6.3 Algorithm of the Elastic Solutions Method

Let us consider the primary initial system of equations (16.77), (16.87), and (16.84) for Problem 2. In the case of the elastic work of the rod $\omega = 0$. Then Eqs. (16.87) and (16.77) give P_0 and P_1, formula (16.84) is used to define strain, and then we find stresses knowing strains under Hooke's law.

In the case of elastic–plastic work, to solve the system of equations (16.77), (16.87), and (16.84) we use the method of successive approximations based on the assumption that ω is a small parameter. As the first approximation, let us assume the elastic solution obtained at $\omega = 0$

$$P_0^{(1)},\ P_1^{(1)},\ \varepsilon^{(1)},\ \sigma^{(1)}.$$

If $|\varepsilon^{(1)}| < \varepsilon_s$, the rod shows no plastic strains and the first approximation is the accurate solution to the elastic problem. Otherwise, there is an elastic–plastic state. Then the condition $|\varepsilon^{(1)}| \geqslant \varepsilon_s$ is used to find the area $\Omega^{(1)}(x)$ where the plastic state has occurred. With known $\varepsilon^{(1)}(x)$ from the diagram $\sigma \sim \varepsilon$, we find $\omega^{(1)}(\varepsilon^{(1)}) = \omega^{(1)}[x]$.

The found value $\omega^{(1)}[x]$ is substituted into Eqs. (16.87) and (16.84) so as to obtain the second approximation.

$$P_0^{(2)},\ P_1^{(2)},\ \varepsilon^{(2)},\ \sigma^{(2)},\ \omega^{(2)}[x].$$

Integration in the interval $[0; l]$ is done for $x \in \Omega^{(1)}(x)$ only. The iteration process is continued until the difference of values obtained from two successive approximations is within the permitted limits. The calculations show that usually two or three approximations are enough.

To conclude, we shall note that the convergence of the method of elastic solutions is not rigorously proven yet.

References

1. A. Il'yushin, *Prikl. matematika i mekhanika* [Applied Mathematics and Mechanics], no. 9, izz. 3,, Moscow, MGU Publ., chapter *Svyaz' s teoriei Sen-Venana, Levi, Mizesa i s teoriei malykh uprugoplasticheskii deformatsii* [Connection with the theory of Saint-Venan, Levy, Mises and the theory of small elastoplastic deformations] (1945), pp. 207–218
2. A. Il'yushin, *Mekhanicheskie svoistva i ispytanie metallov* [Mechanical properties and testing of metals] (OGIZ Publ., Moscow, Leningrad, 1948)
3. V. Molotnikov, *Mekhanika konstruktsii* [Mechanics of structures] (Lan' Publ., SPb., Moscow, Krasnodar, 2012)
4. S. Timoshenko, J. Gud'er, *Teoriya uprugosti* [Theory of elastic strength] (Nauka Publ., Moscow, 1975)

Chapter 17
Additions and Generalizations to the Strain Theory of Plasticity

17.1 Generalizations of Goldenblatt and Prager

The simplicity of the ratios of the strain theory of plasticity

$$\sigma'_{ij} = 2G_s \varepsilon'_{ij}, \quad \sigma_i = \Phi(\varepsilon_i), \quad \sigma_0 = 3K\varepsilon_0 \qquad \text{with } dJ'_2 > 0,$$

$$\sigma'_{ij} = 2G\varepsilon'_{ij}, \quad \sigma_0 = 3K\varepsilon_0 \qquad \qquad \text{with } dJ'_2 < 0 \qquad (17.1)$$

and their formal affinity with the respective ratios of elasticity theory attracted wide attention of engineers and researchers. Another important advantage of the theory is an opportunity to solve multiple applied elastic–plastic problems. Some of these solutions were given in the previous chapter. It was also shown that a simple process of approximate solution was built based on the ratios (17.1) to solve marginal problems—a method of elastic solutions.

After 1928, when the fundamental study of Lode [15] was published, multiple experimental works were done to find the applicability limits of the strain theory of plasticity. It was found and generally admitted that we must know to what extent the conditions of proportional loading are implemented in each element of the body volume to judge about the possibility of using the strain theory. In the case of proportional loading, the laws (17.1) of the strain theory have been experimentally proved.

However, for a non-hardening material, the term of proportional loading has no sense, since when the plasticity condition is met, which binds components of the stress tensor, the stressed state may change only by changing the ratios between stresses. Nevertheless, strain theory is used in the case of ideal plastic material taking the obtained results as approximation.

Experimental studies show that the laws of the strain theory of plasticity remain rather accurate even when loading differs from proportional. Slight divergence from experimental data is found when the main axes of the stress tensor are turned during loading.

As formulas (17.1) show, the Hencky–Nadai law is based on the proportionality of stress and strain deviators. This provision can be replaced by a generalized one: the strain deviator is the function (F) of the stress deviator.

$$\varepsilon'_{ij} = F(\sigma'_{ij}). \tag{17.2}$$

The dependency (17.2) must be invariant relative to coordinates since the material is isotropic in the initial condition. Taking into account the considerations of tensor dimensionality, I. I. Goldenblatt [4] and V. Prager later [13] suggested that in a general case (17.2) might look as follows:

$$\varepsilon'_{ij} = F(J'_2, J'^2_3) \left[P(J'_2, J'^2_3)\sigma'_{ij} + Q(J'_2, J'^2_3)t_{ij} \right], \tag{17.3}$$

where

$$J'_2 = \frac{1}{2}\sigma'_{ij} \cdot \sigma'_{ij}, \quad J'_3 = \frac{1}{3}\sigma'_{ij}\sigma'_{jk}\sigma'_{kl};$$

$$t_{ij} = \frac{\partial J'_3}{\partial \sigma'_{ij}} = \sigma'_{ik}\sigma'_{kj} - \frac{2}{3}J'_2\sigma'_{ij},$$

and the symbols P and Q designate some functions of these arguments.

If we assume $Q = 0$, $P = 1$, and $F = F(J'_2)$ in formula (17.3), the Hencky–Nadai law (17.1) is obtained from the ratios (17.3) as a partial case. We shall also note that the Prager law (17.3) includes not only the second invariant of the stress deviator into the ratios of the link between stresses and strains but also the third one, which is rather significant for some materials.

Both the Hencky–Nadai law and the Prager law are distinctive by the fact that stresses and strains in them are related by finite ratios. Plasticity theories with such links are usually called theories of strain type.

17.2 Tensor–Linear Ratios in Plasticity Theories

Stresses and strains can be linked by some integration–differentiation operations. The simplest class of such relations are tensor–linear relations where tensor matrices are related by some linear operators.

If we remain within the tensor–linear ratios, any plasticity law, as noted by Prager, can be represented as follows:

$$L\left[\left(\sigma'_{ij}\right)\right] = L'\left[\left(\varepsilon'_{ij}\right)\right], \tag{17.4}$$

where (σ'_{ij}), (ε'_{ij}) are matrices of stress and strain deviators, L, L' are linear scalar deviators

$$L\left[\left(\sigma'_{ij}\right)\right] = A \cdot (\sigma'_{ij}) + B\frac{d}{d\lambda}(\sigma'_{ij}) + \int\limits_0^\lambda C \cdot (\sigma'_{ij})d\lambda + \cdots,$$

$$L'\left[\left(\varepsilon'_{ij}\right)\right] = A' \cdot p(\varepsilon'_{ij}) + B' \cdot \frac{d}{d\lambda}(\varepsilon'_{ij}) + \int\limits_0^\lambda C' \cdot (\varepsilon'_{ij})d\lambda + \cdots. \qquad (17.5)$$

Here, the coefficients $A, \ldots, C'$ are constant or some scalar functions from the invariants of stress and strain deviators, and λ is some parameter characterizing the loading sequence.

The ratios (17.4) and (17.5) include [5] multiple variants of plasticity theory as partial cases, in particular, the Hencky–Nadai–Ilyushin strain theory. Indeed, it is sufficient to assume $A = 1$, $B = C = \ldots = 0$, $A' = 2G_s = g(J'_2)$, and the ratios (17.4)–(17.5) turn into the law (17.1) of the strain theory of plasticity.

A partial case of the tensor–linear dependency (17.4) is also the Saint-Venant – Levy – Mises plasticity theory mentioned above (p. 149) in the overview of theories. Assuming in operators (17.5) $A' \neq 0$, $B' \neq 0$ and assuming all other coefficients as zero, we will obtain the condition of coinciding director tensors of strain and stress rates.

We can say that the Prandtl theory [14] also belongs to the type of a tensor–linear link of stresses and strains.

Opposite to the enumerated variants of the theory, an example of the simplest tensor–non-linear ratio between stresses and strains is the Prager theory (17.3), since the dependency (17.3) does not belong to the class (17.4) (due to the multiplier t_{ij}).

Later in 1947, A. A. Ilyushin proved that in the case of simple loading, all possible laws of plasticity based on the ratio (17.4) should coincide with the Hencky–Nadai–Iluyshin law and the latter was general in the class of tensor–linear ratios.

This circumstance caused more interest in strain theory. On the one hand, experimental works to define its applicability limits were continued. On the other hand, objections of theoretical nature were imposed against strain theory.

Chronologically, the first question was: to what extent can the condition of proportional loading be ensured in the deformation of a real body by external forces? As we know (p. 231), the answer to this question is given by the theorem of simple loading. Despite a constraining condition imposed by the theorem on the type of dependency between the intensities of stresses and strains, this condition is, first of all, sufficient, but not necessary. Second, experimental data showed [16] that if the theory soundly described the proportional loading process, it gave satisfactory results for loading close to proportional. Here we have another serious objection against strain theory—the violation of the so-called continuity condition. Let us explain this by the example of experiments with a thin-wall tube.

Fig. 17.1 Possible direction
of additional loading

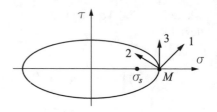

In the case of axial elongation and torsion of the tube, the plasticity condition looks as follows:

$$\sqrt{\sigma^2 + 3\tau^2} = const.$$

In the plane $\sigma \sim \tau$, it is depicted as an ellipse (Fig. 17.1). Let us elongate the tube beyond the yield stress σ_s with no torsion (point M in Fig. 17.1). In the case of additional loading in the direction 1, we have $dJ_2' > 0$ and, as per the law (17.1), we have active loading. The link between gains of stress and strain will be obtained by differentiating the first of formulas (17.1):

$$d\sigma_{ij}' = 2G_s(J_2')d\varepsilon_{ij}'. \tag{17.6}$$

In the case of additional loading in the direction 2, on the opposite, $dJ_2' < 0$, and unloading occurs under the elastic law. In this case, the link between gains of stress and strain is found by differentiating the second of the formulas from (17.1):

$$d\sigma_{ij}' = 2Gd\varepsilon_{ij}'. \tag{17.7}$$

Additional loading 3 is referred to as the *neutral loading*. It can be considered as a limit case of active strain or as a limit case of unloading strain. Since the plastic shear modulus G_s depending on the invariant J_2' is not equal to the elastic shear modulus G, the results (17.6) and (17.7) do not coincide in these limit cases.

It is obvious that switching from active plastic strain to unloading for real materials must be continuous. Thus, the obtained result indicates a serious flaw in strain theory.

17.3   Vector Representation of Tensors

Any tensor having n components can be (as any system of n values) represented in an infinite number of ways in the n-dimensional vector space. If only n independent tensor components are among n components, a respective vector lies in the space having n dimensions. In what follows, we will use geometric terminology.

Let us match the stress deviator (σ'_{ij}) with the stress vector $\mathbf{S}$ with components S_i, $(i = 1, 2, \ldots, 9)$ and match the strain deviator (ε'_{ij}) with the strain vector $Э_{(i)}$, $(i = 1, 2, \ldots, 9)$ under the following law:

$$S_1 = \sigma'_{11}, \ S_2 = \sigma'_{22}, \ S_3 = \sigma'_{33}, \ S_4 = \sigma'_{13}, \ \ldots, \ S_9 = \sigma'_{32};$$
$$Э_1 = \varepsilon'_{11}, \ \ldots, \ Э_9 = \varepsilon'_{32}, \quad (1, 2, 3 \sim x, y, z).$$

Let us note that since stress and strain deviators in the most general case have only five independent components, then the vectors $\mathbf{S}$ and $Э$ are always in the five-dimensional sub-space of the nine-dimensional space. Such five-dimensional spaces were introduced by A. A. Ilyushin.

In the five-dimensional orthogonal space of Ilyushin (S_5), $(Э_5)$ the link between the components of the deviators D_σ, D_ε and the components of the vectors $\mathbf{S}$ and $Э$ is defined by mutually unambiguous linear ratios so that the second invariants of deviators are equal to the squares of the vector moduli $|\mathbf{S}|^2$ and $|Э|^2$, respectively:

$$J'_2 = \sigma'_{ij}\sigma'_{ij} = \sum_{i=1}^{5} S_i^2 = |\mathbf{S}|^2 = \frac{2}{3}\sigma_i^2;$$
$$I'_2 = \varepsilon'_{ij}\varepsilon'_{ij} = \sum_{i=1}^{5} Э_i^2 = |Э|^2 = \frac{3}{2}\varepsilon_i^2.$$

An example of such ratios can be setting the components of the vectors $\mathbf{S}$ and $Э$ in the form of

$$S_1 = \frac{\sqrt{2}}{2}(\sigma_x - \sigma_y), \ S_2 = \sqrt{\frac{3}{2}}(\sigma_z - \sigma_0),$$
$$S_3 = \sqrt{2}\tau_{xy}, \ S_4 = \sqrt{2}\tau_{yz}, \ S_5 = \sqrt{2}\tau_{zx};$$
$$Э_1 = \frac{\sqrt{2}}{2}(\varepsilon_x - \varepsilon_y), \ Э_2 = \sqrt{\frac{3}{2}}(\varepsilon_z - \varepsilon_0), \ Э_3 = \sqrt{2}\varepsilon_{xy} = \frac{\sqrt{2}}{2}\gamma_{xy},$$
$$Э_4 = \sqrt{2}\varepsilon_{yz} = \frac{\sqrt{2}}{2}\gamma_{yz}, \ Э_5 = \sqrt{2}\varepsilon_{zx} = \frac{\sqrt{2}}{2}\gamma_{zx}. \tag{17.8}$$

While loading, the end of the vector $Э$ circumscribes an arch in the space $Э_5$ (Fig. 17.2) that is called *a strain trajectory*. Proportional loading in the five-

Fig. 17.2 Strain trajectory in the space $Э_5$

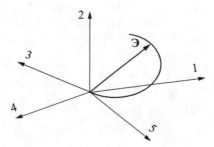

dimensional space corresponds to a straight-line trajectory coming from the reference point.

Since it is assumed in plasticity theory that the link between stresses and strains does not substantially depend on time, we can take the length of the strain trajectory arch as a parameter characterizing the loading sequence:

$$d\Im = \sqrt{d\Im_1^2 + d\Im_2^2 + \ldots + d\Im_5^2};$$

$$\Im = \int_{t_0}^{t} \sqrt{d\Im_1^2 + d\Im_2^2 + \ldots + d\Im_5^2}, \tag{17.9}$$

where $\Im$ is the length of the strain trajectory arch from the moment t_0 to the moment t and $d\Im$ is its differential.

An image of the strain process is an aggregate of the strain trajectories $\Im = \Im$ (t) and the law of conformity between **S** and $\Im$ that permits building a stress vector **S** in each point of the strain trajectory.

Similarly to formulas (17.9) for the space S_5, we can write as follows:

$$ds = \sqrt{dS_1^2 + \ldots + dS_5^2},$$

$$s = \int_{t_0}^{t} \sqrt{dS_1^2 + \ldots + dS_5^2}, \tag{17.10}$$

where s is the length of the arch and ds is an element on the vector trajectory **S** in the space S_5.

In the space S_5, we will call the process image an aggregate of the trajectories **S** = **S**(t) and the law of conformity $\sim$ **S**.

The primary problem of plasticity consists in finding the law that permits building a vector **S** in each point of the trajectory $\Im$ and vice versa.

17.4   Transformations of Rotation and Reflection

Let us at first consider transformation of the strain trajectory in a two-dimensional case. In particular, in the case of axial elongation and torsion of a thin-wall tube as per formulas (17.8), we have

$$S_1 = \sqrt{\frac{3}{2}} (\sigma_x - \sigma_0) = \sqrt{\frac{2}{3}} \sigma_x; \quad S_2 = 0; \quad S_3 = \sqrt{2} \, \tau_{xy};$$

$$\Im_1 = \sqrt{\frac{3}{2}} \, (\varepsilon_x - \varepsilon_0); \quad \Im_3 = \frac{1}{\sqrt{2}} \, \gamma_{xy}.$$

Consequently, the strain trajectory will be a plane curve.

Transformation of the trajectory rotation $\Im = \Im(t)$ *is another trajectory* $\Im' = \Im'(t)$ *obtained from the first one by rotating it as a rigid whole relative to the reference point.*

In the considered case, this transformation is done using the formulas

$$\Im_1' = \Im_1 \cos \alpha + \Im_3 \sin \alpha,$$
$$\Im_3' = -\Im_1 \sin \alpha + ?_3 \cos \alpha,$$

whereas the rotation angle of the trajectory α (Fig. 17.3a) does not depend on the trajectory arch length.

Transformation of the trajectory reflection $\Im = \Im(t)$ *is another trajectory* $\Im' = \Im'(t)$ *obtained by mirroring the trajectory* $\Im$ *in an arbitrary beam coming from the reference point.*

An example of reflection transformation is given in Fig. 17.3b. Here $OA'B'$ is a mirror reflection of the trajectory OAB in the beam representing a bisecting line of the angle $A'OA$. Mathematically, the transformation of reflection is done in a two-dimensional case using the formulas

$$\Im_1' = \Im_1 \cos \alpha + \Im_3 \sin \alpha,$$
$$\Im_3' = -\Im_1 \sin \alpha - \Im_3 \cos \alpha.$$

The results obtained above for two-dimensional trajectories are apparently generalized for the case of a five-dimensional space. The rotation transformation matrix must keep the trajectory arch length, and its determinant must be equal to one. To transform rotation, the matrix determinant must be equal to -1.

The internal geometry of a plane trajectory is well characterized by two parameters: arch length $\Im$ and curvature χ:

$$\chi^2 = \left(\frac{d^2 \Im}{d\Im^2} \right)^2.$$

Fig. 17.3 Transformation of rotation (**a**) and reflection (**b**)

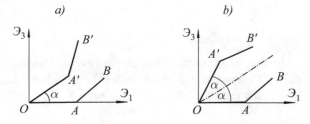

These two parameters fully define the strain trajectory with an accuracy up to the transformation of rotation of reflection.

The internal geometry is fully defined by a natural orthogonal frame [12] built in any point of the trajectory. The block axes $\mathbf{p}_i$, $(i = 1, 2, \ldots, 5)$ satisfy the conditions

$$\mathbf{p}_i\mathbf{p}_j = \delta_{ij}$$

and are calculated using the Frenet–Serret formulas.

17.5   Ilyushin's Isotropy Postulate

If we switch from tensor values to vector ones in the law (17.1) of Hencky–Nadai–Ilyushin, we will obtain

$$\mathbf{S} = 2G_s(э)Э. \tag{17.11}$$

We have already said that the law (17.1) is rather precise in the case of simple loading. The representation of the law as (17.11) indicates its dependency on the vector direction Э. The latter means that the five-dimensional space of Ilyushin is isotropic relative to the rotation of the vector Э. The generalization of this provision onto complicated loadings was called *Ilyushin's isotropy postulate*. For this generalization, the law invariance relative to rotation is supplemented by the invariance of the link (17.11) relative to reflection in all possible planes and directions. As a result, the formulation of the isotropy postulate is given in the following wording:

A link between $\mathbf{S}$ *and* Э *is invariant relative to transformations of rotation and reflection.*

The postulate also permits a different formulation: *the image of the strain process is fully defined by the internal geometry of the strain trajectory,* or else: *the space* $Э_5$ *is isotropic relative to the process image.*

Let us indicate some consequences resulting from the postulate.

Consequence 1 *The link*

$$\mathbf{S} = L(Э)$$

must be vector-linear

Indeed, when transforming the rotation or reflection of the vector Э into Э′, the vector $\mathbf{S}$ is transformed into $\mathbf{S}'$ under the same law. Therefore, the angle α will not look as a transformed dependency

$$\mathbf{S}' = L(Э')$$

only if the operation L is vector-linear relative to the vector $\ni$, e.g.

$$L(\ni) = A\ni + B\frac{d\ni}{d\ni} + \ldots, + \int\limits_{t_0}^{t} C\ni d\ni,$$

where the coefficients A, B, $\ldots$, C are the functions of internal parameters of the trajectory $(\ni, \chi_1, \chi_2, \ldots)$ and any operators of these parameters.

Consequence 2 *The plasticity law*

$$S = L(\ni)$$

has a property of isomorphism.

Let us explain it first using the example of a plane process trajectory. In a plane case, natural axes of the strain trajectory coincide with the direction of the tangent line and normal line to any point of the trajectory. Single vectors $(\mathbf{p}_1, \mathbf{p}_2)$ of the natural benchmark of the strain trajectory are represented by the formulas

$$\mathbf{p}_1 = \frac{d\ni}{d\ni}; \quad \mathbf{p}_2 = \frac{1}{\chi}\frac{d^2\ni}{d\ni^2},$$

and the stress vector can be written as

$$S = S_{p_1}\mathbf{p}_1 + S_{p_2}\mathbf{p}_2, \tag{17.12}$$

whereas the scalar functions S_{p_1}, S_{p_2} (functionally) depend only on the parameters of the internal geometry of the trajectory and therefore are invariant relative to transformations of rotation and reflection.

The vector S can also be decomposed into two components in the direction $\ni$ and $\mathbf{p}_1$ and written as follows:

$$S = S_{\ni}\ni + S_{p_1}\frac{d\ni}{d\ni}, \tag{17.13}$$

where $S_{\ni}$ and S_{p_1} depend only on the internal geometry of the strain trajectory.

On the other hand, the principal law (17.11) can be solved relative to the vector $\ni$ and represented as

$$\ni = \ni_s S + \ni_{p_1}\frac{dS}{ds}. \tag{17.14}$$

The property of the plasticity law (17.11) to be capable of being represented by the space $\ni_5$ as (17.12) or (17.13), or in the space S_5 as (17.14) or in similar types in other spaces formed from S_5 and $\ni_5$ by means of linear transformations was called by A. A. Ilyushin *isomorphism*.

Fig. 17.4 Angular point

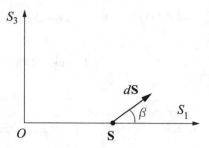

Binomial representation of the dependency (17.13) takes place in the case of a plane strain trajectory. In a general case of a 5-dimensional vector Э, the strain law can be written as

$$\mathbf{S} = \sum_{i=1}^{5} \Lambda_i \mathbf{p}_i, \qquad (17.15)$$

where Λ_i are functionals from the internal parameters of the strain trajectory, $\mathbf{p}_i$ are single vectors of the natural benchmark:

$$\Lambda_i = \Lambda_i(Э, \chi_1, \ldots, \chi_4),$$
$$\chi_i = \chi_i(Э).$$

It has been proved (Ilyushin, [6]) that the law (17.15) in the class of tensor–linear ratios is general. The study of the applicability conditions of the law (17.15) in loading different from simple one and, in particular, in the case of an angular point at the loading trajectory (Fig. 17.4) had [6] the following results.

In an infinitely small vicinity of the angular point, gains of strain and stress vectors are related by the dependency

$$d Э = \frac{1}{N} d\mathbf{S} + \frac{N - P}{N P} \left(\frac{\mathbf{S} d\mathbf{S}}{\mathbf{S}^2} \right) \mathbf{S}, \qquad (17.16)$$

where N does not depend and P significantly depends on the angle β (Fig. 17.4) of the trajectory break. These dependencies must be found experimentally.

17.6 Delay Law

Assume that we know the strain trajectory in a process. In an arbitrary point K, a single vector of the tangential line $\mathbf{p}$ can be built, as well as the stress vector $\mathbf{S}$ under the law (17.11) (Fig. 17.5).

Fig. 17.5 Trace of delay

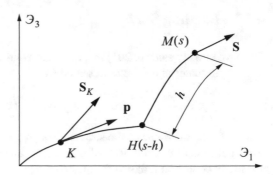

On the strain trajectory, let us consider a precedent point $H(s - h)$ along with the current point $M(s)$ where h is the final segment of the arch called [7] *the trace*. If we conduct full unloading from the point H, the stress vector $\mathbf{S}$ and the elastic strain vector $\Im^e$ turn to zero, and the plastic strain vector $\Im^p$ has the same value as in the point H. The point corresponding to full unloading from the point H is called O_H.

A difference of values of any vector in the point O_H is called the final gain of this vector on the trace h. The vectors of stresses $\mathbf{S}$ and elastic strain $\Im^e$ disappearing in the case of full unloading are called *reversible* and their final gains are deemed equal to the vectors themselves. Vectors $\Im$ and $\Im^p$ are called *irreversible*.

Experimental studies have found [8, 10] that the direction of final gains of vectors relative to the strain trajectory in some point K does not depend on the entire trajectory but only on its segment within the trace h, whereas the trace h different for various materials has a length of about three to ten elastic strains at the yield stress. In any case, it is established that $h < \infty$. This is the *delay law*.

Physically, the delay law means that in any point $\bar{s}$, the material "remembers" the previous history of loading only on the trace h and "forgets" everything that happened beyond the trace on the path s at $s < \bar{s} - h$.

There are alternative opinions as to the delay law. For example, it was believed that the length of the delay trace was evaluated by the dependency $h \simeq e^{-a^2 s}$, $(a = const)$.

The experiments of V. S. Lensky [8, 9, 11] showed that if the strain trajectory had a low curvature ($\chi < 1/h$), the strain vector $\mathbf{S}$ in any point was directed along a tangential line to the trajectory:

$$\mathbf{S} = |\mathbf{S}| \frac{d\Im}{d}.$$

The last condition allows expanding the applicability area of strain-type ratios.

17.7   Loading Surface

We have already said (p. 201) that the yield condition in the case of primary loading of a virgin material can be represented as

$$f(J_1, \ J_2, \ J_3) = k^2, \quad (k = const),$$

where J_i are invariants of the stress tensor. If we assume that the first invariant (J_1) does not affect the conditions of occurrence and development of plastic strain, the plasticity condition looks as follows:

$$f(J_2', \ J_3') = k^2. \tag{17.17}$$

In the deviator space σ', Eq. (17.17) depicts some enclosed surface. In particular, in the five-dimensional space of Ilyushin, this is a five-dimensional sphere Σ^0 shown in Fig. 17.6 that is also called *the initial yield surface (loading)*. The surface Σ^0 confines the area of elastic strains, and the point O corresponds to a non-stressed state of a body.

   If the vector end σ goes to the initial surface and then there is gain $d\sigma$ beyond its surface, plastic strain occurs in the body. The initial surface is deformed and displaced (Fig. 17.6). Since loading can be generated from any stressed state, the loading surface must change so that the end of the vector $\sigma_1 = \sigma + d\sigma$ must belong to the loading surface all the time. A touching point of the stress vector and loading surface is called the *loading point at any moment in time.*

   When the additional loading vector $d\sigma$ is directed beyond the yield surface Σ, there is active loading accompanied by a gain in plastic strain. When directing $d\sigma$ inside Σ, unloading takes place under the elastic law. If the additional loading vector $d\sigma$ is directed along the tangential line to the yield surface, neutral loading *is present.*

   Multiple theoretical and experimental studies (including those ongoing) are dedicated to the type of yield surface. Proportional and non-proportional paths of loading have been implemented as a result of testing tubes in a plane stressed state.

Fig. 17.6 Transformation of the yield surface during loading

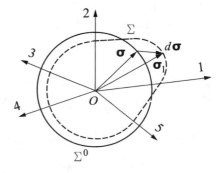

Let us take into account the following results. First of all, the initial yield surface of an isotropic material according to experimental works better complies with the Mises condition than the Tresca condition. Second, in most experimental studies, a convex smooth surface was obtained after preliminary simple loading, but in some cases of proportional plastic loading, we see a trend to forming an area of large curvature near the loading point reminding of blunt angles. The latter was used as a basis for a hypothesis of singular yield surfaces that will be discussed later.

17.8 Drucker Postulate

As shown above (p. 247), the theory of low elastic–plastic strains does not satisfy the continuity condition when switching from active loading to unloading. When analyzing other variants of plasticity theory, we can face such anti-natural and physically unacceptable consequences of various theories.

Therefore, it is desirable to have some universal criterion that must be satisfied as a minimal requirement for plasticity theory. This criterion was formulated by Drucker [2, 3].

Assume that when loading a body beyond the elasticity limits, some stressed state σ_{ij}^* was reached, which is depicted in a non-dimensional space by the loading point M^* corresponding to the stress vector end $\boldsymbol{\sigma}^*$ (see Fig. 17.7). An enclosed stress surface Σ^* going through the point M^* divides the areas of elastic and plastic states of the material. This means that further loading related to the vector end $\boldsymbol{\sigma}^*$ going beyond the area limited by the surface Σ^* leads to additional plastic strain.

Let us consider another stressed state σ_{ij}' corresponding to the loading point M_1 with the vector radius $\boldsymbol{\sigma}'$. Additional loading when switching from the point M^* to the point M_1 is $\boldsymbol{\sigma}' - \boldsymbol{\sigma}^*$. Assume now that we left the point M^* and returned to the same point along a closed path that partially goes beyond the surface Σ^*. This

Fig. 17.7 To the Drucker postulate

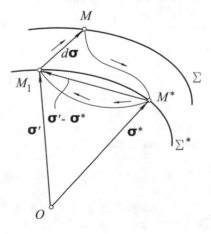

closed path and direction of going around it are shown in Fig. 17.7 with arrows. The
gain of the stress $d\boldsymbol{\sigma}$ will cause a respective gain of strain $d\mathbf{e}$.

The Drucker postulate states that the work of additional stresses on the closed
loading path is not negative:

$$\oint (\sigma_{ij} - \sigma_{ij}^*)de_{ij} \geqslant 0. \tag{17.18}$$

Let us show that formula (17.18) results in a number of important consequences.

Assume that loading is done over a closed path shown in Fig. 17.7. The loading
point M^* corresponding to the stress vector $\boldsymbol{\sigma}^*$ lies within Σ^* or maybe on the
surface as shown in the figure. The path goes from the point M^* to the point M_1
as directed by the arrow for any trajectory fully lying inside Σ^*. Then we impart
gain $d\boldsymbol{\sigma}$ to stress $\boldsymbol{\sigma}'$ in the direction of the point M located on the external side of
the enclosed area Σ^*. Now a new yield surface Σ goes through the point M, so
returning from the point M to the point M^* along any path fully lying inside Σ
means unloading. In this manner, strain is elastic in the sections M^*M_1 and MM^*
and elastic–plastic in the section M_1M and consists of elastic ($d\mathbf{e}^y$) and plastic
($d\mathbf{e}^{\text{II}}$) parts:

$$d\mathbf{e} = d\mathbf{e}^y + d\mathbf{e}^{\text{II}}. \tag{17.19}$$

The work of additional loading on elastic displacements with closed loading path
equals zero. Indeed,

$$\oint (\boldsymbol{\sigma} - \boldsymbol{\sigma}^*)d\mathbf{e}^y = \oint \boldsymbol{\sigma} d\mathbf{e}^y - \boldsymbol{\sigma}^* \oint d\mathbf{e}^y.$$

The first integral in the right part of this equation represents the complete work of
stress in elastic displacement with a closed path and, according to the definition of
the elasticity property, equals zero. The second integral also equals zero due to the
unambiguity of elastic strains. It follows that an irreversible work is done only in
the case of gaining plastic strain. Therefore, from formulas (17.18) and (17.19) we
can write as follows:

$$(\sigma_{ij}' - \sigma_{ij}^*)de_{ij}^{\text{II}} \geqslant 0. \tag{17.20}$$

In the case of continuous active strain from the point M^* to an infinitely close point
M past the intermediate loading, we have a gain of plastic strain during the entire
period of additional loading. If we return from the point M to the point M^* via an
arbitrary path fully lying within Σ, the irreversible work is done again on the plastic
component of strain only. Therefore,

$$d\sigma_{ij}de_{ij}^{\text{II}} \geqslant 0. \tag{17.21}$$

By considering the left part of the condition (17.20) as a scalar product of
the vectors $\boldsymbol{\sigma}' - \boldsymbol{\sigma}^*$ and $d\mathbf{e}^{\text{II}}$, we conclude that these vectors form a sharp angle

Fig. 17.8 Vector of plastic strain gain for a smooth loading surface

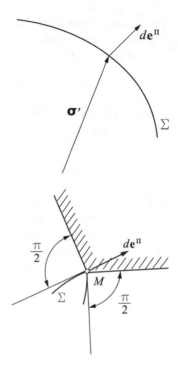

Fig. 17.9 Plastic strain gain for singular loading surface

between them. This means that if the loading surface is smooth and we can draw a single tangential hyper-plane in the point M_1 (Fig. 17.7), the vector de^p must be irreversibly directed along the normal line to the yield surface Σ (Fig. 17.8) irrespectively of the direction of the additional loading vector $d\sigma$. The normality of the vector de^{Π} to the loading surface is called [17] *the gradientality principle,* and the Drucker postulate is sometimes taken as the definition of material hardening.

The supposition of smoothness of the yield surface in the loading point is not compulsory. Moreover, as just stated (p. 257), loading points close to conical are experimentally proved for some materials. If M is a conic point of the loading surface, the Drucker postulate (17.20) means that the vector of plastic strain gain de^{Π} must be within the cone formed by normal lines to the surface Σ in the vicinity of the point M (Fig. 17.9). There are no other restrictions to the direction of vector de^p in the case of a singular yield surface in the Drucker postulate.

The condition (17.18) expresses that the work of any additional impacts on displacements caused by them is positive or equals zero in the case of a closed cycle. On the other hand, the first law of thermodynamics states that the work of a complete system of forces acting on the body is not negative. There are no grounds to believe that the Drucker postulate results from thermodynamics laws, which has been underlined by the author many times. The similarity of formulations is expressed by the author by using the term "quasi-thermodynamic" postulate.

We shall also note that A. A. Ilyushin showed that the Drucker postulate can be true for a specific class of materials and specific loading paths. In particular, if we consider a change of elastic constants of the material in the strain process, the gradientality principle has no sense. Ilyushin believes that the issue of convexity of the loading surface is also speculative.

17.9   On the Applicability Limits of the Strain Theory of Plasticity

The applicability of the ratios of the strain theory of plasticity in the case of loading different from proportional ones is studied in detail by Budiansky [1]. According to Budiansky, we assume consistency with the Drucker postulate as a criterion of validity of using these ratios.

If we consider the material to be non-compressible for simplicity, let us write the principal equations of the Hencky–Nadai–Ilyushin theory

$$\varepsilon'_{ij} = \frac{1}{2G_s}\sigma'_{ij}, \tag{17.22}$$

setting links between deviators of stresses (σ'_{ij}) and full (elastic–plastic) strains (ε'_{ij}).

If the hardening curve is known

$$\tau_o = G_s\gamma_o, \quad \left[\tau_o = \left(\frac{1}{3}\sigma'_{ij}\sigma'_{ij}\right)^{1/2}\right] \tag{17.23}$$

(Fig. 17.10), the secant (plastic) modulus G_s is known in each point K. The elastic components of strain are defined similarly to formula (17.22) where the plastic modulus must be replaced by the shear modulus G:

Fig. 17.10 Hardening diagram and its parameters

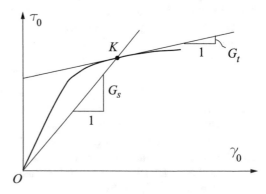

$$\varepsilon_{ij}^{'y} = \frac{1}{2G}\sigma_{ij}'.$$

By subtracting these strains from (17.22), let us write the plastic strain tensor as follows:

$$\varepsilon_{ij}^{\Pi} = e_{ij} = \varepsilon_{ij}' - \varepsilon_{ij}^{'y} = \frac{1}{2}\left(\frac{1}{G_s} - \frac{1}{G}\right)\sigma_{ij}'. \tag{17.24}$$

Let us find gain of plastic strain. By differentiating formula (17.24), we can write as follows:

$$de_{ij} = \frac{1}{2}\left(\frac{1}{G_s} - \frac{1}{G}\right)d\sigma_{ij}' + \frac{1}{2}\sigma_{ij}'d\left(\frac{1}{G_s}\right). \tag{17.25}$$

Due to

$$\frac{1}{G_s} = \frac{\gamma_o}{\tau_o}; \quad d\left(\frac{1}{G_s}\right) = \frac{-\gamma_o d\tau_o + \tau_o d\gamma_o}{\tau_o^2} = \frac{d\tau_o}{\tau_o}\left(\frac{d\gamma_o}{d\tau_o} - \frac{\gamma_o}{\tau_o}\right),$$

the expression (17.25) may look as follows:

$$de_{ij} = \frac{1}{2}\left(\frac{1}{G_s} - \frac{1}{G}\right)d\sigma_{ij}' + \frac{1}{2}\left(\frac{1}{G_t} - \frac{1}{G_s}\right)\frac{\sigma_{ij}'}{\tau_o}d\tau_o, \tag{17.26}$$

where G_t is the tangential modulus (Fig. 17.10) equal to the derivative function $\tau_o(\gamma_o)$ upon its argument.

Let us remind that the strain theory of plasticity suggests a unified hardening diagram $t_o \sim s_o$ (Fig. 17.10). This assumption is true for carbon and low-alloy steel and for titanium alloys. However, there is no unified hardening diagram for high-strength steels, aluminum and magnesium alloys in elongation and compression as well as in elongation and shear. Having this in mind, let us return to the Drucker's postulate.

In the case of no volumetric compression, the requirement of the Drucker postulate (17.21) is written as follows:

$$d\sigma_{ij}'de_{ij} \geqslant 0. \tag{17.27}$$

Let us make an expression in the left part of the condition (17.27). To do it, let us multiply formula (17.26) by $d\sigma_{ij}'$

$$d\sigma_{ij}'de_{ij} = \frac{1}{2}\left(\frac{1}{G_s} - \frac{1}{G}\right)(d\sigma_{ij}'d\sigma_{ij}') + \frac{1}{2}\left(\frac{1}{G_t} - \frac{1}{G_s}\right)\frac{\sigma_{ij}'d\sigma_{ij}'}{\tau_o}d\tau_o. \tag{17.28}$$

Now let use the expression (17.23) to find as follows:

$$dτ_o = \frac{1}{2}\left(\frac{1}{3}σ'_{ij}σ'_{ij}\right)^{-\frac{1}{2}} \cdot \frac{2}{3}σ'_{ij}dσ'_{ij},$$

for example,

$$τ_o dτ_o = \frac{1}{3}σ'_{ij}dσ'_{ij}.$$

By substituting the last expression into formula (17.28), we obtain

$$dσ'_{ij}de_{ij} = \frac{1}{2}\left(\frac{1}{G_s} - \frac{1}{G}\right)\left(dσ'_{ij}dσ'_{ij}\right) + \frac{3}{2}\left(\frac{1}{G_t} - \frac{1}{G_s}\right)(dτ_o)^2. \qquad (17.29)$$

The expression (17.29) will be positive only if

$$G > G_s > G_t,$$

which is always complied with for monotonously hardening materials. Thus, the plasticity law of the strain theory does not contradict the requirement of the Drucker postulate (17.27) for any loadings. Restrictions implied by the postulate can refer only to the type of the loading surface dividing the area of active plastic strain and unloading.

Now let us refer to the second requirement of the postulate (17.20). For a non-compressible material, the plastic strain and loading surface will depend on the history of setting the stress deviator components and this requirement will look as follows:

$$(σ'_{ij} - σ^*_{ij})de^{II}_{ij} \geqslant 0. \qquad (17.30)$$

Let us use the nine-dimensional imaging space. For the plastic strain to satisfy the requirement (17.30), the loading surface in the loading point must have a conic feature.

Indeed, for a smooth loading surface, the vector direction $d\mathbf{e}$ depends only on the loading vector $\boldsymbol{σ}$, but not on $d\boldsymbol{σ}$. In the case of strain theory, in the expression (17.26) of the plastic strain gain, the first addend depends directly on $dσ'_{ij}$, so the direction of the vector $d\mathbf{e}^{II}$ depends on $d\boldsymbol{σ}$. In this manner, it must be suggested that the loading point M (Fig. 17.11) is conical.

By considering the plane projection of the imaging space, let us use $β$ to designate the angle between the cone generatrix and the vector radius $\boldsymbol{σ}$ of the loading point M, $α$ to designate the angle between the vectors $\boldsymbol{σ}$ and $d\boldsymbol{σ}$, and $δ$ to designate the angle between the plastic strain gain vector $d\mathbf{e}^{II}$ and the vector $\boldsymbol{σ}$ (Fig. 17.11). From the Drucker postulate (p. 259), it follows that the vector $d\mathbf{e}^{II}$ must lie within the cone of normal lines to the loading surface in the point M. This condition is reduced to fulfilling the inequation:

Fig. 17.11 Case of a conic
point on the loading surface

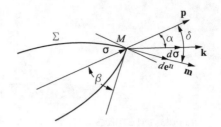

$$\beta + \delta \leqslant \frac{\pi}{2}. \tag{17.31}$$

Let us introduce designations:

$$\sigma = \tau_o \mathbf{p}, \quad d\sigma = \mathbf{k}dq, \quad q\mathbf{e}^{II} = \mathbf{m}dr, \tag{17.32}$$

$$\frac{1}{2}\left(\frac{1}{G_s} - \frac{1}{G}\right) = A, \quad \frac{1}{2}\left(\frac{1}{G_t} - \frac{1}{G_s}\right) = B, \quad N = 1 + \frac{B}{A}.$$

The singular vectors $\mathbf{p}$, $\mathbf{k}$, $\mathbf{m}$ are shown in Fig. 17.11. Taking into account these designations, formula (17.26) is written as a vector equation:

$$d\mathbf{e}^{II} = Ad\sigma + B\frac{\sigma}{\tau_o}d\tau_o.$$

By using designations (17.32) and considering that $\tau_o d\tau_o = \sigma d\sigma = \tau_o dq \cos\alpha$, let us write the previous equation as follows:

$$\mathbf{m}dr = (A\mathbf{k} + \mathbf{p}B\cos\alpha)dq. \tag{17.33}$$

Let us square both parts of Eq. (17.33). We obtain

$$\frac{dq}{dr} = \left[A^2 + (B^2 + 2AB)\cos^2\alpha\right]^{-1/2}.$$

Multiplying Eq. (17.33) by a single vector $\mathbf{p}$ gives

$$\cos\delta dr = (A + B)\cos\alpha dq.$$

Let us find from two last equations:

$$\cos\delta = \frac{N\cos\alpha}{\left[1 + (N^2 - 1)\cos^2\alpha\right]^{1/2}}. \tag{17.34}$$

The inequation (17.31) is equivalent to the following:

$$\cos \delta \geqslant \sin \beta$$

or

$$\frac{N^2 \cos^2 \alpha}{1 + (N^2 - 1) \cos^2 \alpha} \geqslant \sin^2 \beta.$$

The obtained expression can be revised as follows:

$$\frac{N^2}{N^2 + \text{tg}^2\alpha} \geqslant \frac{\text{tg}^2\beta}{1 + \text{tg}^2\beta}.$$

Hence we obtain

$$\text{tg}\, \alpha \leqslant \frac{N}{\text{tg}\, \beta}. \tag{17.35}$$

The inequation (17.35) shows that the strain theory laws do not contradict the Drucker postulate when the additional loading vector $d\sigma$ is directed to the area limited by the cone whose generatrixes make an angle α with a radius vector of the point M. If we assume that plasticity theory does not depend on the loading history, possible loading paths must be limited by the condition $\alpha \leqslant \beta$. By assuming in (17.35) $\alpha = \beta$, we will find the value β_o for which the angle α is the highest:

$$\text{tg}\, \beta_o = N^{1/2}. \tag{17.36}$$

Thus, the practical utility of the obtained result is as follows:

Assume that any problem of plasticity theory is solved using strain theory. This means that for each point of the body, a loading path is built and, therefore, the angle α is known between the point radius vector in the space of the stress deviator and the tangential line to the loading path, and the value of N is calculated for each point, so the formula (17.36) can be used to find the plasticity angle β_o. If it appears that $\alpha < \beta_o$, using strain theory for this problem can be deemed justified in a sense that the requirements of the Drucker postulate are not violated.

References

1. B. Budiansky, A reassessment of deformation theories of plasticity. J. Appl. Mech. **26**(2), 259–264 (1959)
2. D. Drucker, chapter A more fundamental approach to plastic stress-strain relations, in *Proceedings of the First U.S. National Congress of Applied Mechanics, ASME* (ASME, New York, 1951), pp. 487–491
3. D. Drucker, Coulomb friction, plasticity and limit loads. J. Appl. Mech. ASME **21**(1), 71–74 (1954)

4. I. Gol'denblat, *Nekotorye voprosy teorii uprugikh i plasticheskikh deformatsii* [Some questions of the theory of elastic and plastic deformations] (Stroiizdat Publ., Moscow–Leningrad, 1950)
5. A. Il'yushin, *O svyazi mezhdu napryazheniyami i malymi deformatsiyami v mekhanike sploshnykh sred* [On the relationship between stresses and small deformations in continuum mechanics]. PMM [Appl. Math. Mech.] **18**(6), 641–666 (1954)
6. A. Il'yushin, *Plastichnost'. (Osnovy obshchei matematicheskoi teorii)* [Plasticity. (Fundamentals of General mathematical theory)] (Izd-vo AS USSR Publ., Moscow, 1963)
7. A. Il'yushin, *Trudy. T. 2 (1946–1966) / Sostaviteli: E.A. Il'yushina, N.R. Korotkina* [Scientific work. Vol. 2 (1946–1966). Compilers: E.A. Ilyushina, N.R. Korotkina] (Fizmatlit Publ., Moscow, 2004)
8. V. Lenskii, *Ehksperimental'naya proverka zakonov izotropii i zapazdyvaniya pri slozhnom nagruzhenii* [Some new data on the plasticity of metals under complex loading], Izv. An SSSR, OTN, Mech. Mech. Eng. **11**, 15–24 (1958)
9. V. Lenskii, *Nekotorye novye dannye o plastichnosti metallov pri slozhnom nagruzhenii* [Some new data on the plasticity of metals under complex loading], *Izv. AN SSSR, OTN, Mekh-ka i mashinostr* [Izv. An SSSR, OTN, Mech. Mech. Eng.] **4**, 57–64 (1960)
10. V. Lenskii, *Proc. of the First U.S. National Cong. Appl. Mech., ASME*, Moscow, Izd-vo AN SSSR Publ., chapter *Ehksperimental'naya proverka postulatov obshchei teorii uprugo-plasticheskikh deformatsii* [Experimental verification of the laws of isotropy and delay under complex loading] (1961), pp. 58–82
11. V. Lensky, Analysis of plastic behavior of metals under complex loading, in *Proceedings of 2nd Symposium on Naval Structural Mechanics* (1960), pp. 259–268
12. A. Norden, *Kratkii kurs differentsial'noi geometrii. Izd. 2-e.* [A short course in differential geometry. 2nd ed.] (Moscow, Fizmatgiz Publ., 1958)
13. V. Prager, *Problemy teorii plastichnosti* [Problems of plasticity theory] (Fizmatlit Publ., Moscow, 1958)
14. L. Prandtl, Ueber die hart plastischer korper, in *Gottingen Nachrichten* (1920), pp. 74–85
15. Y. Rabotnov, (ed.), *Teoriya plastichnosti: sb. perev. inostr. statei* [Plasticity Theory: Foreign Translation Collection. articles] (IL Publ., Moscow, 1948)
16. A. Zhukov, *O plasticheskikh deformatsiyakh izotropnogo metalla pri slozhnom nagruzhenii* [On plastic deformations of an isotropic metal under complex loading], *Izv. AN SSSR, OTN* [Izvestia of the USSR Academy of Sciences, Department of technical Sciences] **12**, 72–87 (1956)
17. V. Zubchaninov, *Gipoteza ortogonal'nosti i printsip gradiental'nosti v teorii plastichnosti* [The hypothesis of orthogonality and the principle of gradientless in the theory of plasticity]. Izv. RAN. MTT [Izvestiya RAS. Solid state mechanics] **5**, 68–73 (2008)

Chapter 18
Theories of Plastic Yield

18.1 General Ratios

We have already said that plasticity theories of the strain type take the non-linear elasticity theory as the reference point and represent its generalization in the case of a non-elastic material. The link between stresses and strains is usually represented by finite ratios. Unlike it, the theories of plastic yield set links between infinitely low gains of strains and stresses, the stresses themselves, and some parameters of plastic condition.

The following initial provisions are taken in most variants of yield theory.

1. *The body is deemed initially isotropic.*
2. *The volumetric strain is deemed elastic* (14.30)

$$\varepsilon_0 = \frac{\sigma_0}{3K},$$

or

$$d\Theta = \frac{d\sigma_0}{K},$$

where

$$\Theta = \varepsilon_{ii}; \quad \varepsilon_0 = \frac{1}{3}\Theta; \quad \sigma_0 = \frac{1}{3}\sigma_{ii}; \quad K = \frac{E}{3(1-2\nu)}.$$

3. *The gains of the full strain components are presented by the sums of gains of the respective components of elastic and plastic strain*

$$d\varepsilon_{ij} = d\varepsilon_{ij}^y + d\varepsilon_{ij}^{\Pi}. \tag{18.1}$$

The gains of the elastic strain components are related to the gains of stress components by Hooke's law

$$de_{ij}^y = \frac{1}{2G}\left(d\sigma_{ij} - \frac{3v}{1+v}\delta_{ij}d\sigma_0\right). \tag{18.2}$$

4. *The stress deviator D_σ and the deviator of plastic strain $D_{d\varepsilon}^{\Pi}$ are proportional*

$$D_{d\varepsilon}^{\Pi} = D_\sigma d\lambda, \tag{18.3}$$

where $d\lambda$ is some infinitely small scalar multiplier.

Taking into account that plastic strain is not accompanied by changes in the body volume, as per the first of formulas (13.21), the dependency (18.3) can be recorded as follows:

$$d\varepsilon_{ij}^{\Pi} = \sigma_{ij}'d\lambda, \tag{18.4}$$

where σ_{ij}' are the components of the stress deviator D_σ. By calculating the gain of plastic strain work, let us find

$$dA_{\Pi} = \sigma_{ij}d\varepsilon_{ij}^{\Pi} = d\lambda \cdot \sigma_{ij}\sigma_{ij}' = 2d\lambda \cdot \tau_0^2, \tag{18.5}$$

whereas τ_0 is the octahedral tangential stress defined by formula (13.31).

In formula (18.5), we see that the multiplier $d\lambda$ is associated with the gain of the plastic strain work. Since $dA_{\Pi} \geqslant 0$, $d\lambda \geqslant 0$. Having this in mind, according to formula (18.1), we obtain the full gains of the strain components

$$d\varepsilon_{ij} = d\varepsilon_{ij}^y + d\lambda \cdot \sigma_{ij}', \tag{18.6}$$

where the gains of the elastic strain components are defined under Hooke's law (18.2). It is then easy to find the gain of any strain work

$$dA = dA_y + dA_{\Pi}, \tag{18.7}$$

where dA_{Π} is already defined (18.5), and the gain of the elastic strain work can be defined [7] as $dA_y = d\Pi$, where the elastic potential Π is found using the formula

$$\Pi = \frac{\sigma_0^2}{2K} + \frac{\tau_0^2}{2G}. \tag{18.8}$$

For $d\lambda = 0$, Eqs. (18.6) turn into Hooke's law. In a general case, the system of equations (18.6) is not complete, since it contains a non-defined multiplier. Therefore, it is necessary to close the system with an additional ratio.

18.2 Prandtl–Reuss Yield

As an additional ratio, let us take the Huber–Mises yield condition (14.15)

$$\tau_0 = \tau_s,$$

where τ_s is the shear yield stress. Then we have as follows from the condition (18.5):

$$d\lambda = \frac{dA_{\Pi}}{2\tau_s^2}. \tag{18.9}$$

Formula (18.9) shows that the multiplier $d\lambda$ is proportional to the gain of the plastic strain work, and since the latter is defined by the formula $\sigma_{ij}d\varepsilon_{ij}^?$, there is no unique dependence of the gains of the strain components on the stress components and their gains, which takes place in the case of an ideally plastic body.

If the plasticity condition $\tau_0 = \tau_s$ is fulfilled, then $d\tau_0 = 0$ and plastic strain occurs. If $d\tau_0 < 0$, unloading under the elastic law occurs.

For the Mises plasticity condition, Eqs. (18.6) were proposed by Reuss [10] in 1930. For the plane problem, they were offered by Prandtl [9] in 1924.

18.3 Saint-Venant–Mises Yield Theory

In the case of developed strain, the components of its elastic part are low as compared to the plastic strain components, and they can be neglected. With this assumption, the Prandtl–Reuss theory equations go to the equations of the Saint-Venant–Mises plasticity theory

$$d\varepsilon_{ij} = d\lambda \cdot \sigma_{ij}'.$$

Dividing this equation by an infinitely small interval of time dt gives

$$\dot{\varepsilon}_{ij} = \lambda' \sigma_{ij}', \tag{18.10}$$

where

$$\lambda' = \frac{1}{2\tau_s^2} \cdot \frac{dA_?}{dt} = \frac{1}{2\tau_s^2}\sigma_{ij}\dot{\varepsilon}_{ij} = \frac{1}{2\tau_s^2}\sigma_{ij}'\dot{\varepsilon}_{ij}.$$

The latter formula shows that the multiplier λ' is proportional to the plastic strain power and characterizes the dissipation of strain energy. If we remove the strain components from this formula using the dependencies (18.10), we will find

$$\lambda' = \frac{\Gamma}{2\tau_s},$$

where Γ is the intensity of shear strain rates of a non-compressible medium defined by the formula

$$\Gamma = \left(2\varepsilon_{ij}\varepsilon_{ij}\right)^{1/2}.$$

Thus, Eqs. (18.10) can also be written as

$$\frac{\dot{\varepsilon}_{ij}}{\Gamma} = \frac{\sigma'_{ij}}{2\tau_s}. \tag{18.11}$$

These ratios show that strains cannot be unambiguously defined when setting strain rate stresses. The uncertainty of strain rate components related to an uncertainty of the multiplier λ' is necessary for satisfying the conditions of joint strain. On the opposite, for the defined rates of strain $\dot{\varepsilon}_{ij}$, the components of the stress deviator σ'_{ij} are defined unambiguously. In this case, the deviator components defined using formulas (18.11) identically satisfy the Mises plasticity condition.

Equations (18.10) for the case of plane strain for the Tresca plasticity conditions were given by Saint-Venant [9]. In a general case, they are defined by Levy and Mises [9].

18.4   Plastic Yield in Isotropic Hardening

Assume that the loading surface Σ undergoes even expansion during the plastic strain of a material. This hardening is called *isotropic*. Assume that in this case the equation Σ has the following form:

$$f[J_2(D_\sigma), J^3(D_\sigma)] = F(q), \tag{18.12}$$

where F is the ascending function of some parameter q characterizing the extent of material hardening. In a simpler case, we can assume that the function f depends only on the square invariant of the stress deviator, and we can write the dependency (18.12) as

$$\tau_0 = f(q). \tag{18.13}$$

The work of plastic strain can be taken as a measure of material hardening

$$A_? = \int A_{ij} d\varepsilon_{ij}^{\Pi}.$$

The hardening condition (18.13) then looks as follows:

$$\tau_0 = f(A_\text{II}). \tag{18.14}$$

If the universal hardening function $\tau_0 = \tilde{\Phi}(\gamma_0)$ is found from experiments, we can write $\gamma_0 = \Phi(\tau_0)$ as a result of its inversion and represent as follows:

$$dA_\text{II} = \frac{d\Phi(\tau_0)}{d\tau_0}d\tau_0 = \Phi'(\tau_0)d\tau_0.$$

By introducing this representation in formula (18.5) and designating

$$\frac{\Phi'(\tau_0)}{2\tau_0^2} = \Psi(\tau_0),$$

we obtain

$$d\lambda = \Psi(\tau_0)d\tau_0. \tag{18.15}$$

Substituting formulas (18.15) into Eqs. (18.6) allows calculating the gains of the full strain components

$$d\varepsilon_{ij} = d\varepsilon_{ij}^y + \sigma'_{ij}\Psi(\tau_0)d\tau_0. \tag{18.16}$$

The ratios (18.16) are true for active strain $d\tau_0 \geqslant 0$. If $d\tau_0 = 0$, there is neutral loading, and in the case of $d\tau_0 < 0$, unloading occurs under the elastic law. When switching from active loading to neutral one and to unloading, the strain component gains are continuously changed. Therefore, the ratios of yield theory are free from the above distortions of continuity suffered by strain theory.

We shall note that these ratios set an unambiguous dependency of strain component gains on stresses and their gains if the hardening is present. The hardening state has no condition binding the stress components (as in the case of ideal plasticity), and the multiplier λ is well defined by formula (18.15).

18.5   Handelman–Lin–Prager Plasticity Theory

This variant of plasticity theory is sometimes called the simplest yield theory. The following pre-requisites are taken as fundamental.

1. The gain of the strain deviator is well defined by the stress deviator and its gain.
2. The link of the strain deviator gain is linear relative to the stress deviator and its gain.
3. The continuity condition is true.

4. The loading criterion is defined by the condition $dJ_2' > 0$.

Based on two first pre-requisites, we can write as follows:

$$d\varepsilon_{ij}' = A_{ijkl}d\sigma_{kl}', \quad (dJ_2' > 0), \tag{18.17}$$

where the fourth-rank tensor A_{ijkl} depends only on the stress deviator D_σ.

Similarly to formula (18.1), we represent the full strain gain as a sum of gains of its elastic and plastic parts

$$d\varepsilon_{ij} = d\varepsilon_{ij}^y + de_{ij}, \quad (e_{ij} = \varepsilon_{ij}^{\mathrm{II}}). \tag{18.18}$$

The gain of elastic strains is defined by Hooke's law, and we assume as follows for the gain of the plastic strain component:

$$de_{ij} = C_{ijkl}d\sigma_{kl}', \tag{18.19}$$

where the tensor C_{ijkl} depends on the stress deviator.

We can write as follows from the continuity condition in the case of neutral loading:

$$C_{ijkl}d\sigma_{kl}' = 0, \quad dJ_2' = 0. \tag{18.20}$$

The loading neutrality condition gives

$$dJ_2' = d(\sigma_{kl}'\sigma_{kl}') = 2\sigma_{kl}'d\sigma_{kl}' = 0. \tag{18.21}$$

Simultaneous zeroing of two lines relative to $d\sigma_{kl}'$ having forms (18.20) and (18.21) makes us assume that

$$C_{ijkl} = G_{ij}\sigma_{kl}',$$

and represent the plasticity law as

$$de_{ij} = \begin{cases} G_{ij}\sigma_{kl}'d\sigma_{kl}' \equiv \dfrac{1}{2}G_{ij}dJ_2', \ dJ_2' > 0; \\ 0, \hspace{3.5cm} dJ_2' \leqslant 0. \end{cases} \tag{18.22}$$

The tensor analysis shows [1] that the most general form of tensor G_{ij} contained in the law (18.22) will be

$$G_{ij} = G_{ij}[(\sigma_{ij}')] = P(J_2', J_3'^2) + Q(J_2', J_3'^2)J_3't_{ij}. \tag{18.23}$$

The law (18.22)–(18.23) expresses the Handelman–Lin–Prager yield theory. If we assume that the third invariant of the stress deviator poorly affects the ratios of the link between the gains of stress and strain, the ratios (18.22)–(18.23), as a partial case, result in the plastic yield law proposed by Laning

$$de_{ij} = \begin{cases} P(J_2')\sigma_{ij}'dJ_2', & dJ_2' > 0; \\ 0, & dJ_2' \leqslant 0. \end{cases} \tag{18.24}$$

The function $P(J_2')$ included in formula (18.24) must be defined from experiments.

The Laning law (18.24) in the case of simple loading can be integrated. Assume that

$$\sigma_{ij}' = (\sigma_{ij}')^0 \cdot \lambda, \quad dJ_2' = (J_2')^0 \lambda d\lambda,$$

where the upper register "0" indicates some fixed values of respective quantities. By integrating the first of formulas (18.24), we have

$$e_{ij} = \int_0^1 P(\lambda^2 J_2'^0)\lambda\sigma_{ij}'^0 J_2'^0 \lambda d\lambda.$$

If the function $P(J_2')$ is known, the integration result of the last formula will look as follows:

$$e_{ij} = G_s(J_2'^0)\sigma_{ij}'^0.$$

The obtained formula in fact coincides with the law (17.1) of the strain theory of plasticity.

In the conclusion of this paragraph, we shall note that the ratios (18.22)–(18.24) are plasticity laws associated with the Huber–Mises yield condition.

18.6   Yield for Plane Loading Surfaces

Let us assume that the equation of the loading surface Σ going through some loading point M can be written as

$$f(\sigma_{ij}) = k^2, \quad (k = const). \tag{18.25}$$

The function f can depend on stresses, strains, and in any complex way on the loading path or strain path.

If the stress obtains gain $d\sigma_{ij}$, such that

$$df = \frac{\partial f}{\partial \sigma_{ij}}d\sigma_{ij} > 0, \tag{18.26}$$

some plastic strain de^{Π} occurs. Based on the Drucker postulate, the vector de^{Π} is directed to the normal line Σ, so $df \sim \mathbf{n} \cdot d\boldsymbol{\sigma} > 0$ and

$$\dot{e}_{ij}^{\Pi} = \lambda \frac{\partial f}{\partial \sigma_{ij}}. \tag{18.27}$$

Let us specify the form of the multiplier λ included in formula (18.27). Since the plastic strain rate $\dot{e}_{ij}^{\Pi}$ must be proportional to the normal line to the surface Σ of the additional loading vector component $d\boldsymbol{\sigma}$, the gains of the plastic strain components de_{ij}^{Π} must be $\dfrac{\partial f}{\partial \sigma_{kl}} d\sigma_{kl}$. Taking this into account, formula (18.27) can be represented as follows:

$$de_{ij}^{\Pi} = H \frac{\partial f}{\partial \sigma_{kl}} d\sigma_{kl} \cdot \frac{\partial f}{\partial \sigma_{ij}},$$

or else

$$\dot{e}_{ij}^{\Pi} = H \frac{\partial f}{\partial \sigma_{ij}} \left(\frac{\partial f}{\partial \sigma_{kl}} \cdot \dot{\sigma}_{kl} \right), \tag{18.28}$$

where H is the positive hardening function that, as results from the Drucker postulate, can depend on the loading history, strain history but does not depend on the gains $d\sigma_{kl}$ and de_{kl}^{Π}. The latter means that the ratio (18.28) is linearly relative to the gains $d\sigma_{kl}$ and de_{kl}^{Π}.

The indicated linearity of the law (18.28) relative to gains $de_{kl}^{?}$ and $d\sigma_{kl}$ shows, in particular, that the link $de_{kl} \sim d\sigma_{kl}$ does not depend on the angle (β) of the loading trajectory fracture (Fig. 17.4)

$$\mathrm{tg}\,\beta = \frac{dS_3}{dS_1},$$

which is incorrect in a general case. Therefore the law (18.28) must be refined. For example, we can assume that H is a homogeneous function of the zero degree from stress gain.

Let us consider a partial case. Assume that H and f are functions of the second variant of the stress deviator J_2' (or $\sigma_?$ or τ_0, which is the same). By assuming, for example, $f = J_2'$, we have

$$\frac{\partial f}{\partial \sigma_{ij}} = \frac{\partial}{\partial \sigma_{ij}} (\sigma_{ij}' \sigma_{ij}') = 2\sigma_{ij}',$$

as well as

$$\frac{\partial f}{\partial \sigma_{kl}} \dot{\sigma}_{kl} = \frac{\partial J_2'}{dt}.$$

The fact that the second invariant of the stress deviator J_2' differs from the square of stress intensity $\sigma_?^2$ and octahedral tangential stress τ_0^2 by constant multipliers only results in the proportionality of values

$$\frac{\partial J_2'}{\partial t} \sim \sigma_? \dot{\sigma}_{\text{и}} \sim \tau_0 \dot{\tau}_0.$$

By substituting the last results to the law (18.28), we obtain

$$\dot{e}_{ij} = H_1(\sigma_{\text{и}})\dot{\sigma}_{\text{и}}\sigma_{ij}', \tag{18.29}$$

or

$$\dot{e}_{ij} = H_2(\tau_0)\dot{\tau}_0\sigma_{ij}', \tag{18.30}$$

where

$$H_1(\sigma_{\text{и}}) = H(\sigma_{\text{и}})\sigma_{\text{и}}, \quad H_2(\tau_0) = H(\tau_0)\tau_0.$$

By comparing the expressions (18.29) and (18.30) with formula (18.24), we conclude that in the considered case, the law (18.28) results in ratios of the Laning law. Let us remind that in this case for simple loading, the law (18.29) (or (18.30)) can be integrated and is converted into the Hencky–Nadai–Ilyushin law.

18.7 Yield for Some Loading Surfaces

Let us consider a material that finds the *ideal Baushinger effect*. The diagram of the alternating-sign uniaxial stressed state beyond the yield stress of this material is shown [5] in Fig. 18.1. 1 is the yield start in the case of primary loading of the initially isotropic material; 2 is partial unloading with further loading of the same sign; 3 is loading start; 4 is full unloading; 5 is the start of yield in the case of loading of opposite sign. The reduction of the yield stress in point 5 equals hardening in the direction of initial unloading that numerically equals the difference of ordinates of points 3 and 1.

In a general case of loading, when hardening this material in some direction, an equal softening occurs in the opposite direction. The loading surface in each successive moment in time t_1, t_2, t_3 progressively moves as a rigid whole following the loading point as shown in Fig. 18.2. This hardening is called *translational*.

If the initial loading surface equation is represented by formula (18.25)

$$f(\sigma_{ij}) = k^2,$$

Fig. 18.1 Ideal Baushinger effect

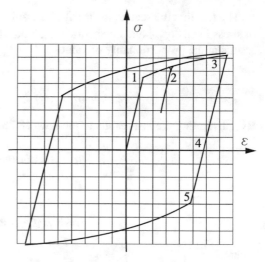

Fig. 18.2 Displacement of loading surface in the case of translational hardening

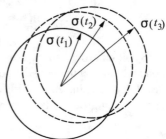

then in the case of its incremental transfer, the equation of the new loading surface will be

$$f(\sigma_{ij} - S_{ij}) = k^2, \tag{18.31}$$

where S_{ij} is the tensor whose components in the stress space are the coordinates of the loading surface center.

It is obvious that the tensor S_{ij} must be related to plastic strains. If we assume J_2' as the function $f(\sigma_{ij})$ and suggest (A. Yu. Ishlinsky) that the tensor components S_{ij} are proportional to the respective components of plastic strain, we obtain

$$S_{ij} = ce_{ij}, \quad (c = const). \tag{18.32}$$

In the case of uniaxial elongation, formulas (18.31) and (18.32) result in linear dependency between stress and plastic strain. We have linear hardening and ideal Baushinger effect.

Fig. 18.3 Kinematic Prager model

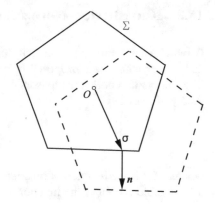

Irrespective of Ishlinsky and approximately at the same time with him, Prager proposed a similar hypothesis later called the hypothesis of kinematic hardening. Let us explain the Prager model by using Fig. 18.3.

Let us have some initial loading surface Σ with the center O corresponding to the non-loaded condition of the material. The surface Σ may not be smooth, and it can have ribs or other specific features. Let us imagine a rigid shell shaped as the surface Σ. In a plane case, this will be a rigid frame shown in Fig. 18.3.

Let us assume that this flat frame can gradually move in the stress plane using a crank of variable length with a fixed end coinciding with the point O. The free end of the crank will be identified with the loading point defined by the vector σ. In the case of no friction between the frame and the crank, the crank rotation will cause the frame to move in the direction of the normal line **n**.

According to the Drucker postulate , the plastic strain vector **e** is proportional to **n**. Consequently, the displacement of the center O characterized by the vector **S** will also be proportional to **e**

$$\mathbf{S} = c\mathbf{e}, \quad (c = const),$$

or in the tensor form

$$S_{ij} = ce_{ij},$$

which coincides with the Ishlinsky theory (18.32).

The Prager model is more common than the Ishlinsky theory. A detailed study of the kinematic hardening law in various sub-spaces is given in the papers by Shield and Zigler [11, 12]. Individual variants of the model are used in solutions of static and dynamic [4] applied tasks.

18.8   Kadashevich–Novozhilov Plasticity Theory

A more general variant of plasticity theory as compared with the Ishlinsky theory and Prager model was proposed by [3] Yu. I. Kadashevich and V. V. Novozhilov. The authors accept that the tensors S_{ij} and $e_{ij}^{?}$ are related by the ratios of the strain theory of plasticity

$$e_{ij} = \frac{1}{2G^*} S_{ij}, \qquad (18.33)$$

whereas G^* is the function of invariants of the tensor S_{ij}. The following is proposed as a possible variant of the function G^*

$$G^* = G^*(J_{2(S_{ij})}').$$

For $G^* = const$, the ratio (18.33) coincides with the Ishlinsky theory. A later paper [8] reveals a physical essence of the tensor S_{ij} by introducing the concept of micro-stresses.

18.9   Singular Loading Surfaces

By studying the applicability limits of the strain theory of plasticity (p. 260), we have found that further loading surfaces can be singular. Here we consider the specifics of a different kind when the initial loading surface defined by the plasticity condition consists of several smooth surfaces forming when crossed by the rib.

The direction of the normal line to the loading surface on the ribs is not defined. As to the direction of the plastic strain vector gain, we only know that it lies in the plane perpendicular to the rib and within the angle limited by normal lines to the surfaces forming the rib. In the case of additional loading with isotropic hardening, in the considered case, the loading surface is expanded keeping similarity, and in the case of translational hardening, the initial surface is displaced in parallel to itself. In both cases, new specifics are not expressed and smooth loading surfaces remain smooth. The simplest example of such a process is obtained if we try to generalize the Saint Venant-Tresca plasticity theory for the case of a hardening material.

Let us assume for simplicity that we know the directions of principal stresses beforehand. Instead of a non-dimensional space of stresses, the process can be considered in a three-dimensional sub-space of principal stresses. For $\sigma_1 > \sigma_2 > \sigma_3$, the maximum tangential stress will be

$$\tau_{max} = \frac{\sigma_1 - \sigma_3}{2},$$

and its constancy condition is written as

$$\sigma_1 - \sigma_3 = \pm 2k, \tag{18.34}$$

and the constant value k is selected such that the plasticity surface goes through the loading point.

In a general case, the ratio between principal stresses during loading can change. In the space of principal stresses, we cannot associate the names of axes with the values of principal stresses without violating the inequation $\sigma_1 \geqslant \sigma_2 \geqslant \sigma_3$. Therefore, let us assume the designations σ_ξ, σ_η, σ_ζ for principal stresses. In various points of the space of principal stresses, the ratio between them is different, and, in this connection, we assign the values of 1, 2, or 3 to the indexes ξ, η, ζ. For this reason, we have six similar conditions (18.34) instead of two conditions:

$$\begin{aligned}
\sigma_\xi - \sigma_\eta &= \pm 2k, \\
\sigma_\eta - \sigma_\zeta &= \pm 2k, \\
\sigma_\zeta - \sigma_\xi &= \pm 2k.
\end{aligned} \tag{18.35}$$

As we already know (p. 185), the conditions (18.35) are equations of six planes that form the loading surface as a hexagonal prism. Let us assume that the loading point belongs to the prism face $f = \sigma_\xi - \sigma_\eta$ corresponding to the first of equations (18.35). Derivatives from the function f will be

$$\frac{\partial f}{\partial \sigma_\xi} = 1, \quad \frac{\partial f}{\partial \sigma_\eta} = -1, \quad \frac{\partial f}{\partial \sigma_\zeta} = 0.$$

Formula (18.28) gives

$$\begin{aligned}
\dot{e}_\xi &= H(\dot{\sigma}_\xi - \dot{\sigma}_\eta), \\
\dot{e}_\eta &= -H(\dot{\sigma}_\xi - \dot{\sigma}_\eta), \\
\dot{e}_\zeta &= 0.
\end{aligned} \tag{18.36}$$

If we assume that the function H depends on the value of the highest tangential stress only, e.g. $H = H(\sigma_\xi - \sigma_\eta)$, the ratios (18.36) can be integrated; we have

$$\begin{aligned}
e_\xi &= h(\sigma_\xi - \sigma_\eta), \\
e_\eta &= -h(\sigma_\xi - \sigma_\eta), \\
e_\zeta &= 0,
\end{aligned} \tag{18.37}$$

where

$$h(\sigma_\xi - \sigma_\eta) = \int H(\sigma_\xi - \sigma_\eta) d(\sigma_\xi - \sigma_\eta).$$

Though the ratios of the yield theory were initial, the adopted assumption of the function form H leads to the final dependencies between plastic strains and stresses as in the strain theory of plasticity. By comparing the dependencies (18.37) with

Fig. 18.4 To singular
loading surfaces

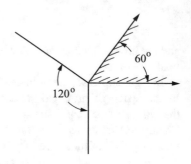

formulas (1.4), we conclude that the plastic strain in the considered assumption represents a pure shift in the plane $\xi O \eta$; the magnitude of this shift in the case of active plastic strain is unambiguously defined by a respective value of tangential stress until $\sigma_\xi > \sigma_\zeta > \sigma_\eta$. As soon as this inequation is violated, another pair from stresses σ_ξ, σ_η, σ_ζ. must be taken as the maximum and minimal principal stresses. Formulas (18.36) will be applied for other indexes, and they can be integrated once again.

For example, assume that additional loading brings us to the loading surface face $\sigma_\xi - \sigma_\eta = \pm 2k$. The integration of ratios in the form of (18.36) will give

$$e_\xi = h(\sigma_\xi - \sigma_\zeta) + e'_\xi,$$
$$e_\eta = e'_\eta,$$
$$e_\zeta = -h(\sigma_\xi - \sigma_\zeta).$$

In these formulas, dashes indicate plastic strains (18.37) acquired by the body during the time when the loading point is on the face $\sigma_\xi - \sigma_\eta = \pm 2k$.

Let us consider the case when the loading point remains on the loading surface rib. Let us assume for example that $\sigma_\xi = \sigma_\eta > \sigma_\zeta$. Then two conditions are fulfilled simultaneously: $\sigma_\xi - \sigma_\zeta = \pm 2k$ and $\sigma_\eta - \sigma_\zeta = \pm 2k$. Figure 18.4 shows two adjacent faces of a prism in the vicinity of the rib sectioned by an octahedral plane. Normal lines to the prism faces form an angle within which possible directions of the plastic strain vector gain are found.

The components of the plastic strain rate whose vector is directed along the normal line to the plane $\sigma_\xi - \sigma_\zeta = +2k$ will be

$$\dot{e}_\xi = H_1(\dot{\sigma}_\xi - \dot{\sigma}_\zeta), \quad \dot{e}_\eta = 0, \quad \dot{e}_\zeta = -H_1(\dot{\sigma}_\xi - \dot{\sigma}_\zeta).$$

In a similar way, the components of the plastic strain rate normal to the second face $\sigma_\eta - \sigma_\zeta = +2k$ will be recorded as follows:

$$\dot{e}_\xi = 0, \quad \dot{e}_\eta = H_2(\dot{\sigma}_\eta - \dot{\sigma}_\zeta), \quad \dot{e}_\zeta = -H_2(\dot{\sigma}_\eta - \dot{\sigma}_\zeta).$$

By adding the respective components, we obtain plastic strain components when positioning the loading point on the prism rib

$$
\begin{aligned}
\dot{e}_\xi &= H_1(\dot{\sigma}_\xi - \dot{\sigma}_\zeta), \\
\dot{e}_\eta &= H_2(\dot{\sigma}_\xi - \dot{\sigma}_\zeta), \\
\dot{e}_\zeta &= -(H_1 + H_2)(\dot{\sigma}_\xi - \dot{\sigma}_\zeta).
\end{aligned}
\tag{18.38}
$$

The value $(H_1 + H_2)$ must be considered as the function of difference $\sigma_\xi - \sigma_\zeta$.

By assuming that plastic strain does not depend on the spherical (hydrostatic) part of the stress tensor, let us apply confining pressure $-\sigma_\xi$ to the body. Then stresses along the axes ξ and η are zeroed, and the confining pressure $-(\sigma_\xi - \sigma_\zeta)$ remains in the direction of the third axis ζ, so we actually have uniaxial compression. Plastic strain in the direction of the axis ζ will be unambiguously defined by this stress. Consequently, as a result of integrating the last of the yield Eqs. (18.38), we must obtain a solution in the form of $e_\zeta^? = -h(\sigma_\xi - \sigma_\zeta)$. Therefore the final result of integration (18.38) can be written as

$$
\begin{aligned}
e_\zeta &= -h(\sigma_\xi - \sigma_\zeta), \\
e_\xi &= \lambda h(\sigma_\xi - \sigma_\zeta), \\
e_\eta &= (1 - \lambda)h(\sigma_\xi - \sigma_\zeta).
\end{aligned}
\tag{18.39}
$$

The parameter λ in the solution (18.39) remains indefinite. Its value lies in the fact the condition $\sigma_\xi = \sigma_\eta$ limits the choice of possible stressed states. To satisfy the strain conformity equations with these restrictions, strains must have a specific freedom. This uncertainty of strain in a singular point of the loading surface results in physically unacceptable consequences. For example, in the case of uniaxial elongation or compression, transverse strains can be whatsoever if the material volume remains constant. This and similar consequences of the strain uncertainty allow looking at the yield theory with a piecewise linear surface of loading as an approximation of a physically more real smooth loading surface. However, in many cases, the calculation results using such piecewise linear approximation give an acceptable error.

The general theory of plastic yield based on the arbitrary linear approximation of the loading surface was developed by Hodge [2]. An original variant of the yield theory with non-associated yield laws is proposed by V. M. Marchenko [6]. One of the recent variants of the plastic yield theory is the paper by V. G. Zubchaninov [13].

References

1. A. Gorshkov, L. Rabinskii, D.V. Tarlakovskii, *Osnovy tenzornogo analiza i mekhanika sploshnoi sredy : uchebnik dlya vuzov* [Fundamentals of tensor analysis and continuum mechanics: textbook for universities] (Nauka Publ., Moscow, 2000)

2. P. Hodge, J.N. Goodier, *Elasticity and Plasticity. The Mathematical Theory of Plasticity* (Wiley, New York, 1958)
3. Y. Kadashevich, S. Pomytkin, Ehndokhronnaya teoriya plastichnosti, obobshchayushchaya teoriyu Sandersa-Klyushnikova [Endochronic theory of plasticity generalizing the theory of Sanders-Klyuchnikova]. Inzh.-stroit. zhurnal [Eng. - builds. journal] **1**, 82–86 (2013)
4. M. Krivosheina, I. Konysheva, M. Kozlova, Razrushenie i uprugo-plasticheskoe deformirovanie anizotropnykh materialov pri dinamicheskom nagruzhenii [Destruction and elastic-plastic deformation of anisotropic materials under dynamic loading]. Mekhan. kompozits. mater. i konstr. [Mehan. the entire effort. Matera. and constr.] **12**(4), 502–513 (2006)
5. A. Kurkin, B. G.P., Metodika modelirovaniya znakoperemennoi uprugo-plasticheskoi deformatsii izotropnogo materiala [Method of modeling alternating elastic-plastic deformation of an isotropic material]', *Zavodskaya laboratoriya. Diagnostika materialov* [Plant laboratory. Diagnostics of materials] (2008)
6. V. Marchenko, Teoriya plasticheskogo techeniya s orientirovannoi poverkhnost'yu nagruzheniya [Plastic flow theory with oriented loading surface]. Uchenye zapiski TSAGI [Scientific notes of the Central Aerohydrodynamic Institute] **7**(5), 98–107 (1976)
7. V. Molotnikov, *Mekhanika konstruktsii* [Mechanics of structures] (Lan' Publ., SPb., Moscow, Krasnodar, 2012)
8. V. Novozhilov, Y. Kadashevich, *Mikronapryazheniya v konstruktsionnykh materialakh* [Microstresses in structural materials] (Mashinostroenie Publ., Leningrad, 1990)
9. Y. Rabotnov (ed.), *Teoriya plastichnosti: sb. perev. inostr. statei* [Plasticity Theory: Foreign Translation Collection. articles] (IL Publ., Moscow, 1948)
10. A. Reuss, Vereintachte berechnung der plastischen foamänderung in der plastizitetstheorie. Zeitsch. angew. Math. und Mech. Bd. 10. H. 2 (1930), pp. 266–274
11. R. Shield, H. Ziegler, On Prager's hardening rule. Zeitsch. ang. Math. Phys. **9a**, 260–276 (1958)
12. H. Ziegler, A modification of Prager's hardening rule. Quart. Appl. **17**(1), 55–65 (1959)
13. V. Zubchaninov, *Problemy prochnosti i plastichnosti* [Problems of strength and plasticity], Moscow, Izd-vo AN SSSR Publ., chapter *Modifitsirovannaya teoriya techeniya i matematicheskie modeli protsessov plasticheskogo deformirovaniya* [A modified theory of yield and mathematical models of the processes of plastic deformation], izz. 71 (2009), pp. 5–19

Chapter 19
Other Variants of Plasticity Theories

19.1 Batdorf–Budiansky Slip Theory

By the mid-1920s, an idea was developed to account for the physical properties of a real body when analyzing the mechanics of non-elastic strain. The first effort to associate the properties of a poly-crystalline aggregation with the properties of its component grains was undertaken in 1938 by Taylor [29]. The study was based on the hypothesis that arbitrary strain not accompanied by changes in volume can be represented as a result of shifts in five planes and directions. Taylor's idea was developed in the paper by Bishop and Hill [5] and later also used by Lin [18]. However, chronologically it is deemed that the completed theory was formulated by Batdorf and Budiansky [2] . The proposed mechanical model of the plastic strain of metals was called the *slip theory by the authors*. The authors used clearly defined facts:

– a real metal represents an aggregate of disorderly oriented crystalline grains;
– the plastic strain of a single grain is caused by a shift in a specific crystallographic plane and in a specific direction (highest density of atom packing).

A normal line to the slip plane with a single vector n and a single vector β defining the slip direction in this plane forms *the slip system*. If the tangential stress $\tau_{n\beta}$ in the plane n in the direction β exceeds the yield stress, the crystal undergoes plastic strain of pure shift $\gamma_{n\beta}$. It is assumed that the value of this strain is a definite function of stress $\tau_{n\beta}$.

Real crystals have several systems of possible slips where the plane and direction of the highest density of atomic packing are similar. For example, there are 12 systems in crystals with a cubic face-centered grating (Fig. 19.1). The figure highlights one of the four planes of the highest density of atom arrangement. Atoms belonging to the highlighted plane are marked by light highlighting. Arrows show directions with the shortest distances between atoms.

Fig. 19.1 Slip systems in a
crystal

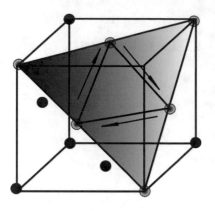

Usually, slip takes place in several systems simultaneously, and shear in each system of slip has a hardening effect on other slip systems. There is no qualitative description (or at least assessment) of these effects of mutual hardening. However, this gap hardly has any value for describing slip phenomena. The thing is that atomic planes are not displaced relative to each other. Such displacements require extremely high stresses. In fact, dislocations are displaced. When one dislocation outcrops to the crystal surface, two parts of it are shifted relative to each other by the value of the Burgers vector.

Thus, the plastic strain mechanism accompanied by hardening appears to be rather complicated. Therefore, Batdorf and Budiansky adopted the simplest scheme. It is supposed that for each grain, there is only one slip system where slip occurs when the tangential stress component reaches the yield stress. A macroscopic effect of plastic strain for a body in general is expressed when slips occur in the chain of plastically deformed grains. If the number of such grains in the considered volume is high, they include a sufficient number of grains for which the normal line to the virtual slip plane will be within the cone with the axis n and the solid angle $d\Omega$ (Fig. 19.2).

Due to the random orientation of grains in the initial condition, the material is deemed to be isotropic. Therefore, the volume of grains having the slip system $n\beta$ is taken as proportional $d\Omega d\beta$. Plastic strain from shifts in the slip system $n\beta$ is hypothetically represented as follows:

$$d\gamma_{n\beta}^{p} = F(\tau_{n\beta})d\Omega d\beta.$$

The plastic strain of the body in general represents a result of imposing an infinitely large number of pure shifts for various slip systems $n\beta$. These shifts must be summed as tensor components. To calculate this sum, let us go over to the components of the strain tensor relative to fixed axes x_1, y_1, z_1 using formulas (13.3)–(13.4) for converting a second-rank tensor. By taking the directions n and β as directions 1 and 2 of the new coordinate system, we must assume all the

Fig. 19.2
Batdorf–Budiansky model

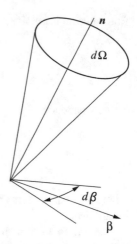

components of the tensor e_{ij} equal zero, except for $e_{12} = \dfrac{1}{2}\gamma_{n\beta}$. We obtain

$$de_{ij}^p = (\beta_{1i}\beta_{2j} + \beta_{2i}\beta_{1j})de_{12}.$$

Having in mind that $\beta_{1i} = n_i$, $\beta_{2i} = \beta_i$, we can record as follows:

$$de_{ij}^p = \frac{1}{2}(n_i\beta_j + n_j\beta_i)F(\tau_{n\beta})d\Omega d\beta.$$

By integration for plastic strain components, we obtain the following formulas:

$$e_{ij}^p = \frac{1}{2}\int_\Omega d\Omega \int_{-\pi/2}^{+\pi/2}(n_i\beta_j + n_j\beta_i)F(\tau_{n\beta})d\beta, \qquad (19.1)$$

where tangential stress is calculated using the formula

$$\tau_{n\beta} = \sigma_{ij}n_i\beta_j. \qquad (19.2)$$

The primary difficulty in practical calculation of strains using formulas (19.1) is that the function $F(\tau_{n\beta})$ differs from zero only where $\tau_{n\beta} > \tau_s$, whereas τ_s is the shear yield stress. Therefore, integrals in formulas (19.1) do not fall upon the entire sphere of the singular radius but only some part of it. An open issue is also the form of the function $F(\tau)$. By considering the uniaxial elongation problem, the paper by Batdorf and Budiansky assumes $\tau_{n\beta} = \sigma_{11}n_1\beta_1$, and the authors obtain the following condition to define the integration area:

$$n_1\beta_1 > \frac{\tau_s}{\sigma_{11}}.$$

The solution is obtained when setting $F(\tau)$ in the form of the following row:

$$F(\tau) = \sum_1^N a_n \left(\frac{\tau}{\tau_s} - 1 \right)^n.$$

The dependency of plastic strain on the elongating stress σ_{11} is obtained as

$$e_{11}^p = \sum_1^N a_n g_n \left(\frac{\sigma_{11}}{2\tau_s} \right),$$

where the parameters g_n are found (for $N = 5$) by time-consuming numerical integration.

According to slip theory, plastic strain starts developing when the tangential stress in any of the slip systems reaches the yield stress. It follows that the initial loading surface complies with the Tresca condition of the maximum tangential stress. Indeed, if $\tau_{max} = \tau_s$, there will always be a group of crystalline grains for which this stress will be tangential in the slip system. For further loading surfaces, the loading point in the Batdorf–Budiansky model will be conic.

As shown in the example of uniaxial elongation, the studies based on formulas (19.1) are extremely complicated even at proportional loading. For a case of complicated loading of Cicala [7], analytical solutions were achieved for the case when a thin-wall tube specimen was elongated beyond the yield stress and then twisted. Rather complicated calculations related to the solution of this task based on the modified slip model are given in detail in Chap. 27. To avoid repetition, we will not reproduce them here and will give only the final results.

Since only one stress component (σ) differs from zero in the first link of the loading trajectory in this problem and only two components (σ and τ) differ from zero in the second link, the process can be imaged in a plane with the coordinates σ and τ (Fig. 19.3). The initial loading surface in the considered two-dimensional cases represents an ellipsis

$$\sigma^2 + 4\tau^2 = const.$$

Fig. 19.3 To the Cicala problem

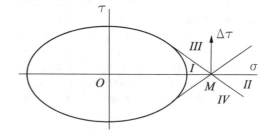

The following loading path is considered: along the X axis from the point O to the point M and then along a beam going from the point M. As a result, it is found that tangential lines to the initial ellipsis drawn from the point M divide the plane $\sigma \sim \tau$ into 4 areas marked by Roman numbers in Fig. 19.3. If stress gains are such that the additional loading trajectory is located in the area I, there is elastic loading. If the additional loading vector is located in the area II, gains of plastic strains are defined by formulas of the strain theory of plasticity. In the case when the additional loading path is located in the area III or IV, the ratios between stresses and strains are obtained from formulas (19.1) in an extremely complicated form. The final result is obtained by the author only for the case of orthogonal additional loading shown in Fig. 19.3 as the vector $\Delta\tau$. For this case,

$$\frac{\Delta\tau}{\Delta\gamma} = G^* = \frac{G}{1 + \frac{3}{2}G\left(\frac{1}{E_s} - \frac{1}{E_t}\right)}. \tag{19.3}$$

The slip theory represented by the Batdorf–Budiansky model was not proved experimentally, which the authors later admitted [3]. Despite the model predicts some quality effects, their quantitative parameters were not rather satisfactory. In Chap. 27 we show that this is not a reason to abandon slip theory in general.

19.2 Two-Dimensional Klyushnikov Model

As said above, the calculation of strain based on the Batdorf–Budiansky slip theory is related to serious mathematical challenges so it is required to use approximated calculations. At the same time, the hypotheses of the Batdorf–Budiansky model are not that reliable to search for precise solutions. To simplify the primary dependencies of slip theory, V.D. Klyushnikov proposed [16], [17] a model of a two-dimensional medium for which $e_{33} = 0$, $\sigma_{33} = \frac{1}{2}(\sigma_{11} + \sigma_{22})$. Then the stress deviator components will be

$$\sigma'_{11} = -\sigma'_{22} = \frac{1}{2}(\sigma_{11} - \sigma_{22}).$$

According to the non-compressibility condition $e^p_{11} + e^p_{22} = 0$, therefore we can write

$$e^p_{11} = -e^p_{22} = \frac{1}{2}(e^p_{11} - e^p_{22}).$$

It is deemed that the material may deform only by shear in planes perpendicular to the plane $x_1 x_2$. In some plane among these planes that make the angle ω with the axis x_1, the tangential stress will be

$$\tau = \frac{1}{2}(\sigma_{11} - \sigma_{22}) \sin 2\omega + \sigma_{12} \cos 2\omega. \tag{19.4}$$

In elements where $\tau > \tau_s$, plastic shift will occur

$$d\gamma = F(\tau)d\omega,$$

causing plastic strain whose components in the axes $x_1 x_2$ will be

$$de_{11}^p = -de_{22}^p = \frac{1}{2}(de_{11}^p - de_{22}^p) = \frac{1}{2}d\gamma \sin 2\omega = \frac{1}{2}F(\tau) \sin 2\omega d\omega,$$
$$de_{12}^p = \frac{1}{2}d\gamma \cos 2\omega = \frac{1}{2}F(\tau) \cos 2\omega d\omega.$$

We will use the two-dimensional space of stresses assuming the following values as coordinates

$$Q_1 = \frac{1}{2}(\sigma_{11} - \sigma_{212}), \quad Q_2 = \sigma_{12}.$$

We will match these stresses to the following coordinates in a two-dimensional space of strains with unit axes i_1, i_2

$$q_1 = \frac{1}{2}(e_{11}^p - e_{22}^p), \quad q_2 = e_{12}^p.$$

In the adopted designations, the initial loading surface is shown by the circumference $Q_1^2 + Q_2^2 = const$, and plastic strains are found using the formulas

$$q_1 = \frac{1}{2}\int F(\tau) \sin 2\omega d\omega,$$
$$q_2 = \frac{1}{2}\int F(\tau) \cos 2\omega d\omega, \tag{19.5}$$

whereas

$$\tau = Q_1 \sin 2\omega + Q_2 \cos 2\omega. \tag{19.6}$$

Assume that additional loading is done at some stressed state. The components of plastic strain will get gains

$$\delta q_1 = \frac{1}{2}\int F'(\tau)\delta\tau \sin 2\omega d\omega,$$
$$\delta q_2 = \frac{1}{2}\int F'(\tau)\delta\tau \cos 2\omega d\omega. \tag{19.7}$$

Fig. 19.4 Proportional
loading according to V. D.
Klyushnikov

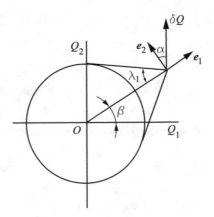

We shall have in mind that in the case of additional loading, integration limits can
also change.

Let us apply the ratios (19.7) to the analysis of proportional loading. In this case,
the loading trajectory is the beam going from the reference point of the plane $Q_1 Q_2$.
Let us designate β as the angle that the loading beams make with the axis Q_1. Then
we can record as follows:

$$Q_1 = Q \cos \beta, \quad Q_2 = Q \sin \beta, \quad \text{and consequently } \tau = Q \sin(2\omega + \beta).$$

Instead of decomposing the vector q by unit axes i_1, i_2, let us introduce the basis
e_1, e_2 related to the trajectory of proportional loading (Fig. 19.4).

Equations (19.5) can be written as follows in the vector form:

$$\mathbf{q} = \frac{1}{2} \int F(\tau)[\mathbf{i}_1 \sin 2\omega + \mathbf{i}_2 \cos 2\omega]d\omega.$$

Let us use the formulas for coordinate conversion

$$\mathbf{i}_1 = \mathbf{e}_1 \cos \beta - \mathbf{e}_2 \sin \beta,$$
$$\mathbf{i}_2 = \mathbf{e}_1 \sin \beta + \mathbf{e}_2 \cos \beta.$$

We obtain

$$\mathbf{i}_1 \sin 2\omega + \mathbf{i}_2 \cos 2\omega$$
$$= \mathbf{e}_1 \sin(2\omega + \beta) + \mathbf{e}_2 \cos(2\omega + \beta).$$

Let us substitute the last expression into the formula expressing the vector q and
make substitution with a variable in the sub-integral expression using the formula

$$2\omega + \beta = \frac{\pi}{2} + \lambda.$$

We have

$$\mathbf{q} = \frac{1}{4} \int F(\tau)(\mathbf{e}_1 \cos \lambda - \mathbf{e}_2 \sin \lambda) d\lambda, \tag{19.8}$$

where $\tau = Q \cos \lambda$ and integration falls upon those values λ for which $\tau > \tau_s$, e.g.

$$\cos \lambda > \frac{\tau_s}{Q}.$$

If we designate $\cos \lambda_1 = \tau_s / Q$, the integration area will be

$$-\lambda_1 \leqslant \lambda \leqslant \lambda_1.$$

Assume that after proportional loading, additional loading $\delta \mathbf{Q}$ is done in an arbitrary direction making the angle α with the unit axis $\mathbf{e}_2$ (Fig. 19.4). By designating the additional loading vector modulus as ε for shortness, we have as follows for its components:

$$\Delta Q_1 = \varepsilon \sin(\alpha - \beta), \quad \Delta Q_2 = \varepsilon \cos(\alpha - \beta).$$

To gain the tangential stress component, we obtain

$$\Delta \tau = \varepsilon \cos(\alpha - \beta - \omega) = \varepsilon \sin(\alpha - \lambda).$$

For the gain of the plastic strain vector, we have

$$\Delta \mathbf{q} = \frac{1}{4} \varepsilon \int F'(\tau) \sin(\alpha - \lambda)[\mathbf{e}_1 \cos \lambda - \mathbf{e}_2 \sin \lambda] d\lambda. \tag{19.9}$$

The simple formula obtained for finding the plastic strain vector gain bears the same challenges as in the more general Batdorf–Budiansky model: in the case of low additional loading, plastic strain is obtained for those elements only where $\Delta \tau > 0$; elastic unloading will occur in elements for which $\Delta \tau < 0$. As per the graphical picture of the process given in Fig. 19.4, the following variants are possible.

1. $\alpha \leqslant -\lambda_1$. In this case $\alpha - \lambda < 0 \ \forall \lambda \in (-\lambda_1; \lambda_1)$, $\Delta \tau < 0$ and elastic unloading takes place. The straight lines $\alpha = \pm \lambda_1$ define the angle inside which the additional loading vector causes elastic unloading in all elements. It is easy to believe that these straight lines touch the initial circumference (see Fig. 19.4).
2. $\alpha \geqslant \lambda_1$; then, $\alpha - \lambda > 0 \ \forall \lambda \in (-\lambda_1; \lambda_1)$ and everywhere $\Delta \tau > 0$—additional loading takes place in all elements. The ratio (19.7) is obtained from formula (19.5) where integration is replaced by varying.

3. $\alpha \in [-\lambda_1; \lambda_1]$. In this case $\Delta\tau > 0$ $\forall\lambda \in (-\alpha; \alpha)$ and integration when calculating the expression (19.9) is done within $-\alpha \leqslant \lambda \leqslant \alpha$.

Thus, in the Batdorf–Budiansky theory and for the simplest two-dimensional model, the problem of integrating plasticity equations for any loading part is so complex that it prevents one from making any qualitative conclusions of the nature of loading surface changes in the case of complex loading trajectories. This conclusion was a source of pessimistic moods of Yu.N. Rabotnov [25] as to the possible progress of plasticity theory. In Part III, we tried to show what can be opposed to these pessimistic conclusions.

19.3 Endochronic Plasticity Theory

Almost in all variants of plasticity theory that we discussed in previous paragraphs of Part II, the concept of a loading surface is used. It is a priori deemed that the loading surface can be defined experimentally. To do it, we must be able to clearly fix the moment of plastic strain, which is impossible in principle. Apparently, any of the existing plasticity theories can be declared unsound as contradicting the data of thin experiments. Therefore, a doubt occurred: "... does the loading surface concept have any real sense and does it have to be used as a basis when building the plasticity theory?" [26, p. 564].

Understanding of justification of this doubt resulted in efforts to build the plasticity theory that is not based on the loading surface concept but directly expresses components of the stress tensor as some functions on loading trajectories. At the brink of the 1950s, A. A. Ilyushin [9, 10] created the theory of the elastic–plastic process of strain of continuous media in the case of complex loading that was later developed in the scientific school created by him (V. G. Zubchaninov [35], R. A. Vasin [34] et al). The same type of theories includes the above-mentioned (p.157) *endochronic theory of plasticity*, which was proposed in 1971 by K. Valanis [32, 33] and is intensively developing today. The evolution of the endochronic theory of plasticity is described in the article by Yu. I.. Kadashevich, and S. P. Pomytkin [14] with a sufficient degree of detail as well as in the book by P. V. Trusov and I. E. Keller [30].

The theory is based on the concept of the so-called internal time. Apparently, the theory name is related to this concept: endo—internal, chronos—time (*Greek*). The determinant functions of the Valanis theory have a comparatively simple structure of hereditary type whose form does not differ from the functions of linear visco-elasticity with the substitution of physical time with the so-called internal time. In the initial variant, the author tried to use the following type of the determinant ratio:

$$\sigma'_{ij} = \int_0^{\mu} L(\mu - \mu')d\varepsilon'_{ij}(\mu'), \tag{19.10}$$

where σ'_{ij} and ε'_{ij} are the deviators of stress and strain tensors and μ is the internal time defined by the ratio

$$d\mu = \frac{d\lambda}{f(\lambda)}, \quad d\lambda = |d\varepsilon'_{ij}|, \quad f(\lambda) > 0, \tag{19.11}$$

where $f(\lambda)$ is some positive function called *the hardening function*. The material is hardening if $df/d\lambda > 0$ and softening if $df/d\lambda < 0$. The requirement of fading memory implies specific requirements to the class of functions that L belongs to, and there must be $dL(\mu)/d\mu < 0$. In particular, this requirement is met by the function

$$L(\mu) = \sum_{i=1}^{N} E_i e^{-\alpha_i \mu}. \tag{19.12}$$

Here E_i, α_i are constant values of the material defined from experiments for complex loading.

The analysis of the ratios (19.10)–(19.12) leads to the conclusion that sources of Valanis's approach are found in the papers by A. A Ilyushin (see, for example, [8]), whereof we already spoke earlier (Chap. 12). We shall also note that before Valanis, in 1969, A. A. Vakulenko [31] introduced the concept of thermodynamic time, which allowed efficiently studying the process of building determinant ratios of non-elastic strains.

Despite the obvious simplicity of the determinant ratios of the Valanis theory, they allowed for a qualitative description of many interesting effects observed in deformation beyond the yield stress. In particular, the model (19.10)–(19.12) qualitatively describes the effects of linear and non-linear hardening, hysteresis and stabilization hysteresis loops in cyclic deformation, "a dive" into stress intensity in the vicinity of the fracture point of the strain trajectory, and some others. Along with that, the quantitative correspondence of experimental data and calculation results in most cases cannot be deemed satisfactory.

A reason for such non-compliance is probably a tempting re-simplification of the model. Striving for simplicity usually turns into the loss of a link with the model. An excessive complication of the model leads to computational and sometimes principal complications. A reasonable compromise here, as advised by R. Bellman, is that "...a scientist, like a pilgrim, must take a straight and narrow path between the Traps of Re-simplification and Swamp of Over-complication" [4].

Critics of the initial variant of the Valanis endochronic theory in multiple publications induced the author and his adherents to create "improved or corrected variants" of the theory. They tried to correct the theory [12, 21, 33] by complicating the type of functionality. However, the implementation of such an approach resulted in the "Swamp of Over-complication" as said above.

The first Valanis modification is done by changing the internal time measure while keeping the initial structure of equations. Ratios for a new measure of internal time were not represented by formulas

$$d\mu = \frac{d\xi}{f(\xi)}, \quad d\xi = \left| d\varepsilon'_{ij} - \frac{1-\alpha}{2G} d\sigma'_{ij} \right|, \quad (19.13)$$

where G is the shift modulus and α is the small parameter of endochronic behavior.

Using a new measure of internal time (19.13) allowed for the partial elimination of disadvantages suffered by a virgin variant of the theory. However, such serious flaws of the initial formulation of the endochronic theory as a violation of the Drucker postulate remained. Furthermore, cyclic rheologic effects not typical of the theory of plastic media were found. The latter effects disappear [21] at $\chi = 1$; however, there is a singularity in ratios and the yield surface occurs. This deprives [16] the endochronic theory of its primary advantage declared at its creation.

Another variant of modifying the endochronic theory of plasticity is proposed by Yu. I. Kadashevich and A. N. Mikhaylov [11]. The cited paper proposes a so-called tensor-parametric representation of determinant functionality

$$\mathbf{S} = \int_0^z L_1(z - z')d\mathbf{R}(z'), \quad dz = \frac{d R}{f(R)}, \quad dR = |d\mathbf{R}|,$$
$$\mathbf{Э} = \int_0^z L_2(z - z')d\mathbf{R}(z'), \quad (19.14)$$

where $\mathbf{R}$ is the auxiliary vector whose form is not specified beforehand; it is believed that this vector characterizes the effect of micro-strains and micro-stresses. Introducing two functionalities defined on various classes of functions L1 and L2 instead of one in Valanis's initial variant allows for a substantial expansion of the opportunities of the theory, but it makes it more complicated. For practical calculations, it is suggested to use a simplified variant of these modifications recorded in a differential form

$$\mathbf{S} + a_1(z)\frac{d\mathbf{S}}{dz} = b_1(z)\mathbf{R} + c_1\frac{d\mathbf{R}}{dz},$$
$$\mathbf{Э} + a_2(z)\frac{d\mathbf{Э}}{dz} = b_2(z)\mathbf{R} + c_2\frac{d\mathbf{R}}{dz}, \quad (19.15)$$

where the functions $a_i(z)$, $b_i(z)$, $c_i(z)$, $(i = 1, 2)$ are defined from experiments. The issue of defining the auxiliary vector $\mathbf{R}$ remains open. In some works, it is suggested to use a vector of plastic strain for this purpose.

By the 1990s, three trends appeared in the development of endochronic theory. Failing to withstand criticism, many foreign researchers abandoned the primary variant (19.10)–(19.12) of the theory and switched to the modification (19.13) believing that $\alpha = 0$ in a limit case and paying no attention that in that limit case they actually repeated the results of A. A. Vakulenko of 1969 [31]

$$\sigma'_{ij} = \tau_0 \frac{d\varepsilon^p_{ij}}{d\mu} + \int_0^\mu L(\mu - \mu') \frac{d\varepsilon'_{ij}}{d\mu'} d\mu',$$

$$\frac{d\mu}{d\lambda} = \frac{1}{m(\lambda, \dot{\lambda})}, \quad d\lambda = \sqrt{d\varepsilon^p_{ij} d\varepsilon^p_{ij}}. \tag{19.16}$$

The second approach is described in the paper by Yu. I. Kadashevich and S. P. Pomytkin [13]. It was proposed to preserve the general variant of endochronic theory but record it in a differential–parametric form paying special attention to a limit case when $\alpha \to 0$. The authors propose determinant ratios in the following form:

$$\frac{1}{2G} \left[\sigma'_{ij} + \alpha\tau \frac{d\sigma'_{ij}}{dR} \right] = \tau \frac{dR'_{ij}}{dR} + \frac{R'_{ij}}{g + \alpha}. \tag{19.17}$$

$$R'_{ij} = \varepsilon'_{ij} - \frac{1 - \alpha}{2G} \sigma'_{ij}, \quad dR = \sqrt{dR'_{ij} dR'_{ij}},$$

$$\varepsilon_{ii} = \frac{\sigma_{ii}}{K}, \quad 0 \leqslant \alpha \leqslant 1, \tag{19.18}$$

$$|\dot{R}| = \sqrt{\frac{dR}{dt} : \frac{dR}{dt}}, \quad \tau = \tau(|R|, |\dot{R}|).$$

τ is the equivalent of the shear yield stress, g is the equivalent of the hardening coefficient, and K is the volumetric compression modulus.

The third approach was developed by V. S. Sarbayev [27]. It is suggested to use the theory of D. Backhouse instead of Valanis's functionality [1]

$$\sigma'_{ij} = \tau(\lambda) \frac{d\varepsilon^?_{ij}}{d\lambda} + \int_0^\lambda L(\lambda, \lambda - \lambda') \frac{d\varepsilon^?_{ij}}{d\lambda'} d\lambda', \tag{19.19}$$

provided that the yield stress $\tau = 0$.

The analysis of the variants of theories built on the ratios (19.16) and (19.19) has shown [14] their identity. For this reason, these ratios are proposed to be called the Vakulenko–Backhouse theory.

In the recent decade, efforts have been made to use the endochronic approach to describe processes of high strains accompanied by volumetric changes.

19.4   On the Methods of Physical Mesomechanics and Synergetics

By the middle of the twentieth century, phenomenologic models of plasticity built at the macroscopic level of the continuum mechanism required expansion of the physical base of the phenomenon and its use in building the theory of no-elastic strains. Electronic microscopy became the breakthrough that physicists made in the microcosm of a deformable solid body. During the next half of the century, plasticity and strength physics was boosted due to studies of occurrence, motion, and changes of the primary type of structural imperfections—dislocations. It was hoped that reliable data on structural faults and their migration in a loaded body will allow making the physical theory of non-elastic deformations at the macro-level. However, soon this period of hopes turned into the period of disappointments: efforts to calculate the "stress–strain" curve failed even for uniaxial compression.

There was a need to search for a new, unconventional approach. One such approach was proposed by V.E. Panin [22, 23], which was later developed into a new domain of solid state physics—physical mesomechanics. An intermediate scale is introduced between micro- and macro-levels—mesoscopic. As a result, the distribution of phenomena accompanying the body loading process looks as follows:

– micro-level: local changes in the crystalline lattice are manifested as the generation of dislocation cores and their displacement in the field of stress gradient;
– meso-level: generation and motion of local zones of non-elastic deformations as stripes of a mesoscopic scale level within individual aggregates of the internal structure; it has been experimentally found that linear displacements of these aggregates are accompanied by their rotation;
– macro-level: generation of a single mains macro-stripe, two parallel macro-stripes as a dipole or two conjugate macro-stripes ending in destruction division of the body into parts.

Local zones of non-elastic strain become concentrators of stresses of various scales and areas of shift instability, and their motion to an equilibrium state is considered as a *synergetic* process. The following [23] synergetic *principles* of this process are formulated.[1]

Principle 1 A shift in a loaded solid body is related to the local loss of shift stability and can be done at the micro-, meso-, and macro-scale levels as a local change in the initial internal structure.

Principle 2 A shift on any scale level can be generated only in the local zone of the stress concentrator of respective scale since in general the structure of a loaded solid body preserves its shift stability under the action of the mean applied loading.

Principle 3 The free surface of the body has the least shift stability in the loaded solid body.

Principle 4 A shift in a continuous medium with a stringent material rotation generates a zone of flexure–torsion in its way, which is a new stress concentrator.

Principle 5 A shift as a relaxation process in a limited elastic–plastic environment with specified boundary conditions generates fading elastic and elastic–plastic self-oscillations.

---

[1] Some of these principles coincide in their essence with the axioms formulated in Chap. 17.

Principle 6 The plastic strain of a solid body develops under the following pattern: primary (basic) stress concentrator—relaxation shift with stringent rotation—further relaxation shift.

Principle 7 Self-organization of shifts in a deformable solid body reflects self-coupling of elastic–plastic rotation modes related to shifts and flexure–torsion zones: for a defined loading axis, the total turn and flexure–torsion in the hierarchy of shifts of all scale levels must equal zero (condition of uniformity preservation). A violation of this condition causes cracks as accommodation rotary modes of strain.

Principle 8 A global loss of shift stability and destruction occur at the place of the stress macro-concentrator and are defined by the development mechanics of macro-stripes of localized strain accommodated by the relaxation process at the meso- and micro-scale levels.

Using the postulated principles, the problem of the elastic–plastic strain of a solid body is included in the problem of self-organization that is the primary subject of synergetics or *non-linear dynamics*. The researcher inevitably comes to a nontrivial problem of building an adequate mathematical model of system behavior in the language of non-linear dynamics. The problems formulated based on such models are called *evolutionary*. The most common feature of evolutionary problems is their ability to describe the process not only in the smooth flow of events but also in the conditions with aggravation characterized by radical changes, the formation of new structures, and properties over rather short time intervals. They say that self-organization in the system takes place by means of passing through dynamic chaos, the disintegration of old structures, and formation of new ones.

The so-called basic equations of synergetics are known [15, 20, 24] and widely discussed in many publications. We will not mention evolution scenarios based on them, conditions of system stability loss, various conditions of going into chaos, and other interesting results. According to P.V. Makarov [19], we ask the question: what synergetics and general properties of basic equations of non-linear dynamics studied by it give us for researching the behavior of stress–strain state in solid bodies during the deformation?

The author proposes the following complete system of equations:

– equations expressing preservation laws:

$$\frac{d\rho}{dt} + \rho \operatorname{div} \mathbf{v} = 0, \quad \rho \frac{dv_i}{dt} = \frac{\partial \sigma_{ij}}{\partial x_j} + \rho F_i, \quad \frac{\partial E}{\partial t} = \frac{1}{\rho} \sigma_{ij} \frac{\partial \varepsilon_{ij}}{\partial t} - q_{i,j}, \quad (19.20)$$

– evolutionary equations of the first group:

$$\dot{\sigma}_{ij} = \lambda(\dot{\theta}^t - \dot{\theta}^P)\delta_{ij} + 2\mu(\dot{\varepsilon}^t_{ij} - \dot{\varepsilon}^P_{ij}),$$
$$\sigma_{ij} = -P\delta_{ij} + s^e_{ij} + s^v_{ij}; \quad -P = \frac{1}{3}\sigma_{ii}, \quad P = f(\rho, E), \quad (19.21)$$

– evolutionary equations of the second group:

$$\dot{\theta}^p = A \frac{\partial}{\partial x_i} B \frac{\partial}{\partial x_i} \theta^p + C(\theta),$$

$$12pt[]\dot{\varepsilon}_{ij}^p = F(\varepsilon_{eff}^p, \sigma_{eff}, S_{ij}, \cdots),$$

(19.22)

for example,

$$\dot{\varepsilon}_{ij}^p = \frac{3}{2} \frac{\varepsilon_{eff}^p}{\sigma_{eff}} S_{ij},$$

whereas

$$\dot{\varepsilon}_{ij}^t = \dot{\varepsilon}_{ij}^e + \dot{\varepsilon}_{ij}^p; \quad \dot{\theta}^t = \dot{\varepsilon}_{ii}^t;$$

$$\dot{\theta}^p = \dot{\varepsilon}_{ii}^p; \quad \dot{\varepsilon}_{ij}^t = \frac{1}{2} \left(\frac{\partial v_i}{\partial x_j} + \frac{\partial v_j}{\partial x_i} \right).$$

Here $\dot{\varepsilon}_{eff}^p$ and σ_{eff} are the second invariants of rates of plastic strains and stresses, respectively; $\mathbf{v}$ is the shift vector; ρ is the medium density; F_i are the defined functions of coordinates and time; λ and μ are Lame coefficients; P is the mean stress; s_{ij}^e and s_{ij}^v are the equilibrium and non-equilibrium parts of shift stress, respectively (in elastic condition $s_{ij}^v = o$); the upper indexes t, e, and p are full, elastic, and plastic strains, respectively; θ is the volumetric strain; E is the energy of strain; q is the dissipative function; A, B, and C are some functions depending on the selection of specific kinetics.

For strain rates from the second of formulas (19.21), the differentiation operation in time in the Jaumann sense is used [28] taking into account the rotation of the axes due to medium strain

$$\frac{Ds_{ij}^e}{Dt} = 2\mu \left(\dot{\varepsilon}_{ij}^e - \frac{1}{3} \dot{\varepsilon}_{kk}^e \delta_{ij} \right),$$

$$\frac{Ds_{ij}}{Dt} = \dot{s}_{ij} - s_{ik} \dot{\omega}_{ij} - s_{jk} \dot{\omega}_{ik},$$

$$\dot{\omega}_{ij} = \frac{1}{2} \left(\frac{\partial v_i}{\partial x_j} - \frac{\partial v_j}{\partial x_i} \right).$$

In a significantly simplified form, the model (19.20)–(19.22) is used [19] to evaluate the stress–strain state and stability of the coal formation roofing having a zone of outcropping. Figure 19.5 depicts a calculation picture of the development of the fracture system in the coal formation roofing under the action of gravity loads obtained by P. V. Makarov at the SB RAS Institute of Strength and Material Strength (Tomsk) using the SKIF cluster Cyberia.

It should be noted that the results obtained by P. V. Makarov are only qualitatively similar to the real situation before the stability loss of rock outcrops in the coal

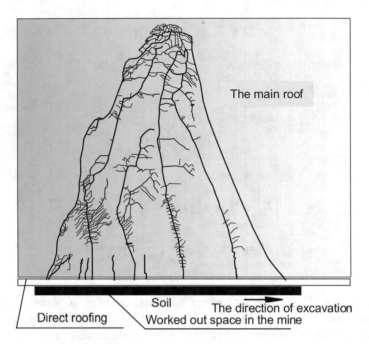

Fig. 19.5 Calculation pattern of fracture evolution in the vicinity of an outcropped coal formation

formation zone. If we mean real geological conditions of a deposit taking into account cleavage fractures and other geological faults, predictions in Fig. 19.5 cannot be adopted even as a rough approximation of reality.

There are also efforts [6] of using synergetic methods in the technology of improving the durability of non-ferrous alloys.

References

1. G. Backhaus, Zur analytischen darstellung des material verhalten in plastiaschen bereich. ZAMM. Bd. **51**, 471–477 (1971)
2. S. Batdorf, B. Budiansky, Splitting tests: an alternative to determine the dynamic tensile strength of ceramic materials, in *NACA TC1871* (1949), pp. 1–31
3. S. Batdorf, B. Budiansky, Polyaxial stress-strain relations of a strain-hardening metal. J. Appl. Mech. **21**(76), 323–326 (1954)
4. R. Bellman, *Dinamicheskoe programmirovanie* [Dynamic programming] (Inostrannaya literatura Publ., Moscow, 1960)
5. J. Bishop, and R. Hill, A theoretical derivation of the plastic properties of a polycrystalline face-centered metal. Philos. Mag. **42**, 1298–1307 (1951)
6. D. Borshchevskaya, povyshenie dolgovechnosti splava amg6m s pozitsii sinergeticheskogo podkhoda [Increasing the durability of the AMg6M alloy from the position of a synergistic approach]. Vestnik KHNADU [Bulletin of the Kharkov National Road University] **54**, 101–106 (2011)

7. P. Cicala, On the plastic deformation. Atti Accad. naz. Lincei. Rend. Cl. Sci. fis. e. nature. **8**, 583–586 (1950)

8. A. Il'yushin, O svyazi mezhdu napryazheniyami i malymi deformatsiyami v mekhanike sploshnykh sred [On the relationship between stresses and small deformations in continuum mechanics]. PMM [Appl. Math. Mech.] **18**(6), 641–666 (1954)

9. A. Il'yushin, *Plastichnost'. (Osnovy obshchei matematicheskoi teorii)* [Plasticity. (Fundamentals of general mathematical theory)] (Izd-vo AS USSR Publ., Moscow, 1963)

10. A. Il'yushin, *Trudy. T. 2 (1946–1966) / Sostaviteli: E.A. Il'yushina, N.R. Korotkina* [Scientific work. Vol. 2 (1946–1966). Compilers: E.A. Ilyushina, N.R. Korotkina] (Fizmatlit Publ., Moscow, 2004)

11. Y. Kadashevich, A. Mikhailov, *o teorii plastichnosti, ne imeyushchei poverkhnosti tekuchesti* [On the theory of plasticity without a flow surface]. DAN SSSR [Reports of the USSR Academy of Sciences] **254**(3), 574–576 (1980)

12. Y. Kadashevich, A. Mosolov, sovremennoe sostoyanie ehndokhronnoi teorii plastichnosti [Current state of endochronous plasticity theory]. Problemy prochnosti [Probl. Strength] **6**, 3–12 (1991)

13. Y. Kadashevich, S. Pomytkin, o vzaimosvyazi teorii plastichnosti, uchityvayushchei mikro-napryazheniya, s ehndokhronnoi teoriei plastichnosti [On the relationship of the theory of plasticity, taking into account micro-stresses, with the endochronous theory of plasticity]. Izv. RAN. MTT [Izv. Russian Academy of Sciences. MTT] **4**, 99–105 (1997)

14. Y. Kadashevich, S. Pomytkin, *Uprugost' i neuprugost' : sb. nauchn. tr.* [Elasticity and inelasticity: a collection of scientific papers], Moscow, Izd-vo MGU Publ., chapter *Ehtapy razvitiya ehndokhronnoi teorii neuprugosti* [Stages of development of the endochronous theory of inelasticity], *Uprugost' i neuprugost' : sb. nauchn. tr. [Elasticity and inelasticity: a collection of scientific papers]* (2011), pp. 232–235

15. G. Khaken, *Sinergetika. Ierarkhiya neustoichivostei v samoorganizuyushchikhsya sistemakh i ustroistvakh* [Synergetics. Hierarchy of instabilities in self-organizing systems and devices] (Mir Publ., Moscow, 1985)

16. V. Klushnikow, Defekty ehndokhronnoi teorii plastichnosti [Defects in the endochronous theory of plasticity]', Izv. AN SSSR. MTT [Izv. USSR ACADEMY OF SCIENCES. MTT] **1**, 176–179 (1989)

17. V. Klyushnikov, *Matematicheskaya teoriya plastichnosti [Mathematical theory of plasticity]* (MGU Publ., Moscow, 1979)

18. T. Lin, A proposed theory of plasticity based on slip, *in* 'Proceedings Second U.S. Nat. Congr. Appl. Mech. New York (1954), pp. 461–468'

19. P. Makarov, *matematicheskaya teoriya ehvolyutsii nagruzhaemykh tverdykh tel i sred* [mathematical theory of evolution of loaded solids and media]. *Fiz. mezomekh-ka* [Physical mesomechanics], issue 3, no. 11, pp. 19–35 (2008)

20. G. Malenetskii, A. Potapov, *Sovremennye problemy nelineinoi dinamiki* [Current problems of nonlinear dynamics] (Editorial URSS Publ., Moscow, 2002)

21. A. Mosolov, *Ehndokhronnaya teoriya plastichnosti* [Endochronic theory of plasticity], *Preprint In-t probl. mekh-ki AN SSSR* [Preprint In-t Probl. Mekh-ki of the USSR Academy of Sciences], no. 353 (1988)

22. V. Panin, *Strukturnye urovni deformatsii tverdykh tel* [Structural levels of deformation of solids]. *Izv. vuzov. Fizika* [University News. Physics], izz. 25. no. 6, pp. 5–27 (1982)

23. V. Panin, *Sinergeticheskie printsipy fizicheskoi mezomekhaniki* [Synergetic principles of physical mesomechanics]', *Fizich. mezomekh-ka* [Physical. masomeh-ka], izz. 6, no. 3, pp. 5–36 (2000)

24. I. Prigozhin, P. Glensdorf, *Termodinamicheskaya teoriya struktury, ustoichivosti i fluktuatsii* [Thermodynamic theory of structure, stability, and fluctuations] (Editorial URSS Publ., Moscow, 2003)

25. Y. Rabotnov, *Polzuchest' ehlementov konstruktsii* [Creep of structural elements] (Nauka Publ., Moscow, 1966)

26. Y. Rabotnov, *Mekhanika deformiruemogo tverdogo tela [Mechanics of a deformable solid]* (Nauka Publ., Moscow, 1988)
27. B. Sarbaev, *ob odnom variante teorii plastichnosti s translyatsionnym uprochneniem* [On a variant of the theory of plasticity with translational hardening]', *Izv. AN SSSR. MTT. [News of the USSR Academy of Sciences. Solid mechanics]*, no. 1, pp. 65–72 (1994)
28. L. Sedov, *Vvedenie v mekhaniku sploshnykh sred* [Introduction to Solid Media Mechanics] (Fizmatlit Publ., Moscow, 1962)
29. G. Taylor, *Stephen Timoshenko 60th Anniversary Volume, McMillan Co., New York, 1938, pp. 218–224.*, McMillan Co., New York, chapter Analysis of plastic strain in a cubic crystal, pp. 218–224 (1938)
30. P. Trusov, *Teoriya opredelyayushchikh sootnoshenii. Ch.2.* [The theory of defining relations. Ch.2]. Perm': Izd-vo Perm. gos. un-ta Publ. (1999)
31. A. Vakulenko, *k teorii neobratimykh protsessov* [On the theory of irreversible processes] *vestn. lgu [lsu bulletin]'*, *Vestn. LGU [LSU Bulletin]*, no. 7, pp. 84–90 (1969)
32. K. Valanis, A theory of viscoplasticity without a yield surfaces. Arch. Mech. Stosow. **23**(4), 517–551 (1971)
33. K. Valanis, '*Obosnovanie ehndokhronnoi teorii plastichnosti metodami mekhaniki sploshnoi sredy [Substantiation of the endochronous theory of plasticity by methods of continuum mechanics]*', Tr. ASME. Teoreticheskie osnovy inzhenernykh raschetov [Work ASME. Theoretical foundations of engineering calculations, **106**(4), 72–81] (1984)
34. R. Vasin, *Itogi nauki i tekhniki. Seriya "Mekhanika deformiruemogo tela"* [Results of science and technology. Series "The Mechanics of deformable bodies"], *McMillan Co., New York,* chapter *Opredelyayushchie sootnosheniya teorii plastichnosti [Defining relations in the theory of plasticity]*, pp. 3–75 (1990)
35. V. Zubchaninov, '*Gipoteza ortogonal'nosti i printsip gradiental'nosti v teorii plastichnosti [The hypothesis of orthogonality and the principle of gradientless in the theory of plasticity]*', Izv. RAN. MTT [Izvestiya RAS. Solid state mechanics], no. 5, pp. 68–73 (2008)

Part III
Development of the Slip Concept in Plasticity Theory

Chapter 20
Problem Setting

20.1 Initial Concepts and Definitions

For plastic materials, we will adopt that the lower yield stress must be associated with shift resistance, as shown later, and the yield stress will mean the maximum tangential stress before the occurrence of plastic strains (now we speak of macro-homogeneous strain). The occurrence of the yield peak can be explained by the fact that in some cases it is more complicated to extract dislocation from the Cottrell cloud formed from homogeneous atoms and vacancies than to move it.

Various phenomena taking place beyond the elastic limit in a general case can be considered as a result of an abrupt rise in the role of diffusion processes leading to a substantial change in the mutual location of Cottrell objects [2] relative to dislocations and, consequently, a slow change in the strength characteristics of a material (occurrence of slip, occurrence and amalgamation of micro-fractures, etc.) Vice versa, strain in reaching the yield stress will be represented as (almost) instantaneous transition from one state of the micro-structure to another one (stable) in the vicinity of the most stressed point, whereas there must be changes in the relative location of molecules.

Consequently, relative displacements of molecules must be equal to or more than distances between molecules. When loading solid bodies, clicks are heard and slip stripes (surfaces) are formed. If we abstract from dynamic phenomena, we must admit as follows.

Axiom 20.1 *Plastic strain in micro-volumes of a solid body is discrete in space and instantaneous in time.*

Discrete displacements occur as a result of shifts of multiple mobile structural imperfections in an elastic body. Their presence in any small volume of a body is deemed reality. It is obvious that defects can be displaced by the value no less than the distance between atoms and can have final power; the usually suggested continuity of plastic strains is an unnatural assumption, which is not always justified

© The Author(s), under exclusive license to Springer Nature Switzerland AG 2021   303
V. Molotnikov, A. Molotnikova, *Theory of Elasticity and Plasticity*,
https://doi.org/10.1007/978-3-030-66622-4_20

even in terms of convenience of computational kind. Relative displacements of atoms in planes with plastic strain (slip) are one thousand times longer than such displacements in volumes with purely elastic strain within the applicability of Hooke's law. At the same time, the total strain is low. It means that the volume where macro-structural changes take place is usually rather low as compared to the volume where purely elastic strain occurs.

Axiom 20.2 *Plastic strain changes only structurally sensitive characteristics of the material.*

Consequence *Elasticity coefficients do not depend on plastic strain.* Structurally sensitive (strength) characteristics are shift and tear resistance. Let us go over to their definition.

20.2   Shift Resistance

Let us consider homogeneous macro-strain of a poly-crystalline body with a disorderly crystal orientation. In each crystal, plastic strain is the result of slips of atomic layers over specific planes and in known directions with the maximum density of atom packing when reaching a specific value by a respective component of tangential stresses. The specified planes in a poly-crystalline material form a fan of slip planes whose opening depends on the stress level.

According to Batdorf and Budiansky [1], let us represent the slip plane going through an arbitrary point as a tangential plane to the half-sphere of a singular radius (Fig. 20.1).

This plane is defined by the normal line $n(\alpha_0, \beta_0)$. It is deemed that [4] slips in this plane occur in the directions l characterized by the angle $\omega_?$. A shift l will occur from local slips along planes with normal lines n enclosed within the solid angle $d\Omega$ in the directions l enclosed within an infinitely low angle $d\omega_0$, which shift will

Fig. 20.1 Graphical representation of slip planes and directions

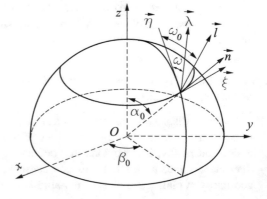

be represented as follows:

$$dy_{nl} = \varphi_{nl} d\omega_0 d\Omega, \qquad (20.1)$$

where φ_{nl} is the slip rate being a function of the directions $n(\alpha_0, \beta_0)$ and $l(\omega_0)$. Using the formulas of converting second-rank tensor components when switching to new axes, we can record as follows based on (20.1):

$$dy_{\nu\lambda} = \varphi_{nl}(n_\nu l_\lambda + n_\lambda l_\nu) d\omega_0 d\Omega, \qquad (20.2)$$

where each of the addends in the brackets is the product of cosines of angles between the respective vectors.

When changing the direction where the shift from these slips is defined into the opposite direction, a respective shift component will change only the sign, e.g.

$$dy_{n(-l)} = -dy_{nl}. \qquad (20.3)$$

Hence,

$$\varphi(\alpha_0, \beta_0, \omega_0 + \pi) = -\varphi(\alpha_0, \beta_0, \omega_0). \qquad (20.4)$$

Summing the respective strain tensor components from shifts (20.3) gives

$$\gamma_{ij} = \frac{1}{2} \iint\limits_R \varphi_{nl}(n_i l_j + n_j l_i) d\omega_0 d\Omega, \quad (i, j = x, y, z), \qquad (20.5)$$

where $l_i, \ldots, n_j$ are the projections of the single vectors l and n onto the respective axis of the Cartesian coordinate system, and R is the area of all positive and negative slips.

If we fix a normal line $\nu(\alpha, \beta)$ from a multitude of slip planes and directions to an arbitrary slip plane and direction $\lambda(\omega)$ in this plane inside the slip direction fan, the tangential stress component in this direction at the moment of slip is called [3] the plastic shift resistance $S_{\nu\lambda}$. In a common case, the shift resistance $S_{\nu\lambda}$ is [3] some operator from the slip rate φ_{nl}, which we will express using a symbolic form $S_{\nu\lambda}\varphi_{nl}$.

If R means an area where slips occur at this point of time, and Γ is the boundary of that area, we have as follows according to the definition of shift resistance:

$$S_{\nu\lambda}\varphi_{nl} = \tau_{\nu\lambda} \qquad \text{when } (\nu, \lambda) \in R. \qquad (20.6)$$

The area where no slips occur at this point in time is defined by the condition:

$$S_{\nu\lambda}\varphi_{nl} > \tau_{\nu\lambda} \qquad \text{when } (\nu, \lambda) \notin R. \qquad (20.7)$$

In the latter conditions, the stress tensor component $\tau_{\nu\lambda}$ is an arbitrary set function of time and is expressed through the known components of the stress tensor under the formula:

$$\tau_{\nu\lambda} = \nu_x\lambda_x\sigma_x + \nu_y\lambda_y\sigma_y + \nu_z\lambda_z\sigma_z + (\nu_x\lambda_y + \nu_y\lambda_x)\tau_{xy}+$$
$$+ (\nu_y\lambda_z + \nu_z\lambda_y)\tau_{yz} + (\nu_z\lambda_x + \nu + x\lambda_z)\tau_{zx}, \tag{20.8}$$

where $\nu_x, \ldots, \lambda_z$ are the projections of the single vectors ν and λ onto the axis of the Cartesian coordinate system $Oxyz$ calculated using the known [5] formulas:

$$\nu_x = \sin\alpha\cos\beta, \; \nu_y = \sin\alpha\sin\beta,$$
$$\nu_z = \cos\alpha, \; \lambda_x = -\sin\omega\sin\beta - \cos\omega\cos\alpha\cos\beta,$$
$$\lambda_y = \sin\omega\cos\beta - \cos\omega\cos\alpha\sin\beta, \; \lambda_z = \cos\omega\sin\alpha, \tag{20.9}$$

$$(\alpha = \widehat{\nu, z}, \quad \beta = \widehat{x, \nu_x Oy}, \quad \omega = \widehat{\lambda, \lambda_z Oy}).$$

Apart from the ratios (20.6)–(20.7), the function φ_{nl} must satisfy the continuity condition, which implies some restrictions on the operator $S_{\nu\lambda}$. It is then believed that in the initial point of time $t = t_0$, the material is isotropic and satisfies the condition

$$\varphi_{nl}\big|_{t=t_0} = 0, \tag{20.10}$$

and the slip rates at the boundary of the Γ slip area are continuous, e.g.

$$\frac{\partial\varphi_{nl}}{\partial t}\big|_\Gamma = 0. \tag{20.11}$$

Based on the notions introduced by us, we can formulate the primary task of mechanics of plastic bodies as follows:

to find such an operator $S_{\nu\lambda}$ for which the slip rate φ_{nl}, defined upon the conditions (20.6)–(20.7) and (20.10)–(20.11), gives dependencies, using formulas (20.5), between the components of stress and strain tensors observed in experiments.

20.3   Slip Synthesis

Single vectors of the axes ξ and η (Fig. 20.1) are defined by the formulas:

$$\vec{\eta} = -\frac{\partial\vec{n}/\partial\alpha_0}{|\partial\vec{n}/\partial\alpha_0|}, \quad \vec{\xi} = \frac{\partial\vec{n}/\partial\beta\beta_0}{|\partial\vec{n}/\partial\beta_0|} \tag{20.12}$$

or in projections:

$$\eta_x = -\cos\alpha_0\cos\cos\beta_0, \quad \eta_y = -\cos\alpha_0\sin\beta_0, \quad \eta_z = \sin\alpha_0,$$
$$\xi_x = -\sin\beta_0, \quad \xi_y = \cos\beta_0, \quad \xi_z = 0.$$

(20.13)

Using formula (20.2), let us calculate the gain of the plastic strain tensor component in the plane with the normal line $\vec{n}$ from all slips in this plane:

$$d\gamma_{n\lambda} = \Phi_{n\lambda}d\Omega,$$

(20.14)

where

$$\Phi_{n\lambda} = \int_{\{L\}} \varphi_{nl}l_\lambda d\omega_{l\lambda}, \quad (\omega_{l\lambda} = \widehat{l\lambda}),$$

(20.15)

whereas λ is a direction in the slip plane with the normal line $\vec{n}$, and $\{L\}$ is the multitude of slip directions in this plane.

The last expression can be represented otherwise as

$$\Phi_{n\lambda} = \Phi_{n\eta}\cos\omega\omega + \Phi_{n\xi}\sin\omega, \quad (\omega = \widehat{\eta\lambda}).$$

(20.16)

Here,

$$\Phi_{n\eta} = \int_{\{L\}} \varphi(\alpha_0, \beta_0, \omega_0)\cos\omega_0 d\omega_0,$$
$$\Phi_{n\xi} = \int_{\{L\}} \varphi(\alpha_0, \beta_0, \omega_0)\sin\omega_0 d\omega_0.$$

(20.17)

From the formula structure (20.16), it is seen that the value $\Phi_{n\lambda}$ in the slip plane is converted as a vector with the components $\Phi_{n\xi}$ and $\Phi_{n\eta}$ along the axes $\vec{\xi}$ and $\vec{\eta}$. Therefore, the value $\Phi_{n\lambda}$ defined by formula (20.15) will be called *the vector slip rate*.

It follows from the above that the components of the plastic strain tensor can be expressed as follows:

$$\gamma_{\nu\lambda} = \frac{1}{2}\iint_{\{D\}} \left[\Phi_{n\eta}(n_\nu\eta_\lambda + n_\lambda\eta_\nu) + \Phi_{n\xi}(n_\nu\xi_\lambda + n_\lambda\xi_\nu)\right]d\Omega,$$

(20.18)

where $\{D\}$ indicates the multitude of all normal lines $\vec{n}$ to slip planes.

Assume that $\vec{\nu}$ indicates a normal line to an arbitrary plane tangential to the half-sphere (Fig. 20.2), and $\vec{\lambda}$ is some direction in this plane. Let us calculate the gain of the plastic strain tensor component $(d\gamma_{\nu\lambda})$ from all slips in planes whose

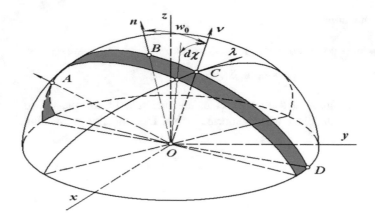

Fig. 20.2 Summing of slips

normal lines belong to an infinitely small element of the half-sphere surface marked in Fig. 20.2 by highlighting. A multitude of normal lines to these slip planes is designated through $\{D'\}$. This surface element is cut on the half-sphere by two planes going through the reference point O. The first of these planes $(OABCD)$ is perpendicular to $\vec{\lambda}$, and the second one is obtained by turning the first one around the axis OA $(\overrightarrow{OA} = \vec{v} \times \vec{\lambda})$ by an infinitely small angle $d\chi$.

The location of the normal line $\vec{n}$ $(\vec{n} \perp \vec{\lambda})$ to an arbitrary plane will be characterized by the angle w_0, counted from the normal line $\vec{v}$, whereas a positive direction of counting the angle w_0 coincides with the clockwise rotation direction if viewed from the end of the vector $\vec{\lambda}$. Then,

$$d\Omega = \cos w_0 dw_0 d\chi, \tag{20.19}$$

and the gain of the plastic strain tensor component from all slips upon the multitude of planes $\{D'\}$ will be

$$d\gamma_{v\lambda} = r_{v\lambda} d\chi, \tag{20.20}$$

where

$$r_{v\lambda} = \int_{\{W\}} \left(\Phi_{nn} \eta_\lambda + \Phi_{n\xi} \xi_\lambda \right) n_v \cos w_0 dw_0, \tag{20.21}$$

whereas the multitude $\{W\}$ being a sub-multitude $\{D'\}$ includes all normal lines $\vec{n}$ satisfying the condition

$$\vec{n} \cdot \vec{\lambda} = 0. \tag{20.22}$$

The value $r_{\nu\lambda}$ defined by formulas (20.21)–(20.22) is called *the tensor slip rate*. By substituting formulas (20.17) into the expression (20.21), we can obtain

$$r_{\nu\lambda} = \int\limits_{\{W\}} \int\limits_{\{L\}} \varphi(\alpha_0, \beta_0, \omega_0)(\eta_\lambda \cos\omega_0 + \xi_\lambda \sin\omega_0) \cos^2\omega_0 d\omega_0 dw_0. \quad (20.23)$$

Let us set the directions of the axes ν and λ using the angles α, β, and ω counted under the same rules as the angles α_0, β_0, and ω_0 for the vectors $\vec{n}$ and $\vec{l}$. Then the condition (20.22) can be represented as

$$\text{tg}\,\omega_0 \sin\alpha_0 \sin(\beta - \beta_0) = \sin\alpha \cos\alpha_0 - \sin\alpha_0 \cos\alpha \cos(\beta - \beta_0). \quad (20.24)$$

For the fixed α, β, and ω, the last formula defines the arch equation used for integration in formula (20.21). Taking into account that $\vec{n} \cdot \vec{\lambda} = \cos w_0$ and $\vec{n} \times \vec{\nu} = \vec{\lambda} \sin w_0$, we obtain the following ratio that will be used in what follows:

$$-\cos^2 w_0 dw_0 = \frac{F_1^2(\alpha, \alpha_0, \omega) \sin\alpha_0 d\alpha_0}{F_2(\alpha, \alpha_0, \omega)}, \quad (20.25)$$

where

$$F_1(\alpha, \alpha_0, \omega) = \cos\alpha \cos\alpha_0(1 + \text{tg}^2\omega) +$$
$$+ \sin\alpha \,\text{tg}\,\omega \sqrt{\sin^2\alpha_0(1 + \text{tg}^2\omega) - \sin^2\alpha}\,,$$
$$F_2(\alpha, \alpha_0, \omega) = \cos\omega(\cos^2\alpha +$$
$$+ \text{tg}^2\omega)^2 \sqrt{\sin^2\alpha_0(1 + \text{tg}^2\omega) - \sin^2\alpha}.$$

20.4 Definition of Principal Strains

Let 1, 2, and 3 indicate the principal axes of the strain tensor at some point in time (Fig. 20.3), and ε_1, ε_2, and ε_3 are the principal (elastic–plastic) strains in this point. Let us select the axis Oz to be coinciding with the principal axis 3 and the axes Ox and Oy to be inclined by the angle $\pi/4$ to two other principal directions. For the full strains ε_z and γ_{xy} in this case, we have as follows:

$$\varepsilon_z = \varepsilon_3,$$
$$\gamma_{xy} = \varepsilon_1 - \varepsilon_2. \quad (20.26)$$

Let us record the strain components as a sum of their elastic and plastic parts by designating both elastic and plastic parts of strain using the upper indexes "y" and "п", respectively:

Fig. 20.3 To the definition of principal strains

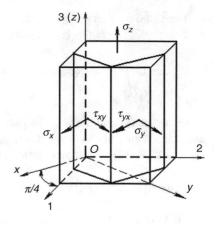

$$\varepsilon_1 = \varepsilon_1^y + \varepsilon_1^{\text{II}},$$
$$\varepsilon_2 = \varepsilon_2^y + \varepsilon_2^{\text{II}},$$
$$\varepsilon_3 = \varepsilon_3^y + \varepsilon_3^{\text{II}}, \qquad\qquad (20.27)$$
$$\gamma_{xy} = \gamma_{xy}^y + \gamma_{xy}^{\text{II}}.$$

Taking into account that plastic strain occurs without changes in the volume ($\varepsilon_1^p + \varepsilon_2^p + \varepsilon_3^p = 0$), we can represent as follows:

$$\varepsilon_{1,2} = \frac{1}{2}(\varepsilon_1^y + \varepsilon_2^y \pm \gamma_{xy}^y \pm \gamma_{xy}^{\text{II}} - \varepsilon_z^{\text{II}}),$$

$$\varepsilon_3 = \varepsilon_z^y + \varepsilon_z^{\text{II}}. \qquad\qquad (20.28)$$

By defining strains under Hooke's law, let us represent formulas (20.28) as follows:

$$\varepsilon_{1,2} = \frac{1}{2}\left\{\frac{1}{E}\left[3\sigma(1-\nu) - (1+\nu)(\sigma_z \mp 2\tau_{xy})\right] + \gamma_{xy} - \varepsilon_z\right\},$$

$$\varepsilon_3 = \frac{1}{E}\left[\sigma_z - \nu(\sigma_x + \sigma_y)\right] + \varepsilon_z, \qquad\qquad (20.29)$$

where σ is the mean stress

$$\sigma = \frac{1}{3}(\sigma_x + \sigma_y + \sigma_z),$$

and E and ν are the Young modulus and the Poisson coefficient of a material in an elastic state.

The dependencies (20.29) show that principal strains will be fully defined if the components of plastic strain γ_{xy} and ε_z are known. To calculate these, let us use the definition of the tensor slip rate. Let $\nu = x'$ and $\lambda = y'$, whereas x' and y' are

Fig. 20.4 Special case of slip

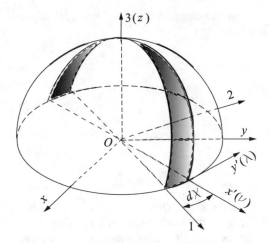

arbitrary orthogonal axes in the planes 1 and 2. With this selection of the directions ν
and λ, the element D' of the half-sphere surface covered by highlighting in Fig. 20.2
will be oriented as shown in Fig. 20.4.

With the selected element D', from formula (20.21), we will find as follows by
simple substitution of indexes:

$$r_{x'y'} = \int_{\{W\}} \Phi_{n\xi} \sin^2 \alpha_0 d\alpha_0. \tag{20.30}$$

Let us introduce designations:

$$r_{x'x'} = -\int_{\{W\}} \Phi_{n\eta} \cos \alpha_0 \sin^2 \alpha_0 d\alpha_0,$$

$$r_{zz} = \int_{\{W\}} \Phi_{n\eta} \sin^3 \alpha_0 d\alpha_0. \tag{20.31}$$

The values defined by formulas (20.31) will be called *normal tensor slip rates*.

Based on formulas (20.18) and (20.13), for the components of plastic strain ε_z
and γ_{xy}, we have as follows:

$$\varepsilon_z = \frac{1}{4} \iint_{\{D\}} \Phi_{n\eta} \cos \alpha \alpha_0 \sin^2 \alpha_0 d\alpha_0 d\beta_0,$$

$$\gamma_{xy} = \frac{1}{2} \iint_{\{D\}} \left[\Phi_{n\xi} \cos 2\beta_0 - \Phi_{n\eta} \sin 2\beta_0 \right] \sin^2 \alpha_0 d\alpha_0 d\beta_0. \tag{20.32}$$

From the comparison of sub-integral expressions in formulas (20.30)–(20.32), it follows that[1]

$$\varepsilon_z = \frac{1}{4} \int\limits_0^\pi r_{zz} d\beta_0,$$

$$\gamma_{xy} = \frac{1}{2} \int\limits_0^\pi (r_{x'x'} \sin 2\beta_0 + r_{x'y'} \cos 2\beta\beta_0) d\beta_0. \qquad (20.33)$$

The following is proved.

Lemma *If the directions ν and λ are located in the plane of the principal axes* (1, 2) *of the strain tensor and the third principal axis coincides with the symmetry axis of the half-sphere, principal strains are defined through the three* $(r_{x'y'}, r_{x'x'},$ *and* $r_{zz})$ *components of the tensor slip rate using formulas* (20.29) *and* (20.33).

References

1. S. Batdorf, B. Budyanskii, *Mekhanika: Sb. Perev., No. 5(33)* (Mechanics: a collection of translations, no. 5(33)) (Inostrannaya literatura Publ., Moscow, 1955). *Zavisimost' mezhdu napryazheniyami i deformatsiyami dlya uprochnyayushchegosya metalla pri slozhnom napryazhennom sostoyanii* (The relationship between stress and deformations for the hardening of metals under complex stress state), Mekhanika: sb. perev., no. 5(33) (Mechanics: a collection of translations), pp. 120–127
2. A. Kottrell, *Teoriya Dislokatsii* (Dislocation theory) (Mir Publ., Moscow, 1969)
3. M. Leonov, V. Molotnikov, K teorii deformatsii metallov s yarko vyrazhennym predelom tekuchesti (On the theory of deformations of metals with bright expressed yield strength). Izv. AN Kirg. SSR (Izv. Academy of Sciences of Kyrghyz. SSR) **6**, 3–10 (1974)
4. M. Leonov, V. Molotnikov, B. Rychkov, *XIII International Congr. According to the Theory. and Appl. Mekh-ks (IUTAM)* (IUTAM Publ., Moscow, 1972). The development of the concept the slip theory of plasticity, pp. 31–32
5. A. Lyav, *Matematicheskaya teoriya uprugosti*(Mathematical theory of elasticity) (ONTI NKTP SSSR Publ., Moscow, 1935)

---

[1]In formulas (20.33), sub-integral functions can turn zero at some segments of the interval $0 \leqslant \beta_0 \leqslant \pi$.

Chapter 21
Strain Specifics of Plastic Bodies

21.1 Elongation Diagram of a Plastic Material Specimen

Figure 21.1 gives a typical elongation diagram for a specimen made of plastic material. Here we can find the following characteristic features.

The area OA of elastic strain in some section AB is followed by a low non-elastic strain that is most frequently called pre-yield (or micro-yield) strain; its value is within $10^{-5} \ldots 10^{-4}$.

Then a loading drop can be observed (to the point C); the peak forming thereat is referred to as the yield drop [14]. It is followed by the area CD (or BE) corresponding to the occurrence and growth of micro-yield strain. Hereafter we will refer to the strain process on the yield plateau as yield. During yield, the specimen strain is characterized by high heterogeneity. A detailed study of yield development in a specimen of low-carbon steel at various testing rates is made, for instance, in papers by Yu. N. Rabotnov and his colleagues [10, 12].

It has been found [12] that plastic strain occurs at first in a small area as compared with the specimen length, which is in the place of stress concentrations in the cylindrical specimen, most frequently in the vicinity of one of the fillets. When plastic strain occurs, stress starts dropping from σ_v (point B, Fig. 21.1) to some value σ_s (point C, Fig. 21.1). Simultaneously with the stress drop, strain in the area of the initial plastic strain quickly rises to some value ε_0 (Fig. 21.2, [12], curve 1), which is constant for the defined elongation rate. At that time, the material is partially unloaded in the entire remaining volume of the specimen, which can be noticed on curve 2 (Fig. 21.2) built upon indications of the strain sensor attached in the middle of the specimen.

Then the plastic strain area grows; the interface of elastic and plastic zones moves along the specimen as a front. t^* is used in Fig. 21.2 to designate the moment when plastic strain occurs in the initial plasticity area and t^{**} is the moment corresponding to the plastic strain front approaching the middle of the specimen, and T_0 is the time while the plastic strain front covers the entire working length of the specimen. After

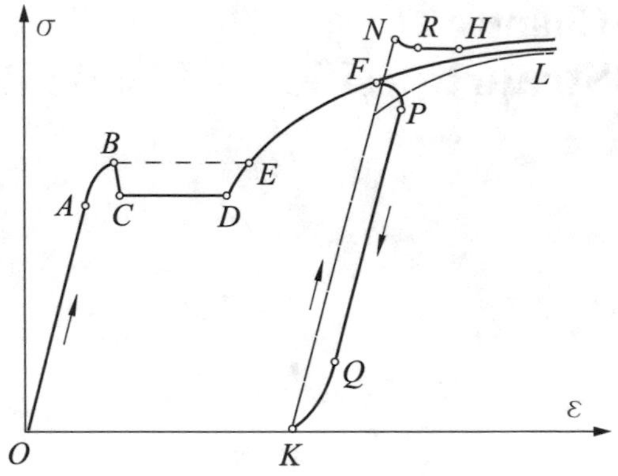

Fig. 21.1 Typical elongation diagram of a specimen made of plastic material

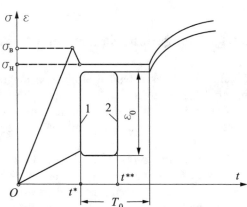

Fig. 21.2 Distribution diagram of plastic strain in specimen elongation

the front covers the entire length of the specimen, a hardening section appears in the "stress– strain" curve (DEF, Fig. 21.1), and strain becomes homogeneous along the entire length of the specimen working length.

Depending on the specimen material and testing conditions, there can be no yield drop in the elongation diagram [10]. In this case, yield occurs at almost no constant stress σ_V. This case corresponds to the line BE in Fig. 21.1.

Experiments [12] with a changed rate v of grip displacements of a testing machine with strain on the yield plateau show that the stress value σ_s, at which the yield occurs, is defined by the value v only at the considered moment in time and does not depend on the preceding strain process. If the grip displacement rate is changed incrementally in the yield process, the yield stress σ_s also changes incrementally (Fig. 21.3), and in the case of the same rates v, the values σ_s are the same. It is also characteristic that the hardening section is defined in these conditions only by the rate value (v_0) of grip displacement at the moment (t_1) of hardening start and does not depend on the history of v changes in previous moments ($t < t_1$).

In the case of unloading from some point (for example, from the point F, Fig. 21.1) of the hardening section within a small area FP, significant deviations from Hooke's law [20, 23] are observed in unloading: despite the decreased

Fig. 21.3 Dependency of yield on the strain rate

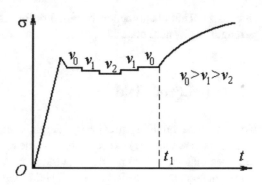

deforming stress, unloading is accompanied by strain growth. This is followed by an area (PQ) (Fig. 21.1) of unloading when the material almost follows Hooke's law, whereas the tangential modulus in this area has almost no difference from the elastic modulus in the area OA. Finally, at some stress (point Q, Fig. 21.1), a small drop of the tangential modulus starts, which drop increases after that. In the case of full unloading (only K, Fig. 21.1), this modulus can be 30...40 % lower than the elastic modulus. The described phenomenon is known as the Bauschinger effect.

There are various opinions in the interpretation of the described effects. Some researchers [5, 20, 23] believed that the unloading diagram was almost linear, its deviation from the parallelism of the curve OA was a consequence of changes in the elastic modulus during plastic strain. To substantiate this conclusion, a thesis is used stating that if the deviations found are deemed a result of plastic strain changes during unloading, this contradicts the plastic strain definition as being residual after full unloading. Other researchers [9, 13, 21] adhere to an alternative point of view and consider the described effects as a result of changes in plastic strain during unloading.

To find the truth, let us consider as follows. If we start elongating the specimen again after full unloading, the initial section of the secondary loading diagram will be almost linear, that is, almost parallel to OA (Fig. 21.1). If loading of the opposite sign is done after unloading, the tangential modulus in the point K (Fig. 21.1) significantly differs [18, 23] from the elastic one. Hence, the tangential modulus depends on the direction strain, which means that the material behavior near the point K is not elastic. Therefore, the observed deviation from Hooke's law in unloading is a result of changes in plastic strain during unloading. This is supported by the fact that the value of the above effects in unloading significantly depends on time factors [23]. We shall also note that the comment of the author [20] as to the previously mentioned contradiction reveals an imperfection of defining plastic strain as residual. If we define plastic strain as adopted by us (see p. 309) as defined by Rice [17], there is no indicated contradiction.

Let us return to Fig. 21.1. If loading of the same sign is done after full unloading (trajectory $FPQK$), yield plateaus in the repeated loading diagram are not observed (Fig. 21.1, dash line KL). However, if the specimen is subject to normalization or low-temperature annealing after full unloading, in the case of repeated loading of the same sign, the yield drop and the yield plateau are found again (line $KNRH$,

Fig. 21.1). This phenomenon is called deformational aging in the literature on the strength of physical materials.

21.2 Delay of Yield

Assume that elongating stress $\sigma(t)$ is created in the specimen at the moment of time t. If it appears that $\sigma(t)$ exceeds some value σ_*, which is constant for this material, the specimen yield does not occur at the moment t, but only after some time $\Delta t(\sigma)$. This phenomenon is called *delay of yield,* and the time period $\Delta t(\sigma)$ *is the yield time of delay.* It depends on the type of the loading function $\sigma(t)$. When studying the delay of yield, $\sigma(t) = const$ is usually adopted [2] (rectangular loading impulse). In this case, the stress σ is applied to the specimen instantaneously.

An exemplary type of dependency between the time of yield delay Δt and the instantaneously applied maximum tangential stress τ_m in an elongated specimen is shown in Fig. 21.4. Here τ_p^∞ designates the maximum tangential stress in specimen elongation causing the instantaneous occurrence of yield, and τ_y means the highest maximum tangential stress when the specimen yield is not observed yet, however long we may be waiting for it.

The yield delay has been a subject matter of multiple studies since the middle of the previous century. Most theoretical studies are done in terms of dislocation theory. These studies were started in the pioneering works by Cottrell [6, 8]. The condition of yield occurrence is represented [7] as

$$\frac{1}{\tau_0} \int_0^t \Phi(\sigma, T^0)d\tau = 1, \tag{21.1}$$

Fig. 21.4 Delay of yield

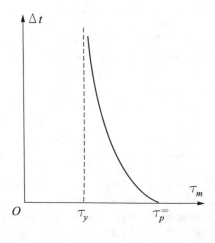

where τ_0 is the constant value, T^0 is the absolute temperature, t is the moment of yield occurrence, and Φ is some function.[1]

The following empirical representations were proposed for the function Φ:

$$\Phi = (|\sigma|/\sigma_*)^n; \tag{21.2}$$

$$\Phi = \begin{vmatrix} a(|\sigma|/\sigma_* - 1)^n & \text{at } \sigma > \sigma_*, \\ 0 & \text{at } \sigma < \sigma_*; \end{vmatrix} \tag{21.3}$$

$$\Phi = A\exp(|\sigma|/b). \tag{21.4}$$

The parameters n, a, A, and b in these expressions are temperature functions. Formula (21.2) proposed by Yekobori [1] gives the final time of yield delay at $\sigma = \sigma_*$ and therefore is contradictory. The equation of Campbell 21.4 also gives [16] incorrect results for stress close to σ_*. The function (21.3) whose use in the yield conditions (21.1) satisfactorily coordinates [16] with the experiment remains non-contradictory.

Let us shortly consider experimental studies of yield delay. A rather comprehensive overview of these studies is contained in materials with a pronounced yield plateau. Let us call the highest value of the maximum tangential stress preceding the formation of yield drop or plateau as the yield stress of such materials.

A sufficiently complete overview of these researches is contained in the article by Yu.V. Suvorova [19]. It should be noted that the number of experiments intended to study this phenomenon is rather limited. Apparently, this is related to the complexity of setting experiments and with a wide scatter of experimental data. Among the existing experimental results, the most reliable are experiments by Clark, Wood, and Gendrikson [2, 22], as well as Lomakin E. V. [11]. Let us also note that as of now, we do not know any experimental works in the field of studying the yield delay in the conditions of complex stressed state. All known experiments belong to uniaxial elongation or compression. Characteristics of yield delay given in the literature are defined by various authors for various rates of loading, and therefore their analysis and comparison are rather problematic.

21.3   Yield Stress and Loading Rate

We already noted (p. 333) that this book discusses materials with a pronounced yield plateau. Let us agree that the yield point of such materials is the highest value of the maximum tangential stress preceding formation of a yield drop or a yield plateau.

During elongation tests, the effect of yield delay is expressed by the fact that the yield stress of a plastic material is very sensitive to the loading rate (to be more

---

[1]This book considers athermal plasticity at normal temperatures.

Fig. 21.5 Influence of the
loading rate on the yield
stress

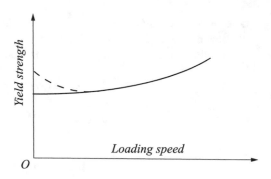

precise, to duration and sequence of applying loads or loading history). Fig. 21.5
shows a specific characteristic form of the dependency between the yield stress
and loading rate. The figure shows that as the loading rate falls down, the yield
stress decreases. However, in the case of very low loading rates, sometimes strength
properties of the material are restored, and in the case of rather low rates [4], the
yield stress may rise with the loading rate declining. The latter can be explained by
the fact that the process of picking up dislocations from blocking atmospheres at
such rates will be slow, since the cloud is not "dispersed" and will "congeal again"
by means of diffusion processes occurring in the body.

On the other hand, in the case of rather high loading rates, accompanying inertia
forces must be taken into account, and wave processes occurring in the specimen
must be considered. In this connection, let us set the interval $[T_1, T_2]$ of time
variance within which the loading process can be deemed independent of these
effects before the onset of yield.

Let us consider that the minimal time (T_1) until the yield stress is $5 \ldots 10$ times
higher than the time of an elastic wave running through the specimen. In this case,
the change of the deforming loading during a single wave run through the specimen
will be low. Therefore, elastic disturbances caused by these load changes will also
be low. Moreover, as a result of multiple waves running during the loading time, the
stress and strain states of the specimen in general can be deemed homogeneous.

If we set a tenfold wave running through the specimen, before the strain reaches
the value of ε_s, e. g. strains at the time of yield start, we can define that the maximum
displacement rate of the specimen end must be

$$v_{\max} = 0.1\varepsilon_s c_0, \tag{21.5}$$

where c_0 is the propagation rate of an elastic wave defined in a known manner [15].
By assuming that $\varepsilon_s \approx 0.2\%$ and taking into account that the propagation rate of
elastic disturbances (c_0) in steel approximately equals $5 \cdot 10^3$ m/s [15], we find that
$v_{\max} \approx 1$ m/s, which, for a specimen length of $l_0 = 100$ mm, corresponds to the
maximum strain rate

$$\dot{\varepsilon} = \frac{v_{max}}{l_0} \approx 10 \; s^{-1}.$$

Then the lower boundary T_1 of the interval $[T_1, T_2]$ will be equal to

$$T_1 = \frac{\varepsilon_s}{\dot{\varepsilon}_{max}} \approx 2 \cdot 10^{-4} s.$$

This means that in the case of loading at the strain rate below $10 \; s^{-1}$, wave processes cannot be considered. Such processes are sometimes called quasistatic. The maximum duration of T_2 loading to the yield stress will be defined by such minimal strain rate for which the restoration of the mechanical properties of the material over the loading time can be neglected. Unfortunately, we do not know experimental data for the restoration time of mechanical properties at normal temperatures. However, according to existing data [3, 13] for zinc, copper, and some alloys, we can approximately believe that the loading time until yield must not exceed $200\ldots400$ min. It can be easily calculated that the strain rate $\dot{\varepsilon}$ must be at least $0.2 \cdot 10^{-6} \; s^{-1}$.

In this manner, we will consider only such loadings when the strain rate is within $10^{-8} \ldots 10 \; s^{-1}$, which corresponds to the loading duration until yield of $2 \cdot 10^{-4} \ldots 2 \cdot 10^3$ s.

References

1. T. Ekobori, *Fizika i mekhanika razrusheniya i prochnosti tverdykh tel* (Physics and mechanics of fracture and strength of solids) (Metallurgiya Publ., Moscow, 1971)
2. J. Hendrickson, D. Wood, The effect of rate of stress application and temperature on the upper yield stress of annealed mild steel. Trans. ASM. **50**, 498–516 (1958)
3. R. Khonikomb, *Plasticheskaya deformatsiya metallov* (Plastic deformation of metals) (Mir Publ., Moscow, 1972), 408p.
4. J. Klepaczko, The strain rate behaviour of iron in pure shear. Int. J. Silod Struct. **5**, 533–548 (1969)
5. V. Klushnikov, *Fiziko-matematicheskie osnovy prochnosti i plastichnosti: ucheb. posobie* (Physical and mathematical foundations of strength and plasticity: textbook. Stipend) (MGU Publ., Moscow, 1994)
6. A. Kottrell, *Effect of Solute Atoms on the Behavior of Dislocations* (Strength. of Solids Publ., Bristol, 1947)
7. A. Kottrell, *Teoriya dislokatsii i plasticheskoe techenie v kristallakh: per. s angl* [Dislocation theory and plastic flow in crystals: TRANS. from English.) (GNTIL po chern. i tsv. metallurgii Publ., Moscow, 1958)
8. A. Kottrell, *Teoriya dislokatsii* (Dislocation theory) (Mir Publ., Moscow, 1969)
9. M. Leonov, Y. Klyshevich, Z. Sulaimanov, *Prosteishaya zadacha plasticheskogo techeniya* (The simplest task plastic flow). Izv. AN Kirg. SSR **6**, 4–13 (1971)
10. E. Lomakin, *Rasprostranenie uprugo-plasticheskikh voln v malouglerodistykh stalyakh* (Propagation of elastic-plastic waves in low-carbon steels). Izv. AN SSSR Mekhan. tv. tela **5**, 152–160 (1970)

11. E. Lomakin, *Zapazdyvanie tekuchesti v uglerodistykh stalyakh* (Delayed fluidity in carbonaceous steels) (MGU Publ., Moscow, 1971)
12. E. Lomakin, A. Mel'shanov, *Povedenie malouglerodistoi stali pri rastyazhenii* (Behavior of low carbon steel at stretching). Izv. AN SSSR, Mekhan. tv. tela **4**, 150–158 (1971)
13. F. Makklintok, A. Argon, *Deformatsiya i razrushenie materialov* (Deformation and destruction of materials) (Mir Publishers, Moscow, 1970)
14. B. Petukhov, *Teoriya zuba tekuchesti v malodislokatsionnykh kristallakh* (Theory of tooth fluidity in low-dislocation crystals). J. Tech. Phys. **71**(11), 42–47 (2001)
15. Y. Rabotnov, *Soprotivlenie materialov* (Strength of materials) (Gosfiztekhizdat Publ., Moscow, 1962)
16. Y. Rabotnov, *Model' uprugo-plasticheskoi sredy s zapazdyvaniem tekuchesti* (Model of elastic-plastic medium with delay fluidity). Prikladn. mekhan. i tekhn. fizika **3**, 45–54 (1968)
17. D. Rais, *O strukture sootnoshenii mezhdu napryazheniyami i deformatsiyami pri plasticheskom deformirovanii metallov, zavisyashchem ot vremeni* (On the structure of relations between stresses and deformations for time-dependent plastic deformation of metals). Appl. Mech. Proc. Am. Soc. Mech. Eng. **3**, 67–74 (1970)
18. G. Stepanov, *Uprugo-plasticheskoe deformirovanie materialov pod deistviem impul'snykh nagruzok* (Elastic-plastic deformation of materials under by pulsed loads) (Naukova Dumka Publ., Kiev, 1979)
19. Y. Suvorova, *Zapazdyvanie tekuchesti v stalyakh* (Retardation of fluidity in steels). Prikl. mekhka i tekhn. fizika **3**, 55–62 (1968)
20. G. Talypov, *Plastichnost' i prochnost' stali pri slozhnom nagruzhenii* (Ductility and strength of steel under complex loading) (Izd-vo LGU Publ., Leningrad, 1968)
21. A. Vakulenko, *Issledovaniya po uprugosti i plastichnosti* (Studies on Elasticity and Plasticity: Collection of Scientific), vol. 8 (LGU Publ., Leningrad, 1971). *Problemy reologii plasticheskikh sred* (Problems of Rheology of Plastic Media), pp. 76–97
22. D. Wood, D.S. Clark, *The Influence of Temperature upon the Time Delay for Yielding in Annealed Mild Steel* (American Society for Metals Publication, New-York, 1951)
23. A. Zhukov, *Slozhnoe nagruzhenie i teoriya plastichnosti izotropnykh metallov* (Complex loading and plasticity theory of isotropic metals). Izv. AN SSSR, OTN **8**, 81–92 (1955)

Chapter 22
Axioms of the Inelastic Body Model

22.1 Deformational Softening

We have already said that (p. 317) the yield stress of materials studied here substantially depends on the loading rate and it goes down when the rate decreases. The lowest value of the yield stress is called the yield stress (τ_y). It is believed that the elastic limit is defined only by the type of the stressed state and, unlike the stress yield stress, does not depend on the loading rate.

22.2 Initial Shear Resistance

According to Cottrell's [1] representations of the picking up of dislocations from blocking atmospheres, we can suggest the following mechanism of deformational softening. In case stress exceeds the elastic limit, some dislocations are liberated from Cottrell clouds. Other dislocations can be displaced together with the surrounding atmosphere. Successive liberation of dislocations from atmospheres leads to changes in the structure and, consequently, mechanical properties of the material. When the yield stress occurs, liberated dislocations start moving. Since displacements of free dislocations require much lower stress than for smoothening of dislocations from blocking atmospheres, the start of plastic yield is accompanied by the drop of material resistance to deformation. This results in the following.

© The Author(s), under exclusive license to Springer Nature Switzerland AG 2021   321
V. Molotnikov, A. Molotnikova, *Theory of Elasticity and Plasticity*,
https://doi.org/10.1007/978-3-030-66622-4_22

Fig. 22.1 Theoretic
elongation diagram in the
plastic zone of the specimen

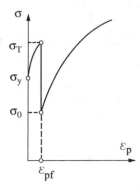

Axiom 22.1 *Initial shear resistance is less than the yield stress.*[1]

Due to this axiom, the theoretic diagram of "stress–plastic strain" for this element of the material where plastic strain occurs will look as shown in Fig. 22.1, where σ_y and σ_T indicate normal stresses corresponding to the elastic limit and yield stress, ε_{pf} (prefluidity) is the non-elastic strain of pre-yield, ε_p is the plastic strain (p. 309), and σ_0 is the initial shear resistance.

Due to low strains of pre-yield, they may not be considered, and the strain is deemed elastic until yield occurs. In this connection, the observed deformational softening preceding the yield will also be called elastic softening. Since pre-yield strains are primarily a result of diffusion processes rather than slipping, we will not consider them unless specifically agreed.

It has been established [3] that when a specimen material will go to a plastic state, the nature of the softening curve for the deformed part of the specimen does not depend on the history of the previous deformation and is defined by plastic strain and its rate at a given time. However, such phenomena as the Bauschinger effect make us accept as follows.

Axiom 22.2 *Any local slip in a poly-crystalline body changes mechanical properties (almost) in all directions.*

Seger [4] considered plastic strain of crystals in terms of dislocation theory and found out that the rate of plastic strain ($\dot{\gamma}$) was a function (V) of the applied stress τ and absolute temperature (T^o) and did not depend on the stress variance rate:

$$\dot{\gamma} = V(\tau, T^o).$$

---

[1]The latter is true for a sufficiently pure metal with relatively large grains. If the material contains relatively many impurities, these diffusion processes of changing the mutual arrangement of blocking clouds and dislocations "disguise," and there can be no drop in the resistance of the material to deformation.

Based on this result, we can suggest that for a poly-crystalline body, the strain rate $(\dot{\gamma}_{nl})$ in the slip plane n in the direction l at time t depends only on the tangential stress component (τ_{nl}) in this direction at this time, but it does not depend on the rate of this component.

Since plastic strain is a consequence of local slips characterized by the slip intensity φ_{nl}, the following results from the above.

Axiom 22.3 *In the case of deformation beyond the yield stress, the shear resistance is an integration–differentiation operator of slip intensity and does not depend on the rate of elastic strains (stresses).*

This axiom is a generalization of the anisotropy postulate of M. Ya. Leonov [2].

22.3 Function of Elastic Softening

Due to the elastic anisotropy of crystalline grains and their chaotic arrangement, internal forces in solid bodies are characterized by some irregularity even in the case of macro-ordinary strain. Due to this irregularity of stressed state, there are always crystals in a poly-crystalline body whose tangential stress components in the shear plane and direction exceed the nominal value, and shears may occur at some loading in these crystals. However, for these slips to cause a macroscopic effect, it is insufficient to have shears occurred in a single grain only; in the case of loading, elastically deformed neighboring elements may return this grain to the initial state.

During the macro-uniform loading of a poly-crystalline body, the maximum number of grains (their relative volume) where shears may occur depends on the multitude of directions where high tangential stresses occur, and the latter depends on the type of the stressed state. An increased number of planes and directions where high stresses occur at the same maximum tangential stress increases the relative volume of grains where plastic strains may occur. This results in the following.

Axiom 22.4 *In case any components of tangential stresses rise in all planes, the increment of the shear resistance from elastic strains only is negative.*

Let us represent that the proportional loading of a body element occurs beyond the elastic limit at an infinitely low speed. A respective maximum tangential stress at the yield stress is designated as S_m^∞. The value S_m^∞ being the highest yield stress will be called *instantaneous yield stress*. The following is assumed.

Axiom 22.5 *In the case of deformation of a plastic material with an infinitely high speed, first slips occur in planes and directions including the planes with the maximum tangential stress in the direction of its action.*

Let us suggest that the considered materials have the same instantaneous yield stresses at uniaxial elongation and compression. For such materials, normal hydrostatic stresses do not have effects on plastic strains, so the yield condition in the case of their instantaneous deformation does not depend on the first invariant of the stress tensor, and it can be represented as

$$F(J_2, J_3) = 0,\tag{22.1}$$

where F is the function of the second (J_2) and third (J_3) invariants of the stress tensor. These invariants can be expressed through the maximum (τ_m) and octahedral (τ_i) tangential stresses. Therefore, the condition (22.1) can be rewritten as follows:

$$S_m^\infty = f(\tau_i),\tag{22.2}$$

where f is the function set for a specific material defining the effects of the stressed state type on the condition of yield occurrence with an infinitely high rate. Since stresses are unambiguously related to elastic strains before the fulfillment of the yield condition (22.2), we will state below that the function f defines the effects of elastic strain on the instantaneous yield stress, and we will call it the *function of elastic softening.*

Due to Axiom 22.4, the function $f(\tau_i)$ defines the initial shear resistance in the case of instantaneous material strain. From Axiom 22.3, it follows that this function is decreasing.

By changing the stressed state type, we can use formula (22.2) to find the function $f(\tau_i)$ in some area. Assume that S_p^∞ and S_k^∞ designate the instantaneous yield stress at elongation and torsion of a thin-wall tubular specimen, respectively. Octahedral tangential stresses corresponding to these two types of stressed state will be

$$\tau_i^{(1)} = \frac{2\sqrt{2}}{3} S_p^\infty \quad \text{and} \quad \tau_i^{(2)} = \sqrt{\frac{2}{3}} S_k^\infty.\tag{22.3}$$

The stresses S_p^∞ and S_k^∞ give the values of the elastic softening function in two points: $\tau_i = \tau_i^{(1)}$ and $\tau_i = \tau_i^{(2)}$. By using the latter, we will approximate f by the linear function assuming that

$$f(\tau_i) = T - k\tau_i, \quad (T, k - const),\tag{22.4}$$

where

$$k = \frac{S_k^\infty - S_p^\infty}{\tau_i^{(1)} - \tau_i^{(2)}}, \quad T = S_k^\infty \left(1 + k\sqrt{\frac{2}{3}}\right),\tag{22.5}$$

whereas $\tau_i^{(1)}$ and $\tau_i^{(2)}$ are defined by formulas (22.3).

The yield condition in the case of instantaneous loading (22.2) can be written otherwise as

$$S_m^\infty = \psi(\tau_i, \tau_m),\tag{22.6}$$

where τ_m is the maximum tangential stress, and the function $\psi(\tau_i, \tau_m)$ is set by the formula

$$\psi(\tau_i, \tau_m) = \frac{f(\tau_i) - m\tau_m}{1 - m}, \quad (m - const), \tag{22.7}$$

since the conditions (22.6)–(22.7) and (22.2) are identical. The function ψ defined by formula (22.7) will be called *a generalized function of elastic softening*.

References

1. A. Kottrell, *Effect of Solute Atoms on the Behavior of Dislocations* (Strength of Solids Publ., Bristol, 1947)
2. M. Leonov, *Osnovnye postulaty teorii plastichnosti* (The basic postulates of the theory of plasticity).DAN SSSR [Reports of the USSR Academy of Sciences], no. 1, vol. 199 (1971), pp. 51–54
3. E. Lomakin, A. Mel'shanov, *Povedenie malouglerodistoi stali pri rastyazhenii* (Behavior of low carbon steel at stretching). Izv. AN SSSR, Mekhan. tv. tela, No. 4 (1971), pp. 150–158
4. D. Mak Lin, *Mekhanicheskie svoistva metallov* (Mechanical properties of metals). (Metallurgiya Publ., Moscow, 1965)

Chapter 23
The Fluidity at the Finite Speed of Loading

23.1 Yield Strength at the Final Loading Speed

Let us consider the loading of a body element beyond the elastic limit at any final rate. In this case, the body will have diffusion processes even before yield, which will cause low non-elastic strains of pre-yield (see, for example, [1, 6, 7]). These diffusion processes may significantly decrease the shear resistance of the considered materials due to changes in the arrangement of dislocations in the cloud of Cottrell [2, 3, 5].

Let us designate the time to reach the yield stress (t_y) until the moment (T_0) of reaching the yield stress as T_1. During the time T_1, the material will have diffusion processes that will cause structural softening. As a result, the yield stress of the material will go down to some value $S_m < S_m^\infty$.

If the stress $\tau_m(t)$ exceeding the yield stress τ_y was applied at the moment t, which acted for the time dt, this stress will give some change in the yield stress as a result of structural softening. Let us assume that for the studied materials, the following conclusion is true.

Hypothesis *Structural softening is defined only by maximum tangential stress.*

This hypothesis expands formula (22.2) to the case of loadings differing from the uniaxial elongation. Based on the suggestion formulated in the hypothesis, the contribution of the stress $\tau_m(t)$ over the time dt to the change in the yield stress can be represented as

$$dS_m = -F[\tau_m(t)]dt,$$

where F is a yet unknown function that we will call *the aging function*. We will find as follows for the entire time interval T_1 :

© The Author(s), under exclusive license to Springer Nature Switzerland AG 2021
V. Molotnikov, A. Molotnikova, *Theory of Elasticity and Plasticity*,
https://doi.org/10.1007/978-3-030-66622-4_23

$$S_m = S_m^\infty - \int_{t_y}^{T_0} F[\tau_m(t)]dt.$$ (23.1)

If the function F is known, the latter formula gives the yield condition at loading with an arbitrary rate. In a partial case of uniaxial elongation, it coincides with the Cotrlell–Campbell [2] condition (21.3). Indeed, if we assume in (21.3) $\tau_0 = S_m^\infty$, $t = T_0$, and

$$\Phi = \begin{vmatrix} 0 & \text{at} & \tau < t_y, \\ F[\tau_m(\tau)] + \dfrac{1}{2}S_m\delta(\tau - T) & \text{at } t_y \leqslant \tau \leqslant T_0, \end{vmatrix}$$

by directly checking, we can make sure that formula (21.3) gives (23.1). (In the latter formula, the symbol δ designates the Dirac function.)

23.2 Defining the Aging Function

As per the definition (Sect. 22.1), when reaching (T_0), the moment of yield stress (S_m), we have

$$S_m = \tau_m(T_0).$$

Substituting the yield conditions into the latter equation (23.1) gives

$$S_m^\infty - \tau_m(T_0) = \int_{t_y}^{T_0} F[\tau_m(t)]dt.$$ (23.2)

Assume that the stress $\tau_m > \tau_y$ is applied instantaneously and before the moment (T_0) of yield occurrence, it remains constant. Then we find as follows from formula (23.2):

$$S_m^\infty - \tau_m = F[\tau_m(t)]\Delta t,$$ (23.3)

where $\Delta t = T_0 - t_y$ is the time of yield delay. From formula (23.3), we obtain

$$F[\tau_m] = \frac{S_m^\infty - \tau_m}{\Delta t(\tau_m)}.$$ (23.4)

In this manner, in case the dependency $\Delta t \sim \tau_m$ is known, formula (23.4) defines the aging function.

Let us assume that the dependency chart [4] between the yield delay time and the instantaneous applied stress (Fig. 21.4) is obtained from the experiments. For analytical representation, we will approximate this dependency by the formula

$$\tau_m = D \cdot \exp\left[-\eta(\Delta t)^N\right] + \tau_y, \tag{23.5}$$

where η, D, N, and τ_y are constant values defined as coefficients of approximation of the dependency $\Delta t \sim \tau_m$ using formula (23.5). Assuming in formula (23.5) $\Delta t = 0$, $\tau_m = S_m^\infty$, we obtain $D = S_m^\infty - \tau_y$, e.g.

$$\tau_m = \left(S_m^\infty - \tau_y\right) \exp\left[-\eta(\Delta t)^N\right] + \tau_y,$$

or

$$\Delta t = \left[\frac{1}{\eta} \ln \frac{S_m^\infty - \tau_y}{\tau_m - \tau_y}\right]^{1/N}. \tag{23.6}$$

Let us find for the aging function:

$$F[\tau_m(t)] = \frac{S_m^\infty - \tau_m(t)}{\left[\dfrac{1}{\eta} \ln \dfrac{S_m^\infty - \tau_y}{\tau_m(t) - \tau_y}\right]^{1/N}}. \tag{23.7}$$

23.2.1 Example

Assume that the proportional loading of the specimen is done under the constant loading rate v, e.g. $\tau_m = \tau_y + vt$. Then, by substituting the function (23.7) into the yield condition (23.1), we obtain the following dependency between the yield stress (S_T) and loading rate:

$$v = \frac{1}{S_m^\infty - S_T} \int_{\tau_y}^{S_T} \frac{\left(S_m^\infty - x\right)dx}{\left[\dfrac{1}{\eta} \ln \dfrac{S_m^\infty - \tau_y}{\tau - \tau_y}\right]^{1/N}}. \tag{23.8}$$

Circles in Figs. 23.1 and 23.2 indicate experimental data obtained by Gendrikson and Wood [4] in experiments for elongation of samples made of annealed steel containing 0.17% of carbon. Continuous lines in these figures depict diagrams built using formulas (23.8) and (23.6). The following constant values are adopted: $S_m^\infty = S_p^\infty = 240\,\text{MPa}$, $N = 0.379$, $\eta = 3.767\ s^{-1}$, and $\tau_y = 150\,\text{MPa}$. As figures show, the dependencies (23.8) and (23.6) well approximate experimental data.

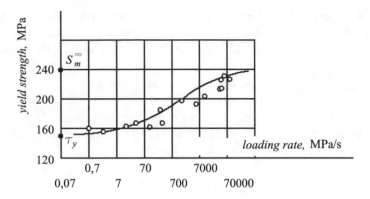

Fig. 23.1 Dependency of the steel yield stress on the loading rate

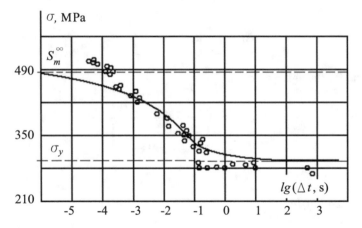

Fig. 23.2 Time of yield delay in the case of instantaneous application of loading

23.3   Components of Deformational Softening

Assume that diffusive structural changes cause material softening equal in all planes and directions. This softening will be called isotropic. Let us represent the component of isotropic deformational softening by some function of the following argument:

$$\Sigma = \int_{t_y}^{T_0} F[\tau_m(t)]dt. \tag{23.9}$$

Let us call the value Σ defined by formula (23.9) as *the primary argument of isotropic aging.*

Then, as a result of aging, material softening occurs in some direction λ, and softening with the equal modulus but opposite in the sign occurs in the opposite direction $(-\lambda)$. Such aging will be called anti-isotropic, and the value of softening that occurs in this case will be called the *anti-isotropic component of aging*. It is then believed that the anti-isotropic component of aging can be represented as a function of the following argument:

$$\Sigma_{v\lambda} = \frac{\gamma_{v\lambda}}{\gamma_m}\Sigma, \tag{23.10}$$

where Σ is the primary argument of isotropic aging, and γ_m and $\gamma_{v\lambda}$ are the maximum plastic shear and the component of plastic strain in the axes v and λ, respectively.

23.4 Almost Simple Strain

Let us consider such loadings for which the maximum plastic shear (γ_m) for continuous plastic strain is almost a monotonous function of time. This strain will be called *almost simple*.

The requirement of the continuity of plastic strain is associated with the fact that stops in loading and keeping the plastically deformed material under constant load or in the case of its partial or full loading is accompanied by material "recovery" caused by diffusion processes.

An exemplary form of such an elongation diagram with recovery at partial loading is given in Fig. 23.3. Effects found in recovery are usually low. Moreover, since these effects are caused not by slip but by diffusion processes, apparently, their accounting and analytical description cannot belong to the subject matter of the theory of strain of plastic materials developed here.

It is assumed that in the case of at least simple strain, softening as a result of strain aging can be represented as a sum of isotropic and anti-isotropic components of aging. By designating this softening as $U_{v\lambda}$, let us represent it as

Fig. 23.3 Recovery effect at partial unloading

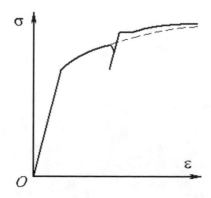

$$U_{\nu\lambda} = A\Sigma_{\nu\lambda} + B\Sigma, \tag{23.11}$$

where A and B are some values that we will deem constant in the first approximation.

Due to Axiom 22.1, the values A and B being a part of the representation of softening $U_{\nu\lambda}$ under formula (23.11) must satisfy the inequation

$$A + B > 1.$$

This condition must be taken into account when designing a shear resistance operator.

References

1. I. Borodin, A. Maier, Y. Petrov, A. Gruzdkov, Maksimum predela tekuchesti pri kvazistaticheskoy i vysokoskorostnoy plasticheskoy deformatsii metallov [Maximum yield strength at quasi-static and high-speed plastic deformation of metals]. Fizika tverdogo tela **56**(12), 2384–2393 (2014)
2. J. Campbell, K. Marsh, The effect of grain size on the delayed yielding of mild steel. J. Theoretical, Experimental and Applied Physics. **7**(78), 933–952 (1962)
3. T. Ekobori, *Fizika i mekhanika razrusheniya i prochnosti tverdykh tel* (Physics and mechanics of fracture and strength of solids). (Metallurgiya Publ., Moscow, 1971)
4. J. Hendrickson, D. Wood, The effect of rate of stress application and temperature on the upper yield stress of annealed mild steel. Trans. ASM. V. **50**, 498–516 (1958)
5. A. Kottrell, *Teoriya dislokatsii* (Dislocation theory). (Mir Publ., Moscow, 1969)
6. Y. Rabotnov, *Elementy nasledstvennoy mekhaniki tverdykh tel* (Elements of hereditary mechanics of solids). (Nauka Publ., Moscow, 1977)
7. N. Selyutina, *Razrusheniye i plasticheskoye deformirovaniye konstruktsionnykh materialov pri udarno-volnovykh nagruzkakh* (Destruction and plastic deformation of structural materials under Shock-Wave loads). (SPbSU Publ., Sankt-Petersburg, 2016)

Chapter 24
Specimen Elongation with Yield Drop

24.1 Original Assumption

Let us remind (p. 333) that during elongation tests of specimens (Fig. 24.1) of structural materials such as low-carbon steels and some other bodies, the following situation is observed (Fig. 21.1). After elastic strain, there is a short non-linear section of the diagram followed by an abrupt drop of the force acting on the sample revealing the so-called yield drop [3]. In some time, the force preserves almost the constant value; the respective section of the diagram is called the yield plateau followed by a hardening section in the diagram. Given below the physical phenomena are drawn to describe the graphic images [2] described above, and a simplified mathematical model is constructed.

24.2 Occurrence of Non-elastic Strain

When the yield plateau starts to form, plastic strain develops extremely unevenly. At first, a small zone (as compared to the specimen length) goes to a plastic condition. As the strain grows, this zone expands, and the stress–strain state of the plastic zone material becomes almost homogeneous everywhere except for the vicinity of the boundary between the plastic and elastic zones. After the yield plateau ends, the entire working part of the specimen experiences homogeneous plastic strain. After that moment, the diagram shows the final hardening that was hidden on the yield plateau creating an illusive image of ideal plasticity.

The fact that plastic strain is not localized in some volume after the drop occurs (similarly to what takes place in the formation of a "neck") but rather distributed across the specimen proves the material hardening from plastic strain immediately after the yield drop occurs.

V. Molotnikov, A. Molotnikova, *Theory of Elasticity and Plasticity*,
https://doi.org/10.1007/978-3-030-66622-4_24

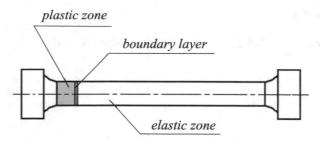

Fig. 24.1 Propagation of plastic strain in the elongation of a specimen of plastic material

We have already said (p. 322) that in the case of loading the specimen beyond the elastic limit, some diffusion processes occur in the material before yield, which cause low non-elastic strain of pre-yield. These diffusion processes can significantly decrease the shear resistance by changing the location of dislocations in the Cottrell cloud.

Let us use the yield condition (23.1) represented as follows:

$$\sigma_y = \sigma_\infty - \int_{t_y}^{T_0} F[\sigma(t)]dt, \tag{24.1}$$

where σ_y is the normal stress at the yield stress of the specimen material in elongation at some constant rate of displacement of pull test machine grips, σ_∞ is the same but for an infinitely high rate of loading, t_y is the moment of reaching the elastic limit, T_0 is the moment of yield occurrence, F is the aging function, and t is time or any other monotonously growing parameter.

24.3   Origins of Boundary Layer Theory

As earlier (p. 309), we will represent a non-elastic strain of the elongated specimen by a consequence of the displacement of structural imperfections in the body such as dislocations blocked by the Cottrell cloud [1] forming from foreign atoms and vacancies. It is harder to pick up a dislocation from a cloud and other obstacles than to move it. Hence it follows that the initial shear resistance (σ_0) is lower than the upper yield stress (see Fig. 22.1 and Axiom 22.1).

At the boundary of the element where plastic strain occurred and a neighboring elastic element, there is a high gradient of the density of structural imperfections. The material of this boundary is an obstacle for the displacement of dislocations moved in the plastic zone. The latter dislocations have pressure upon dislocations located in the boundary layer. As a result of this pressure, there are some changes related to the reconfiguration of the position of dislocations and other structural defects. Therefore, the boundary of the elastic part of the specimen and the plastic zone must be considered as a material layer whose strength properties differ from the properties of an elastic material remotely located from the boundary of the elastic and plastic zones. The strength σ_b of the boundary layer[1] can be deemed the function of the density of dislocations in the plastic zone. The latter is related to the plastic strain. Since plastic strain in the element that has become plastic reaches the value of macro-strain ε_0 over a short period of time at the end of the yield plateau in these test conditions, the boundary layer strength σ_b can be finally deemed to be dependent on the value ε_0, e.g.

$$\sigma_b = p(\varepsilon_0), \qquad (24.2)$$

where p is some positive function of this argument.

Since the boundary layer thickness is indefinitely low, there is a high frequency of stress variance in the yield process (as compared to the own frequency of the test machine). If the yield stress substantially exceeds the initial shear resistance, the formation of the initial yield drop will be accompanied by a jerk that will cause low-frequency converging oscillations mainly defined by the machine design. In a steady-state process, the "physical" yield stress means some average value between the upper yield stress and initial shear resistance; its calculation under the defined upper yield stress and shear resistance is a complex task of dynamics that we will consider in a simplified setting.

24.4 Simplified Model of Non-elastic Strain Growth

As established above, three zones exist in the specimen at the same time whose material has various strength properties and various degrees of deformation: a zone of elastic material where stresses are related to strain by Hooke's law, a plastic zone whose strain properties are described by its relations, and a boundary layer where stress cannot exceed its strength (24.2). If we know the material properties in all three zones, we can build a pattern of yield propagation in a sample. Let us consider the first stage of this task.

Assume that Δx (Fig. 24.2) means the length of a specimen element that is the first to go to the plastic state. To simplify studies, we will adopt that the specimen length has a low "obliquity" of strength properties so that the material strength from

Fig. 24.2 To the definition of the plastic strain rate

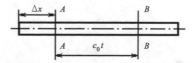

---

[1]Boundary layer strength means the stress when the bearing capacity of this layer is depleted.

one end of the working part of the specimen increases to the other end. Then the element Δx will be located near the end of the specimen with the lower material strength. The origin of the plastic strain in the element is accompanied by the drop of stress in this element from the upper yield stress σ_y to the initial shear resistance σ_0 and then by its rise as per the law (24.1).

This stress drop is of impulse nature, but the diagram shows the rise "blurred" due to the inertia of mobile parts of the loading device. As a result of the stress rise, an elastic wave of unloading will propagate through the specimen at the rate c_0 of sound propagation in the material. Let us use $A - A$ (Fig. 24.2) to designate the boundary of the plastically deformed element Δx and the elastic part of the specimen. Assume that after $t = 0$, there is a moment in time when the stress drops in the element from σ_y to σ_0. As a result of the unloading elastic wave, the cross-section $A - A$ will move for the time dt to the distance

$$du = \frac{\sigma_y - \sigma(t)}{E} c_0 dt,$$

where E is the Young modulus.

For the time t, the elastic wave front will reach the cross-section $B - B$ (Fig. 24.2) distanced from $A - A$ to the distance $c_0 t$, and the displacement of the section $A - A$ will be

$$u(t) = \frac{c_0}{E} \int_0^t [\sigma_y - \sigma(\xi)]d\xi, \quad \left(t < \frac{2l}{c_0}\right), \tag{24.3}$$

whereas l is the length of the specimen working part. The displacement of the section $A - A$ will cause the plastic strain of the element Δx:

$$\varepsilon = \frac{u(t)}{\Delta x} = \frac{c_0}{E \cdot \Delta x} \int_0^t [\sigma_y - \sigma(\xi)]d\xi. \tag{24.4}$$

Let us make an assumption that in the case of uniaxial loading, the link between stress $\sigma(t)$, plastic strain ε, and its rate $\dot{\varepsilon}$ can be expressed by the formula:

$$\sigma(t) = k_0 + k_1\varepsilon(t) + k_2\dot{\varepsilon}(t), \quad (k_0, k_1, k_2 - const). \tag{24.5}$$

By substituting strain (24.4) and its derivative into this formula, we obtain the following integral equation relative to the function $y(t) = \sigma_y - \sigma(t)$:

$$y(t) = \frac{c_1}{c_2} - \frac{c_3}{c_2} \int_0^t y(\varsigma)d\varsigma, \tag{24.6}$$

where

$$c_1 = \sigma_y - k_0; \quad c_2 = 1 + \frac{k_2}{\Delta x\sqrt{E\rho}}; \quad c_3 = \frac{k_1}{\Delta x\sqrt{E\rho}};$$

and ρ is the specimen material density.

A solution to Eq. (24.6) is the function

$$y(t) = \frac{c_1}{c_2} \exp\left[-\frac{c_3}{c_2}t\right],$$

e.g.

$$\sigma(t) = \sigma_y - \frac{\Delta x \sqrt{E\rho}(\sigma\sigma_y - k_0)}{\Delta x \sqrt{E\rho} + k_2} \exp\left[-\frac{k_1 t}{\Delta x \sqrt{E\rho} + k_2}\right]. \tag{24.7}$$

Hence we have

$$\sigma_0 = \sigma(t)|_{t \to 0} = \sigma_y - \frac{\Delta x E(\sigma_y - k_0)}{\Delta x E + k_2 c_0}. \tag{24.8}$$

The latter formula defines the value of the stress rise at the start of specimen yield. The stress σ_0 is the lower (physical) yield stress of the material. As shown in formula (24.8), this stress depends on the length Δx of the initial area of plastic strain that, in its turn, depends on the geometric shape of the sample and the homogeneity and isotropic nature of its material, etc., and so it is a random value if the specimen is rather long.

24.5   Definition of the Plastic Zone Growth Rate

Substituting the solution (24.7) into formula (24.4) gives as follows:

$$\varepsilon(t) = \frac{\sigma_y - k_0}{k_1}\left[1 - \exp\left(-\frac{k_1 c_0 t}{\Delta x E + c_0 k_2}\right)\right].$$

By assuming $\varepsilon(t) = \varepsilon_0$ here, we will find the time t_0, when the entire element Δx acquires plastic strain ε_0 equal to the strain in the end of the yield plateau in the elongation diagram:

$$t_0 = \frac{\Delta x E + c_0 k_0}{k_1 c_0} \ln \frac{\sigma_y - k_0}{\sigma_y - k_0 - \varepsilon_0 k_1}. \tag{24.9}$$

Then we will obtain the following formula for the rate c_p of plastic strain propagation along Δx:

$$c_p = \frac{\Delta x}{t_0} = \frac{\Delta x k_1 c_0}{(\Delta x E + c_0 k_2)\ln[(\sigma_y - k_0)/(\sigma_y - k_0 - \varepsilon_0 k_1)]}. \tag{24.10}$$

24.6 Steady-State Yield

After plastic strain in the element Δx, low-frequency oscillations caused by the formation of the yield drop converge. Before the formation of another boundary layer, the stress reaches the upper yield stress and is then reduced in a zigzag manner causing waves of elastic unloading. These waves partially propagate into the testing device and are partially reflected from specimen fillets. Rapid calculations can be used to show that these wave processes significantly fade after the fourfold reflection of the wave from fillets. If V is the speed of relative displacement of tester grips, the mutual distancing of grips during the fourfold reflection of the wave will be

$$\Delta l = V \Delta t, \tag{24.11}$$

whereas

$$\Delta t = \frac{4l}{c_0}.$$

Let us use Δx_1 to designate the length of another element where plastic strain occurred. Since the relative strain of this element after going to the plastic state is ε_0, we have

$$\varepsilon_0 = \frac{\Delta l}{\Delta x_1}. \tag{24.12}$$

Formulas (24.11)–(24.12) allow getting a condition when wave process will fade before a new rise of stress:

$$\Delta x_1 = \frac{4Vl}{\varepsilon_0 c_0}. \tag{24.13}$$

24.7 Building an Elongation Diagram

The mechanism of the yield drop and plateau formation described above is based on the fact that the initial shear resistance is below the upper yield stress. In accordance with the considered model, the true stress diagram in those elementary volumes of the material with stress drop looks as shown in Fig. 24.3a. After the stress drop, hardening occurs. However, it will be hidden in the "stress–strain" diagram for the entire specimen until the observed yield plateau reaches the limit value. This is when the entire specimen will go to the plastic state and the hardening visible in the diagram is continued hardening shown in Fig. 24.3a with a dashed line.

For the specimen in general, the stress diagram on the yield plateau represents a number of rises shown in Fig. 24.3b by a dashed line; a continuous line reflects the

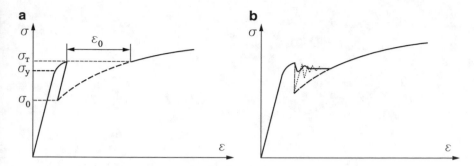

Fig. 24.3 To building the plastic material elongation diagram

dynamic properties of the "sample—loading device" system rather than the strain properties of the specimen material, which creates an illusion of ideal plasticity. Having this in mind, it is logical to represent the yield process on the plateau as having constant stress equal to the upper yield stress and to depict the hardening process with a single curve in both visible and hidden zones of the diagram.

References

1. A. Kottrell, *Teoriya dislokatsii* (Dislocation theory). (Mir Publ., Moscow, 1969)
2. V. Molotnikov, A. Molotnikova, Razvitie neuprugoi deformatsii pri standartnykh ispytaniyakh stal'nykh obraztsov (Development of inelastic deformation at standard tests of steel samples), in *V Polyakhovskie chteniya. Mezhdunar. nauchn. konf. po mekh-ke : tez. dokl.* (Development of inelastic deformation at standard tests of steel samples). (Izd-vo SPBGU Publ., Saint Petersburg, 2009), pp. 56–59
3. B. Petukhov, Teoriya zuba tekuchesti v malodislokatsionnykh kristallakh [Theory of tooth fluidity in low-dislocation crystals]. J. Tech. Phys. **71**(11), 42–47 (2001)

Chapter 25
Building a Shear Resistance Operator

25.1 General Form of the Shear Resistance Operator

Let us remind that due to Axiom 22.3 (see p. 323), the shear resistance $S_{\nu\lambda}$ in the most general case is some integration–differentiation operator from slip intensity (φ_{nl}). As per Axiom 22.2, any local slip in the model of a poly-crystalline body changes mechanical properties (almost) in all directions. Hence, the operator $S_{\nu\lambda}\varphi_{nl}$ must contain the integration operation in the directions n and l. It must also contain an integration–differentiation operation for the time (t), since time effects (aging, recovery, and (based on Axiom 22.1) plastic strain rate) have a significant effect on the dependency between stress and strain beyond the yield stress. Moreover, as given above (Axiom 22.3), shear resistance must depend on elastic strains (but not their rates!); for example, it must explicitly depend on the stress tensor components.

Based on the above, let us adopt that plastic shear resistance is a function (Φ) from the invariants (τ_i, τ_m) of the stress deviator, the primary argument of isotropic aging (Σ), the components of the aging tensor ($\Sigma_{\nu\lambda}$), and some operator ($L_{\nu\lambda}\varphi_{nl}$) from slip intensity, as well as from the rate of this operator:

$$S_{\nu\lambda}\varphi_{nl} = \Phi\left(\tau_i, \tau_m, \Sigma, \Sigma_{\nu\lambda}, L_{\nu\lambda}\varphi_{nl}, \frac{\partial}{\partial t}L_{\nu\lambda}\varphi_{nl}\right). \tag{25.1}$$

For low plastic strains in the conditions of almost simple strain (p. 331), let us represent the function Φ from formula (25.1) as

$$S_{\nu\lambda}\varphi_{nl} = \psi(\tau_i, \tau_m) + \Psi(\tau_i, \ldots)\left[\left(1 + \varepsilon\frac{\partial}{\partial t}\right)L_{\nu\lambda}\varphi_{nl}\right] - U_{\nu\lambda}, \tag{25.2}$$

where ψ is the generalized function of elastic softening, ε is a small parameter, $U_{\nu\lambda}$ is defined by formula (23.11), and Ψ is a function from arguments not depending on the directions n and l.

© The Author(s), under exclusive license to Springer Nature Switzerland AG 2021
V. Molotnikov, A. Molotnikova, *Theory of Elasticity and Plasticity*,
https://doi.org/10.1007/978-3-030-66622-4_25

Using the results of slip synthesis (p. 306), let us choose the operator $L_{\nu\lambda}\varphi_{nl}$ as follows:

$$L_{\nu\lambda}\varphi_{nl} = a\varphi_{\nu\lambda} + b\Phi_{\nu\lambda} + gr_{\nu\lambda} + c\gamma_{\nu\lambda}, \quad (a, b, c, g - const). \tag{25.3}$$

The first addend here is proportional to the displacement intensity of dislocations $\varphi_{\nu\lambda}$ in those planes and the directions where the shear resistance $S_{\nu\lambda}$ is defined. The second addend is proportional to the vector slip intensity and changes as a projection to the direction λ of shear in the direction l for $n = \nu$, e.g. as a shear component in the plane with the normal line ν from slip in the same plane. The third addend is proportional to the tensor intensity of slip, e.g. to the shear component $(\gamma_{\nu\lambda})$ in the plane with the normal line ν from all slips in planes parallel to the direction where the shear resistance is sought. The last addend is proportional to the strain component $\gamma_{\nu\lambda}$ from slips in all planes and directions.

To describe the strain of hardening materials in the studies [2, 5], the operator $L_{\nu\lambda}\varphi_{nl}$ contained only two addends (either $a\varphi_{\nu\lambda}$ and $c\gamma_{\nu\lambda}$ or $b\Phi_{\nu\lambda}$ and $c\gamma_{\nu\lambda}$), since setting the operator $L_{\nu\lambda}\varphi_{nl}$ in a more general form (for example, as (25.3) significantly simplifies the solution of some partial problems. Furthermore, the papers given in this paragraph in the expression of shear resistance (25.2) did not take into account aging and the effects of the slip rate ($\varepsilon = A = B = 0$) and also assumed $\psi \equiv \Psi$. The shear resistance dependency on the rate of the operator $L_{\nu\lambda}\varphi_{nl}$ was first introduced in the article [3].

25.2 Boundary Condition

The operator (25.3) can also be written as follows:

$$L_{\nu\lambda}\varphi_{nl} = \int_{\{L\}} \varphi_{\nu l} F_1(\omega_{l\lambda}) d\omega_{l\lambda} + gr_{\nu\lambda} + c\gamma_{\nu\lambda}, \quad (\omega_{l\lambda} = \widehat{l\lambda}), \tag{25.4}$$

where

$$F_1(\omega) = \frac{a}{2} [\delta(\omega) - \delta(|\omega| - \pi)] + b\cos\omega.$$

In the latter formula, δ is the Dirac function, and the function $F_1(\omega)$ being the core of the integral operator (25.4) characterizes the sensitivity of the material to strain anisotropy.

If we adopt the continuity postulate [1], the function F_1 must satisfy the condition

$$F_1(0) = \infty \qquad \text{at } a \neq 0. \tag{25.5}$$

When the integration–differentiation operator (25.2) is used and when the condition (25.5) is fulfilled, there must be as follows at the boundary Γ of the slip area:

$$\frac{\partial \varphi_{nl}}{\partial t}\Big|_{\Gamma} = 0, \qquad (25.6)$$

which coincides with formula (20.11).

The continuity postulate (25.6) helps defining the slip area boundary. When the operator $L_{\nu\lambda}\varphi_{nl}$ contains no addend giving a feature of (25.5) type (for example, at $a = 0$), the slip area boundary is defined from the condition (20.7), e.g.

$$S_{\nu\lambda}\varphi_{nl} > \tau_{\nu\lambda} \qquad \text{at } (\nu, \lambda) \notin R.$$

25.3 Special Cases

Let us assume in formula (23.11) $A\Sigma_{\nu\lambda}, B\Sigma - const$. In this case, the operator (25.2) analytically expresses as follows:

- shear resistance of a plastic material is defined by the slip intensity (φ_{nl}) and its rate ($\dot{\varphi}_{nl}$) only at the considered moment of time.

This provision will be called the generalized anti-isotropy postulate. It expands the anti-isotropy postulate formulated by M.Ya. Leonov [1] to materials for which the shear resistance depends on the slip rate. Its applicability is related not only to the material properties but also to the nature of body loading.

The necessary condition for the material to satisfy the formulated provision is the requirement of the independent nature of the hardening curve from the loading history until the start of yield. The experiments show [4] that some materials (low-carbon steels) satisfy this requirement.

If we consider that the following equation is fulfilled at the moment (T_0) of slip start

$$A + B = 1 \qquad \text{at } \varphi_{nl}\Big|_{t=T_0} \equiv 0, \qquad (25.7)$$

it follows from formulas (25.2) and (23.1) that the initial shear resistance will be equal to the yield stress. This means that the operator (25.2) will describe a hardening plastic body in a general case whose yield stress depends on the strain duration before the yield start. This body can be represented to consist of an infinitely large number of disorderly oriented threads for each of which the non-elastic properties are described by Eq. (23.1) and operator (25.2) at $A + B > 1$. The non-simultaneity of the yield onset in these threads and the effect of their interaction in macro-homogeneous loading of a body can be taken into account if we consider A and B to be the functions of time satisfying the condition (25.7) at $t = T_0$.

If the yield stress of a hardening material is not sensitive to loading time until the yield onset, then $\Sigma = 0$; the yield stress then coincides with the elastic limit.

Finally, if hardening curves obtained for various test rates can be aligned by parallel transfer, then for such material $\varepsilon = 0$.

References

1. M. Leonov, *Osnovnye postulaty teorii plastichnosti [The Basic Postulates of the Theory of Plasticity]*, DAN SSSR [Reports of the USSR Academy of Sciences], no. 1, vol. 199 (1971), pp. 51–54
2. M. Leonov, K. Rusinko, Analiticheskoe issledovanie ehffekta Baushingera pri kruchenii [Analytical study of the effect Bauschinger at torsion], in *Deformatsiya neuprogogo tela : sb. nauchn. tr. [Deformation Neuprugogo Body: Collection of Scientific Work]* (Izd-vo Ilim Publication, Frunze, 1971), pp. 15–30
3. M. Leonov, V. Molotnikov, B. Rychkov, The development of the concept the slip theory of plasticity, in *XIII International Congress According to the Theoretical and Applied Mekh-ks (IUTAM)* (IUTAM Publication, Moscow, 1972), pp. 31–32
4. E. Lomakin, A. Mel'shanov, *Povedenie malouglerodistoi stali pri rastyazhenii [Behavior of Low Carbon Steel at Stretching]*. Izv. AN SSSR, Mekhan. tv. tela, no. 4 (1971), pp. 150–158
5. K. Rusinko, Nekotorye voprosy deformatsii plastichnykh i khrupkikh tel [Some of the issues of deformation in ductile and brittle solids], Dr. philos. sci. diss. Novosibirsk, Siberian branch of the USSR Academy of Sciences Publ., 1971, 241 p., PhD Thesis, Institute of Mechanics of the USSR Academy of Sciences, 1971

Chapter 26
Full Bauschinger Effect

26.1 Secondary Yield Stress

For the considered materials, the initial section of the secondary loading diagram (after preloading beyond the yield stress, unloading, and changing the loading sign to the opposite) has a form of a smooth rounded curve, and defining the position of the point $(Q$, Fig. 21.1) on it (and on the unloading curve), starting with which plastic strain starts decreasing, causes serious complications. The stress corresponding to this point and characterizing the Bauschinger effect is called [2] the secondary yield stress. Most researchers define the value of the secondary yield stress at some quantity of "tolerance" for plastic strain increment. As shown by L.M. Kachanov [1], the secondary yield stress value substantially depends on the tolerance since hardening in the beginning of the secondary loading diagram for most materials is huge. In this manner, finding the secondary yield stress upon the tolerance is significantly conditional due to arbitrariness in selecting the tolerance.

On the other hand, as noted above, for plastic materials, an increment of non-elastic strain of an opposite sign is observed experimentally irrespective of the plastic strain quantity preceding unloading already in the case of full unloading. Since this increment usually exceeds the size of the order of elastic hysteresis, it may not be taken into account, and it can be believed that the opposite-sign plastic strain starts after changing the loading sign to the opposite one. This assumption eliminates arbitrariness in defining the tolerance when fixing the value of the secondary yield stress.

If we measure the Bauschinger effect by the ratio between the secondary yield stress and the yield stress of the material, according to the above assumption, this ratio equals zero. This Bauschinger effect will be called full.

26.2 Proportional Primary Loading

Let us consider the proportional loading of the body element beyond the yield stress. Assume that N and L designate a normal line to the maximum tangential stress plane and the direction of its action at some point in time t^0 until which proportional loading took place. Starting with the moment t^0 characterized by the stress τ_i^0, we will do proportional unloading of the element in the direction $(-L)$ opposite to L, $(d\tau_i/dt < 0)$. The previously described (see p. 333) behavior of steel specimens in elongation shows that the "freezing" of slips does not occur immediately in unloading in the specimen, since the unloading process is at first accompanied by a plastic strain increment of the same sign. Consequently, unloading is accompanied by a change in the slip area and rate.

Finally, at some moment t^* at the stress τ_i^*, the slip intensity rate $(\partial\varphi_{nl}/\partial t)$ will turn zero. The latter means that at the moment t^*, the "freezing" of slips will occur, and in the case of further unloading, the material will follow Hooke's law.

Based on the definition (20.6), the shear resistance equals the corresponding component of tangential stress: $S_{v\lambda}\varphi_{nl} = \tau_{v\lambda}$. By substituting into this equation the representation (25.2), in the direction of the maximum tangential stress τ_m^* ($\tau_m^* = \tau_{NL}^*$), we will find as follows at the moment t^{*1}:

$$\tau_{NL}^* = \psi(\tau_i^*, \tau_m^*) + \Psi(\tau_i^*, \dots)L_{NL}\varphi_{nl}^* - A\Sigma_{NL} + B\Sigma. \tag{26.1}$$

Hereinafter all values and functions marked with an asterisk are defined at the time moment t^*.

26.3 Proportional Loading of an Opposite Sign

Due to the antisymmetry (20.4) of slip intensity upon the argument l ($\varphi_{nl} \equiv -\varphi_{n(-l)}$), the integral operator $L_{v\lambda}$ has the property of

$$L_{v\lambda}\varphi_{nl} = -L_{v(-\lambda)}\varphi_{nl}. \tag{26.2}$$

For the same reason, a similar property is possessed by the aging tensor component:

$$\Sigma_{v\lambda} = -\Sigma_{v(-\lambda)}. \tag{26.3}$$

Using the condition of the equation of shear resistance to the tangential stress and properties (26.2)–(26.3) in the direction $(-L)$ for the material with full Bauschinger effect, we can write as follows:

---

[1]It is taken into account that at the moment $t = t^*$, $\partial L_{v\lambda}\varphi_{nl}/\partial t = 0$.

Fig. 26.1 To building the function Ψ

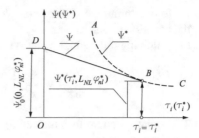

$$0 = \psi(0, 0) - \Psi(0, \ldots)L_{NL}\varphi_{nl}^* + A\Sigma_{NL} - B\Sigma. \tag{26.4}$$

It follows from formulas (26.1) and (26.4) that

$$\Psi^* = \frac{\tau_{NL}^* - \psi\left(\tau_i^*, \tau_{NL}^*\right) + (A + B)\Sigma}{L_{NL}\varphi_{nl}^*}, \quad \left[\Psi^* = \Psi\left(\tau_i^*, L_{NL}\varphi_{nl}^*\right)\right], \tag{26.5}$$

$$\Psi_0 = \frac{\psi(0, 0) + (A - B)\Sigma}{L_{NL}\varphi_{nl}^*}, \quad \left[\Psi_0 = \Psi\left(0, L_{NL}\varphi_{nl}^*\right)\right]. \tag{26.6}$$

The latter formulas define the function Ψ in the strain of a body element in two directions: in the direction L to the stress $\tau_i = \tau_i^*$ and in the opposite direction $-L$. For other directions, the function Ψ is not defined yet. For some direction L_1 differing from L and $-L$, we can find the value of the function Ψ similar to the above, if we know the secondary yield stress in the given direction $(-L_1)$.

The function Ψ must thus depend on the directions N and L. If we depict the calculated values of the function Ψ in the plane $\Psi \sim \tau_i$ (Fig. 26.1), formulas (26.5) and (26.6) give the values of the function Ψ in two points: at $\tau_i = \tau_i^*$ and $\tau_i = 0$. For other directions, the values Ψ can also be calculated and shown on the figure plane. By connecting the acquired points, we find the graphical representation Ψ in the plane $\Psi \sim \tau_i$ that will also reflect the dependency of the function Ψ on the directions (N and L). In the first approximation, we will approximate the function Ψ of a straight line coming through the points $\tau_i = \tau_i^*$ $\tau_i = 0$:

$$\Psi = \Psi_0 - \frac{\tau_i}{\tau_i^*}\left(\Psi_0 - \Psi^*\right), \quad (0 \leqslant \tau_i \leqslant \tau_i^*). \tag{26.7}$$

In Fig. 26.1, the point B corresponds to the value of the function Ψ at the moment t^* when loading in the initial direction (L); the y-coordinate of the point D represents the value Ψ at the moment of slip start when changing the loading sign to the opposite one. The dash line ABC schematically shows the geometric place of the points B whose x-coordinates equal the stress τ_i^* at the moment of freezing slips in unloading and y-coordinates equal the Ψ values at this moment. A continuous line BD depicts the approximation (26.7).

To determine the dependency $\Psi^* \sim \tau_i^*$ (line ABC, Fig. 26.1), we will require the equation of derivatives $\partial \Psi / \partial \tau_i$ and $\partial \Psi^* / \partial \tau_i^*$ in point $\tau_i = \tau_i^*$. Then we obtain the following differential equation:

$$\frac{\partial \Psi^*}{\partial \tau_i^*} = \frac{\Psi^*}{\tau_i^*} \left[1 - \frac{\psi(0,0) + (A-B)\Sigma}{\Psi^* L_{NL} \varphi_{nl}^*} \right]. \tag{26.8}$$

Taking into account formula (26.5), the overall integral of this differential equation is written as follows:

$$\ln \frac{R\tau_i^*}{\Psi^*} = \int \frac{[\psi(0,0) + (A-B)\Sigma]d\tau_i^*}{\tau_i^*[\tau_m^* - \psi(\tau_i^*, \tau_m^*) + (A+B)\Sigma]}, \tag{26.9}$$

where R is the integration constant value.

In the case of proportional loading, there is the equation

$$\tau_m^* = q\tau_i^*, \quad (q - const) \tag{26.10}$$

(for pure shear, $q = \sqrt{3/2}$ and for uniaxial stressed State, $q = 3\sqrt{2}/4$). Taking into account the latter dependency, the right part of formula (26.9) can be integrated. By substituting into (26.9) the expression for the function ψ according to (22.7) and (22.4), after calculating the integral, we obtain as follows:

$$\Psi^* = R\tau_i^* \left[\frac{\tau_i^*}{(k+q)\tau_i^* - T + (1-m)(A+B)\Sigma} \right]^{\frac{T+(1-m)(A-B)\Sigma}{T-(1-m)(A+B)\Sigma}}, \tag{26.11}$$

$$\left(\tau_i^* > [T - (1-m)(A+B)\Sigma]/(k+q) \right).$$

Taking into account the ratios (26.10) and formulas (22.4), the dependency (26.11) can be rewritten as follows:

$$\Psi^* = R\tau_i^* \left[\frac{\tau_i^*}{\tau_m^* - f(\tau_i^*) + (1-m)(A+B)\Sigma} \right]^{\frac{T+(1-m)(A-B)\Sigma}{T-(1-m)(A+B)\Sigma}}, \tag{26.12}$$

$$\left(\tau_i^* > [T - (1-m)(A+B)\Sigma]/(k+q) \right).$$

By combining the dependencies (26.7) and (26.12), we can write as follows:

$$\Psi = \begin{vmatrix} \Psi_0 - \frac{\tau_i}{\tau_i^*}(\Psi_0 - \Psi^*) & \text{at } 0 \leqslant \tau_i \leqslant \tau_i^*, \\ \Psi^* & \text{at } \tau_i = \tau_i^* \end{vmatrix} \tag{26.13}$$

in

$$\left(\tau_i^* > [T - (1 - m)(A + B)\Sigma]/(k + q)\right).$$

Let us note that in the expression (25.3) for the integral operator $L_{\nu\lambda}$, each addend contains a multiplier as the constant value (a, b, c, g). When substituting into the shear resistance (25.2) the functions Ψ using formulas (26.12) and (26.13), we can re-designate $a_1 = aR, \ldots, c_1 = cR$, where $a_1, \ldots, c_1$ are new constant values. Not to implement such redesignations, below we adopt the integration constant value $R = 1$.

26.4 Function Ψ in Almost Simple Strain

Formulas (26.13) define the function Ψ in proportional loading to the stress τ_i^* with further unloading and proportional loading in the opposite direction. In what follows, we will show that using the expressions (26.13) for the function Ψ gives high-quality diagrams in alternating-sign loading. Moreover, as said above, when building formulas (26.13), the dependency of the function Ψ on the directions of further additional loading is indirectly taken into account. This allows suggesting that for the case of almost simple strain, the function Ψ can also be adopted in the form (26.13) if in the expressions (26.5) and (26.6), the value of the operator $L_{NL}\varphi_{nl}^*$ at the time t^* is replaced by the current value of this operator $L_{NL}\varphi_{nl}$ in the direction of maximum tangential stress at the current moment in time t. In the case of such replacement, the expression (26.12) remains the same, and only the upper formula (26.13) changes where the function Ψ_0 now looks as follows:

$$\Psi_0(0, L_{NL}\varphi_{nl}) = \frac{\psi(0, 0) + (A - B)\Sigma}{L_{NL}\varphi_{nl}}.$$

With the function Ψ known, the shear resistance operator (25.2) is fully defined, which allows testing it in specific partial problems.

References

1. L. Kachanov, Ob ehksperimental'nom opredelenii posleduyushchikh poverkhnostei na-gruzheniya i ehffekta Baushingera [About experimental determination of the following loading surfaces and the Bauschinger effect], in *Issledovaniya po uprugosti i plastichnosti : sb. nauchn. tr. Vyp. 8.* (Studies elasticity and plasticity: Collection of scientific. Tr. vol. 8) (Izd-vo LGU Publ., Leningrad, 1971) pp. 108–112
2. G. Talypov, *Plastichnost' i prochnost' stali pri slozhnom nagruzhenii* (Ductility and strength of steel under complex loading) (Izd-vo LGU Publ., Leningrad, 1968)

Chapter 27
Non-elastic Uniaxial Elongation–Compression

27.1 Calculating Slip Intensity

Assume that only the expression $\sigma_z(t)$ differs from zero from all stress tensor components. Taking into account that the tangential stress component $\tau_{\nu\lambda}$ is defined using formula (20.8), and the shear component $\gamma_{\nu\lambda}$ is calculated [6] using the formula:

$$\gamma_{\nu\lambda} = 2(\nu_x\lambda_x\varepsilon_x + \nu_y\lambda_y\varepsilon_y + \nu_z\lambda_z\varepsilon_z) + (\lambda_x\nu_y + \lambda_y\nu_x)\gamma_{xy} + \\ + (\lambda_y\nu_z + \lambda_z\nu_y)\gamma_{yz} + (\lambda_z\nu_x +_x \nu_z)\gamma_{xz}, \tag{27.1}$$

and for uniaxial elongation–compression, we have

$$\tau_m = \frac{\sigma_z}{2}, \ \tau_{\nu\lambda} = \frac{\sigma_z}{2}\sin 2\alpha\cos\omega, \ \gamma_{\nu\lambda} = \frac{3}{2}\varepsilon_z\sin 2\alpha\cos\omega, \ \tau_i = \frac{\sqrt{2}}{3}\sigma_z.$$

According to the definition, plastic strain resistance in the slip area equals the corresponding component of tangential stress: $S_{\nu\lambda}\varphi_{nl} = \tau_{\nu\lambda}$. By substituting the previous formulas into this equation, as well as the representations (25.2) and (25.3), we obtain[1]

$$0.5\sigma_z(t)\sin 2\alpha\cos\omega = \psi(t) + \Psi(t)\left[\left(1 + \varepsilon\frac{\partial}{\partial t}\right)L_{\nu\lambda}\varphi_{nl}\right] - \\ - A\Sigma\sin 2\alpha\cos\omega - B\Sigma, \tag{27.2}$$

$$(\psi(t) = \psi[\tau_i(t), \tau_m(t)], \ \Psi(t) = \Psi[\tau_i(t), L_{NL}\varphi_{nl}(t)]),$$

---

[1]Due to the symmetry of stress and strain states in the considered case of uniaxial elongation, the components $\tau_{\nu\lambda}$ and $\gamma_{\nu\lambda}$, as well as the function φ, do not depend on the variable β.

© The Author(s), under exclusive license to Springer Nature Switzerland AG 2021
V. Molotnikov, A. Molotnikova, *Theory of Elasticity and Plasticity*,
https://doi.org/10.1007/978-3-030-66622-4_27

or

$$0.5\sigma_z(t) \sin 2\alpha \cos \omega = \psi(t) + \Psi(t)\Big\{a[\varphi_{\nu\lambda}(t) + \varepsilon\dot\varphi_{\nu\lambda}(t)]+$$

$$+ b[\Phi_{\nu\lambda} + \varepsilon\dot\Phi_{\nu\lambda}]+$$

$$+ g(r_{\nu\lambda} + \varepsilon\dot r_{\nu\lambda}) + 1.5c[\varepsilon_z(t) + \varepsilon\dot\varepsilon_z(t)] \sin 2\alpha \cos \omega\Big\}-$$

$$\tag{27.3}$$

$$- A\Sigma \sin 2\alpha \cos \omega - B\Sigma,$$

where the point above the symbol designating the function means, as earlier, the differentiation of this function in time.

Taking into account the designations (20.17), formula (20.16) defining the slip vector intensity in the considered case of uniaxial elongation can be represented as

$$\Phi_{\nu\lambda} = 2\int_{-\Omega(\alpha,t)}^{\Omega(\alpha,t)} \varphi(\alpha, \omega_0, t) \cos(\omega - \omega_0)d\omega_0, \tag{27.4}$$

whereas $\Omega(\alpha, t)$ is the boundary value of the angle ω_0, defining the opening of the slip fan in the plane with the normal line $\nu(\alpha, \beta)$. By substituting formulas (20.17) into the definition of the tensor intensity of slip (20.21), we will find

$$r_{\nu\lambda} = 2\int_{-W}^{W} dw_0 \int_{-\Omega(\alpha_0,t)}^{\Omega(\alpha_0,t)} \varphi(\alpha_0, \omega_0, t)(\eta_\lambda \cos \omega_0 + \xi_\lambda \sin \omega_0) \cos^2 w_0 dw_0,$$

$$\tag{27.5}$$

where the interval $[-W, W]$ belongs to the arch (20.24).

27.2   Calculation of the Integral (27.5)

To calculate the curvilinear integral (27.5) along the section $[-W, W]$ of the arch (20.24), let us use the coordinate α_0 as the integration variable. From Eq. (20.24), let us find

$$\cos(\beta - \beta_0) = \frac{\Theta_1(\alpha, \alpha_0, \omega)}{\sin \alpha_0(\cos^2 \alpha + \mathrm{tg}^2\omega)},$$

$$\tag{27.6}$$

$$\sin(\beta - \beta_0) = \frac{\Theta_2(\alpha, \alpha_0, \omega)}{\sin \alpha_0(\cos^2 \alpha + \mathrm{tg}^2\omega)},$$

which designates

$$\Theta_1(\alpha, \alpha_0, \omega) = \sin \alpha \cos \alpha \cos \alpha_0 + H \operatorname{tg} \omega,$$

$$\Theta_2(\alpha, \alpha_0, \omega) = \sin \alpha \cos \alpha_0 \operatorname{tg} \omega +$$

$$+ \cos \alpha \sqrt{\sin^2 \alpha_0 (\cos^2 \alpha + \operatorname{tg}^2 \omega) - \sin^2 \alpha \cos^2 \alpha_0},$$

whereas

$$H = H(\alpha, \alpha_0, \omega) = \sqrt{\sin^2 \alpha_0 (1 + \operatorname{tg}^2 \omega) - \sin^2 \alpha}.$$

Taking into account the latter formulas and expressions (20.9), let us calculate

$$\eta_\lambda \cos \omega_0 + \xi_\lambda \sin \omega_0 = \frac{\cos \omega}{\sin \alpha_0} (\cos \omega_0 \sin \alpha + H \sin \omega_0). \tag{27.7}$$

Now, by calculating the product of single vectors $\vec{n}$ and $\vec{v}$, and by attracting the expressions (27.6), after some identical conversions, we will find

$$\cos w_0 = \frac{\cos \alpha \cos \alpha_0 (1 + \operatorname{tg}^2 \omega) + H \sin \alpha \operatorname{tg} \omega}{\cos^2 \alpha + \operatorname{tg}^2 \omega}. \tag{27.8}$$

Let us express the differential dw_0 via $d\alpha_0$. To do it, let us use the equation

$$\sin w_0 \vec{\lambda} = \vec{n} \times \vec{v}.$$

Hence we have

$$\sin w_0 = \frac{\sin \alpha \cos \alpha_0 \operatorname{tg} \omega - H \cos \alpha}{\cos \omega (\cos^2 \alpha + \operatorname{tg}^2 \omega)}.$$

By differentiating the latter formula, we obtain

$$\cos w_0 dw_0 = -\frac{H \sin \alpha \sin \alpha \alpha_0 \operatorname{tg} \omega + T \sin \alpha_0}{H \cos \omega (\cos^2 \alpha + \operatorname{tg}^2 \omega)} d\alpha_0, \tag{27.9}$$

where

$$T = T(\alpha, \alpha_0, \omega) = \cos \alpha \cos \alpha_0 (1 + \operatorname{tg}^2 \omega).$$

Based on the obtained ratios (27.7)–(27.9), formula (27.5) is converted into

$$r_{\nu\lambda} = 4 \int_{\alpha^*}^{\alpha} d\alpha_0 \int_{-\Omega(\alpha_0,t)}^{\Omega(\alpha_0,t)} \varphi(\alpha_0\omega_0, t) K(\omega, \omega_0, \alpha, \alpha_0) d\omega_0, \tag{27.10}$$

which designates

$$K(\omega, \omega_0, \alpha, \alpha_0) = \left[\frac{T(\alpha, \alpha_0, \omega) + H \sin \alpha \, \mathrm{tg} \, \omega}{\cos^2 \alpha + \mathrm{tg}^2 \omega} \right]^2 \times$$

$$\times \frac{\cos \omega_0 \sin \alpha + H \sin \omega_0}{H(\alpha, \alpha_0, \omega)}, \tag{27.11}$$

where α^* is the boundary of slip planes on the half-sphere of a single radius ($\alpha^* > \pi/4$), whereas based on the condition (25.6) of the rate continuity of the function φ at the boundary of the slip area and the initial condition (20.10), the dependency $\Omega(\alpha, t)$ is found from the condition[2]

$$\varphi(\alpha, \pm\Omega, t) + \varepsilon\dot\varphi(\alpha, \pm\Omega, t) = 0, \tag{27.12}$$

and the function $\alpha^*(t)$ is obtained from the last dependency at $\Omega = 0$, e.g.

$$\alpha^*(t) = \frac{1}{2} \arcsin \left\{ \frac{\psi[t] - B\Sigma}{0.5\sigma_z(t) + A\Sigma\Sigma - 1.5c\Psi[t][\varepsilon_z(t) + \varepsilon\dot\varepsilon_z(t)]} \right\}. \tag{27.13}$$

27.3 Solving the Integral Equation

The substitution of the results (27.4) and (27.10) into Eq. (27.3) gives

$$\frac{1}{2}\sigma_z(t) \sin 2\alpha \cos \omega = \psi[t] + \Psi[t] \{(a[\varphi(\alpha, \omega, t) + \varepsilon\dot\varphi(\alpha, \omega, t)] +$$

$$+ 2b \int_{-\Omega(\alpha,t)}^{\Omega(\alpha,t)} [\varphi(\alpha, \omega_0, t) + \varepsilon\dot\varphi(\alpha, \omega_0, t)] \cos(\omega - \omega_0) d\omega_0 +$$

$$+ 4g \int_{\alpha^*}^{\alpha} d\alpha_0 \int_{-\Omega(\alpha_0,t)}^{\Omega(\alpha_0,t)} [\varphi(\alpha_0, \omega_0, t) + \varepsilon\dot\varphi(\alpha_0, \omega_0, t)] \times \tag{27.14}$$

$$\times K(\omega, \omega_0, \alpha, \alpha_0) d\omega_0 + \frac{3}{2}c[\varepsilon_z(t) + \varepsilon\dot\varepsilon_z(t)] \sin 2\alpha \cos \omega\} -$$

$$- A\Sigma \sin 2\alpha \cos \omega - B\Sigma.$$

Let us introduce designations:

---

[2]This condition is violated in the paper [4], which makes its results incorrect; this error is rectified in the article [5].

$$\overline{\varphi}(\alpha, \omega, t) = \varphi(\alpha, \omega, t) + \varepsilon \frac{\partial}{\partial t} \varphi(\alpha, \omega, t),$$

$$\overline{\varepsilon}_z(t) = \varepsilon_z(t) + \varepsilon \frac{d}{dt} \varepsilon_z(t); \tag{27.15}$$

$$\overline{\lambda}_0(t) = \frac{\psi[t] - B\Sigma}{\Psi[t]}, \quad \overline{\lambda}_1(t) = \frac{\sigma_z(t) + 2A\Sigma}{2\Psi[t]} - \frac{3}{2} c \overline{\varepsilon}_z(t). \tag{27.16}$$

Equation (27.14) will be then written as

$$a\overline{\varphi}(\alpha, \omega, t) + 2b \int_{-\Omega(\alpha,t)}^{\Omega(\alpha,\omega,t)} \overline{\varphi}(\alpha, \omega_0, t) \cos(\omega - \omega_0) d\omega_0 +$$

$$+ 4g \int_{\alpha^*(t)}^{\alpha} d\alpha_0 \int_{-\Omega(\alpha_0,t)}^{\Omega(\alpha_0,t)} \overline{\varphi}(\alpha_0, \omega_0, t) K(\omega, \omega_0, \alpha, \alpha_0) d\omega_0 =$$

$$= \overline{\lambda}_1(t) \sin 2\alpha \cos \omega - \overline{\lambda}_0(t). \tag{27.17}$$

Thus, relative to the function $\overline{\varphi}$, we obtained an integral equation of a compact form. However, the analytical solution of this equation in case when all three constant values (a, b, and g) differ from zero is problematic due to the complexity of the core (27.11).

27.4 Study of the Tensor Intensity of Slips

To get a quality representation of the behavior of tensor intensity of slips ($r_{\nu\lambda}$) at uniaxial elongation, let us consider Eq. (27.17) at $a = \varepsilon = 0$. In this case, due to the symmetry of the problem, the slip vector in an arbitrary plane will coincide in the direction with $\overrightarrow{\eta}$ ($\omega = 0$). By assuming in formulas (27.4) and (27.11) $\omega = 0$ and taking into account that in the considered case $\Phi_{\nu\lambda} = \Phi_{\nu\lambda} \equiv \Phi(\alpha)$, $r_{\nu\lambda} \equiv r(\alpha)$, from (27.17) at $a = 0$ we obtained the following integral equation for the vector intensity of slips $\Phi(\alpha)$:

$$2b\Phi(\alpha) + 4g \int_{\alpha^*}^{\alpha} \Phi(\alpha_0) \frac{\sin \alpha \cos^2 \alpha_0 d\alpha_0}{\cos^2 \alpha \sqrt{\cos^2 \alpha - \cos^2 \alpha_0}} = \overline{\lambda}_1 \sin 2\alpha - \overline{\lambda}_0, \tag{27.18}$$

$$(\alpha \leqslant \alpha_0 \leqslant \alpha^*).$$

The expression (27.18) is an integral equation of Volterra Type 2 with a singular core. By assuming the constant value g to be low, let us select the function $\Phi_0(\alpha) = \overline{\lambda}_1 \sin 2\alpha - \overline{\lambda}_0$ as the zero approximation to solve it. By substituting the last function to Eq. (27.18), we will find the first approximation

Fig. 27.1 Dependency of the tensor intensity of slips on the opening angle of the slip plane fan

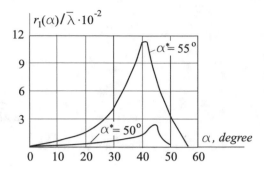

$$\Phi_1(\alpha) = \frac{g}{b} r_1(\alpha) + \frac{1}{2b}\left(\bar{\lambda}_1 \sin 2\alpha - \bar{\lambda}_0\right), \qquad (27.19)$$

where

$$r_1(\alpha) = -4\bar{\lambda}_1 \sin\alpha \left\{ \frac{2(2\cos^2\alpha + \cos^2\alpha^*)\sqrt{\cos^2\alpha + \cos^2\alpha^*}}{3\cos^2\alpha} - \right.$$

$$\left. -\frac{\bar{\lambda}_0}{\bar{\lambda}_1}[D(\cos\alpha) - D(v, \cos\alpha)]\right\},$$

$$D(k) = \frac{K(k) - E(k)}{k^2}, \quad D(v, k) = \frac{F(v, k) - E(v, k)}{k^2},$$

$$v = \arcsin\left(\frac{\cos\alpha^*}{\cos\alpha}\right);$$

$K(k)$ and $E(k)$ are full elliptical integrals [11], and $F(v, k)$ and $E(v, k)$ are full elliptical integrals of the first and second type, respectively. The second and further approximations for the function $\Phi(\alpha)$ are rather big due to the complexity of quadratures, and therefore we do not give them here. Figure 27.1 shows charts of the function $r_1(\alpha)/2\bar{\lambda}_1$ built at $\alpha^* = 50°$ and $\alpha^* = 55°$.

The above study shows the complexity of finding the tensor intensity of slips and, therefore, calculation of strains if the coefficient at tensor intensity in the shear resistance operator differs from zero. Hereinafter in this chapter, we will consider the case when $g = 0$. The problem becomes even simpler if we assume $a = 0$ and $g = 0$. This case for a hardening plastic material is studied in detail in the articles and theses of Blinov and Rusinko [1, 9, 10].

27.5   Determinant Equations in Uniaxial Elongation

For $g = 0$ from formula (27.7), we obtain the following integral equation relative to the function $\overline{\varphi}(\alpha, \omega, t)$:

$$a\overline{\varphi}(\alpha, \omega, t) + 2b \int_{-\Omega(\alpha,t)}^{\Omega(\alpha,t)} \overline{\varphi}(\alpha, \omega_0, t) \cos(\omega - \omega_0) d\omega_0 = \\ = \overline{\lambda}_1(t) \sin 2\alpha \cos \omega - \overline{\lambda}_0(t), \tag{27.20}$$

whereas the core of the last integral equation is confluent. Let us represent the solution of this equation as

$$a\overline{\varphi}(\alpha, \omega, t) = \overline{\lambda}_1(t) \sin 2\alpha \cos \omega - 2b\overline{\Phi}(\alpha, t) \cos \omega - \overline{\lambda}_0(t), \tag{27.21}$$

which designates

$$\overline{\Phi}(\alpha, t) = \int_{-\Omega(\alpha,t)}^{\Omega(\alpha,t)} \overline{\varphi}(\alpha, \omega_0, t) \cos \omega_0 d\omega_0. \tag{27.22}$$

By substituting the function $\overline{\varphi}$ as per (27.21) into the right part of the expression (27.9), let us find as follows after calculation of the integral:

$$\overline{\Phi}(\alpha, t) = \frac{\overline{\lambda}_1(t)(\Omega + 0.5 \sin 2\Omega) \sin 2\alpha - 2\overline{\lambda}_0(t) \sin \Omega}{a + b(2\Omega + \sin 2\Omega)}. \tag{27.23}$$

Based on the condition (27.2), the function $\overline{\varphi}$ at the boundary of the slip area ($\omega = \pm\Omega$) turns to zero. Hence we obtain the dependency between α and Ω:

$$\sin 2\alpha \cos \Omega = \frac{\overline{\lambda}_0(t)}{\overline{\lambda}_1(t)} \left[1 + \frac{b}{a}(2\Omega - \sin 2\Omega) \right]. \tag{27.24}$$

As per formula (20.5), plastic strain in elongation will be

$$\varepsilon_z = \int \int_R \varphi_{nl} l_z n_z d\omega_0 d\Omega, \quad (d\Omega = \sin \alpha_0 d\alpha_0 d\beta_0), \tag{27.25}$$

where R is the slip area. By differentiating equation (27.25) in time, we will find

$$\dot{\varepsilon}_z = \int \int_R \dot{\varphi}(\alpha_0, \omega_0, t) l_z n_z d\omega_0 d\Omega. \tag{27.26}$$

Taking into account the dependencies (20.9) and designations (27.15) from the latter two formulas, it follows that

$$\bar{\varepsilon}_z(t) = \int\limits_{R'} \int \int \bar{\varphi}(\alpha_0, \omega_0, t) \cos \omega_0 \cos \alpha_0 \sin^2 \alpha_0 d\omega_0 d\alpha_0 d\beta_0. \qquad (27.27)$$

To calculate the integral in the right part of formula (27.27), let us successively replace variables assuming at first that

$$\alpha_0 = \frac{\pi}{4} + v, \qquad (27.28)$$

and then

$$\cos 2v = F(\Omega), \quad F(\Omega) = \frac{\bar{\lambda}_0/\bar{\lambda}_1}{\cos \Omega} \left[1 + \frac{b}{a}(2\Omega - \sin 2\Omega) \right]. \qquad (27.29)$$

By using formula (27.22), we will obtain

$$\bar{\varepsilon}_z(t) = \frac{2}{3} \cdot \frac{\bar{\lambda}_1(t)}{a} J(u), \qquad (27.30)$$

where

$$J(u) = \frac{3\pi}{4} \int_0^u F(\Omega) \cdot \frac{F(\Omega)(\Omega + \frac{1}{2}\sin 2\Omega) - 2(\bar{\lambda}_0/\bar{\lambda}_1)\sin \Omega}{1 + \frac{b}{a}(2\Omega + \sin 2\Omega)} \times$$

$$\times \frac{F'(\Omega)d\Omega}{\sqrt{1 - F(\Omega)}}, \qquad (27.31)$$

whereas u is the positive square of Eq. (27.24) for $\sin 2\alpha = 1$, e.g.

$$\frac{\cos u}{1 + \frac{b}{a}(2u - \sin 2u)} = \frac{\bar{\lambda}_0}{\bar{\lambda}_1}. \qquad (27.32)$$

The latter formula and the ratio (27.24) show that the parameter u equals the opening of the slip fan in the plane where the maximum tangential stress acts. The function $J(u)$ can be easily tabulated for various values of the ratio b/a.

By substituting formula (27.30) into the definition $\bar{\lambda}_1$ according to (27.16), we will find

$$\bar{\lambda}_1(t) = \frac{1}{1 + \frac{c}{a}J(u)} \cdot \frac{\sigma_z(t) + 2A\Sigma}{2\Psi[t]}. \qquad (27.33)$$

From formula (27.16) and condition (27.32), we can obtain

$$\frac{\frac{1}{2}\sigma_z(t) + A\Sigma}{\psi[t] - B\Sigma} = \frac{1}{\delta(u)} \left[1 + \frac{c}{a}J(u) \right], \qquad (27.34)$$

where

$$\delta(u) = \frac{\cos u}{1 + \frac{b}{a}(2u - \sin 2u)}. \tag{27.35}$$

Using the result (27.34) and formulas (27.33) and (27.30), we will obtain the following equation for the definition of deformation:

$$\varepsilon_z(t) + \varepsilon \dot{\varepsilon}_z(t) = \frac{2}{3} \cdot \frac{\psi[t] - B\Sigma}{\Psi[t]} \cdot \frac{J(u)}{a\delta(u)}. \tag{27.36}$$

When the dependency $\sigma_z \sim t$ is known, the ratio (27.34) defines the function $u(t)$. Then Eq. (27.36) is a linear differential equation of the first order relative to the strain component $\varepsilon_z(t)$. The initial conditions for this equation will be

$$t = t_p : \quad \varepsilon_z(t_p) = 0,$$

where t_p is the moment in time corresponding to the occurrence of plastic strain. By solving equation (27.36) with the indicated initial conditions, we obtain as follows:

$$\varepsilon_z(t) = \frac{2}{3\varepsilon a} \exp\left[-\frac{1}{\varepsilon}(t - t_p)\right] \int_{t_p}^{t} \frac{\psi[\tau] - B\Sigma}{\Psi[\tau]} \cdot \frac{J[u(\tau)]}{\delta[u(\tau)]} \times$$

$$\times \exp\left[\frac{1}{\varepsilon}(\tau - t_p)\right] d\tau. \tag{27.37}$$

When the plastic strain rate $\dot{\varepsilon}_z(t)$ is defined, the system of Eqs. (27.34) and (27.36) defines the dependency $\sigma_z \sim \varepsilon_z$ in a parametric form (via the parameter u).

Note 1 The calculations show that the link $\sigma_z(\varepsilon_z, \dot{\varepsilon}_z)$ established by formulas (27.34)–(27.36) can be approximated with sufficient accuracy by formula (24.5). Therefore, the coefficients (k_0, k_1, and k_2) of formula (24.5) must be considered as constant values of approximation of the dependency $\sigma_z \sim \varepsilon_z$ in this paragraph by the function (24.5).

Note 2 From the obtained solution (27.34)–(27.36), it follows that for a medium whose non-elastic properties are described by the model (25.2)–(25.3), even in the case of uniaxial elongation, there is no strain theory since the connection between stress and plastic strain depends on the loading history.

27.6 Plastic Strain in Loading and Compression

27.6.1 *Increment of Non-elastic Strain in Loading*

Assume that the rod is elongated beyond the yield stress to some stress $\sigma_z = \sigma_z^o$, $\sigma_z^o = 3\tau_i^o/\sqrt{2}$, where τ_i^o is the octahedral tangential stress at the moment t^o. At $t > t^o$, stress starts decreasing under some law so that

$$\sigma_z(t) < \sigma_z^o, \quad \dot{\sigma}_z(t) < 0 \qquad \text{at} \qquad t > t^o. \tag{27.38}$$

For some time, strain will continue rising. Since neither the shear resistance $S_{\nu\lambda}$ nor the tangential stress component $\tau_{\nu\lambda}$ in the considered case changes its sign, the ratios of the previous paragraph remain in force. In particular, formulas (27.34) and (27.37) are true. Due to the conditions (27.38), formula (27.34) shows that in the case of unloading, the parameter u decreases.

Let us divide the integration interval in the right part of formula (27.37) into two segments: $t_i \leqslant \tau \leqslant t^o$ and $t^o \leqslant \tau \leqslant t$. By assuming $t_i = 0$, let us write

$$\varepsilon_z(t) = \varepsilon_z(t^o) \exp\left[-\frac{t - t^o}{\varepsilon}\right] +$$

$$+ \frac{2}{3a\varepsilon} \exp\left(-\frac{t}{\varepsilon}\right) \int_{t^o}^{t} \frac{\psi[\tau] - B\Sigma}{\Psi[\tau]} \times \frac{J[u(\tau)]}{\delta[u(\tau)]} e^{\tau/\varepsilon} d\tau. \tag{27.39}$$

In this manner, formulas (27.34) and (27.39) define the dependency between stress and strain in unloading. So it is necessary to find the moment of time t^* after which strain in unloading will follow Hooke's law. This moment can be found from the condition $\dot{\varepsilon}_z(t^*) = 0$. Using Eqs. (27.36) and (27.39), this condition gives the following equation to find the moment t^*:

$$\frac{2}{3a} \cdot \frac{[t^*] - B\Sigma}{\Psi[t^*]} \cdot \frac{J[u(*)]}{\delta[u(t^*)]} = \varepsilon_z(t^*) \exp\left[-\frac{t^* - t^o}{\varepsilon}\right] +$$

$$+ \frac{2}{3\varepsilon a} \exp\left(-\frac{t^*}{\varepsilon}\right) \int_{t^o}^{t^*} \frac{\psi[\tau] - B\Sigma}{\Psi[\tau]} \cdot \frac{J[u(\tau)]}{\delta[u(\tau)]}$$

$$\times \exp\left(\frac{\tau}{\varepsilon}\right) d\tau. \tag{27.40}$$

27.6.2 *Strain in Compression*

In the direction of slips going at elongation and unloading, based on Eqs. (26.2), (26.3), and (27.2), we have

$$L_{v\lambda}\varphi_{nl}(t^*) = \frac{\sigma_z(t^*) + 2A\Sigma}{2\Psi[t^*]} \sin 2\alpha \cos \omega - \frac{\psi[t^*] - B\Sigma}{\Psi[t^*]}, \tag{27.41}$$

where t^* still means the moment in time when elastic unloading occurs.

When changing the loading sign to the opposite of the initial one in the direction of newly occurring slips, the shear resistance will be

$$S_{v(-\lambda)}\varphi_{nl} = \psi + \Psi\left[\left(1 + \frac{\partial}{\partial t}\right)\left(L_{v(-\lambda)}\varphi_{nl} - L_{v\lambda}\varphi_{nl}(t^*)\right)\right] +$$

$$+ A\Sigma_{v\lambda} - B\Sigma, \quad t \geqslant t_c,$$

whereas t_c is the moment in time corresponding to the start of applying compressive loading. By substituting the expression (27.41) into this ratio and taking into account that in the slip direction in compression, $S_{v(-\lambda)} = \tau_{v(-\lambda)}$, we again come to an integral equation (27.20) relative to the function $\overline{\varphi}(\alpha, \omega, t) = \varphi(\alpha, \omega, t) + \varepsilon\dot{\varphi}(\alpha, \omega, t)$:

$$a\overline{\varphi}(\alpha, \omega, t) + 2b \int_{-\Omega(\alpha,t)}^{\Omega(\alpha,t)} \overline{\varphi}(\alpha, \omega_0, t) \cos(\omega - \omega_0) d\omega_0 = \tag{27.42}$$

$$= \overline{\lambda}_2(t) \sin 2\alpha \cos \omega\omega - \overline{\lambda}_3(t),$$

where

$$\overline{\lambda}_2(t) = \frac{\sigma_z(t) - 2A\Sigma}{2[t]} + \frac{\sigma_z(t^*) + 2A\Sigma}{2\Psi[t^*]} - \frac{3}{2}c\left[\Delta\varepsilon_z(t) + \varepsilon\Delta\dot{\varepsilon}_z(t)\right],$$

$$\overline{\lambda}_3(t) = \frac{\psi[t] - B\Sigma}{\Psi[t]} + \frac{\psi[t^*] - B\Sigma}{\Psi[t^*]}, \tag{27.43}$$

whereas $\Delta\varepsilon_z$ is the plastic strain increment in compression.

By solving equation (27.42) and repeating calculations similar to (27.21)–(27.36) (in these formulas, $\overline{\lambda}_1$ must be now replaced with $\overline{\lambda}_2$ and $\overline{\lambda}_0$ with $\overline{\lambda}_3$), we will obtain the following ratios:

$$\left[1 + \frac{c}{a}J(u)\right]\overline{\lambda}_2(t) = \frac{\sigma_z(t) - 2A\Sigma}{2\Psi[t]} + \frac{\sigma_z(t^*) + 2A\Sigma}{2\Psi[t^*]},$$

$$\delta[u(t)]\overline{\lambda}_3(t) = \frac{\psi[t] - B\Sigma}{\Psi[t]} + \frac{\psi[t^*] - B\Sigma}{\Psi[t^*]}. \tag{27.44}$$

Here the functions $J(u)$ and $\delta(u)$ are still defined by formulas (27.21) and (27.35). It follows from the dependencies (27.41) and (27.44) that

$$\frac{1 + ca^{-1}J[u(t)]}{\delta[u(t)]} = \frac{\varepsilon^{-1}\sigma_z(t) - A\Sigma + \Psi[t]Q_1}{\psi[t] - B\Sigma + \Psi[t]Q_0},\tag{27.45}$$

$$\Delta\varepsilon_z(t) = \frac{2}{3a\varepsilon}\exp\left[-\varepsilon^{-1}(t - t_c)\right] \times$$
$$\times \int_{t_c}^{t} \frac{\psi[\tau] - B\Sigma + \Psi[\tau]Q_0}{\Psi[\tau]\delta[u(\tau)]} \cdot J[u(\tau)]\exp\left[\varepsilon^{-1}(\tau - t_c)\right]d\tau,\tag{27.46}$$

where

$$Q_1 = \frac{0.5\sigma_z(t^*) + A\Sigma}{\Psi[t^*]}, \quad Q_0 = \frac{\psi[t^*] - B\Sigma}{\Psi[t^*]}.\tag{27.47}$$

The latter formulas define a link between stress and strain in compression after pre-elongation beyond the yield stress and unloading. Let us note that the parameter u defining the area of new slips changes in these formulas starting with zero, e.g. $u(t_c) = 0$. Strain $\varepsilon_z(t)$ is found as a difference between the strain $\varepsilon_z(t^*)$ at the start of elastic unloading and strain $\Delta\varepsilon_z(t)$. Calculations under formulas (27.45)–(27.46) are easily implemented with a computer, for example, using Mathcad.

27.7 Strain Creep and Stress Relaxation

If, starting with some moment in time t^o, the stress σ_z remains constant

$$\sigma_z(t) = \sigma_z(t^o) = const \ (\dot{\sigma}_z(t) = 0) \qquad at \qquad t \geqslant t^o,\tag{27.48}$$

it follows from formula (27.34) that

$$u(t) = u(t^*) = const \quad (t \geqslant t^o),\tag{27.49}$$

e.g. the parameter u characterizing the slip area also becomes constant. In this case, the right part of Eq. (27.36) is constant, and from the equation we obtain the law of plastic strain change due to creep:

$$\varepsilon_z(t) = \frac{2}{3} \cdot \frac{[\psi[t^*] - B\Sigma]J[u(t^o)]}{a\Psi[t^o]\delta[u(t^o)]}\left[1 - e^{-(t-t^o)/\varepsilon}\right] + \varepsilon_z(t^o)e^{-(t-t^o)/\varepsilon}.\tag{27.50}$$

It follows from formula (27.50) that for $t \to \infty$, strain asymptotically tends to the following limit value:

$$\varepsilon_z(t)\big|_{t\to\infty} = \frac{2[\psi[t^o] - B\Sigma]J[u(t^o)]}{3a\Psi[t^o]\delta[u(t^o)]}.\tag{27.51}$$

The result (27.51) can be used when defining the model parameters.

Assume that after elongation at some rate at the moment t_*, full (elastic–plastic) strain is fixed ε_z^p. Plastic strain ε_z at $t > t_*$ will change, since otherwise the equation $S_{\nu\lambda} = \tau_{\nu\lambda}$ is violated. Plastic strain increment will change the shear resistance and, therefore, a respective component of tangential stress. When the stress (σ_z) decreases, the elastic strain ($\Delta\varepsilon_z^y$) decrements under the law

$$\Delta\varepsilon_z^y = \frac{1}{E}\left[\sigma_z^* - \sigma_z(t)\right] \quad (t \geqslant t_*), \tag{27.52}$$

where σ_z^* is the deforming stress at the moment t_*, and E is the Young modulus. For the strain to remain constant, we must demand equations (in modulus) of plastic strain increment ($\Delta\varepsilon_z$) and changes in elastic strain (27.52), e.g.

$$\Delta\varepsilon_z = \varepsilon_z(t) - \varepsilon_z(t_*) = \frac{1}{E}[\sigma_z^* - \sigma_z(t)]. \tag{27.53}$$

Hence, the plastic strain rate will be

$$\dot{\varepsilon}_z(t) = -\frac{1}{E}\frac{d}{dt}\sigma_z(t) \quad (t \geqslant t_*). \tag{27.54}$$

Since in this case of relaxation, after pre-elongation beyond the yield stress, the shear resistance sign and the plastic strain rate sign do not change, the ratios (27.34)–(27.36) are true . By substituting formulas (27.53)–(27.54) into the right part of Eq. (27.36), we obtain

$$\varepsilon_z(t_*) + \frac{\sigma_z^* - \sigma_z(t)}{E} - \frac{\varepsilon}{E}\frac{d}{dt}\sigma_z(t) = \frac{2}{3} \cdot \frac{\psi[t] - B\Sigma}{\Psi[t]} - \frac{J[u(t)]}{a\delta[u(t)]}. \tag{27.55}$$

The latter equation together with formulas (27.34)–(27.36) represents a non-linear differential equation of the first order relative to the function $\sigma_z(t)$, whereas the initial conditions will be

$$\sigma_z(t)\big|_{t=t_*} = \sigma_z^*. \tag{27.56}$$

The integration of equation (27.55) for the initial condition (27.56) is easily done by numerical methods using a computer.

It should be noted that dependencies obtained in previous paragraphs for the case of uniaxial homogeneous stressed state are true also for the case of pure shear ($\tau_{xy} \neq 0$), if we replace $\sigma_z/2$ with τ_{xy} and $3\varepsilon_z/2$, $3\dot{\varepsilon}_z$ with γ_{xy} and $\dot{\gamma}_{xy}$, respectively, and instead of the function $J(u)$, defined by formula (27.31), we use its representation for pure shear given in [3].

27.8   Examples of Building Diagrams in an Uniaxial Stressed State

In Fig. 27.2, based on formulas (27.34)–(27.36), a family of elongation diagrams (curves 1–5) is built for various loading rates. Loading rates (in MPa/s) for each curve are given on the placard placed in the figure. The yield stress for each loading rate was calculated using formula (23.8) for those constant values that were used in the example (p. 329) of the description of experiments of Hendrickson and Wood [2].

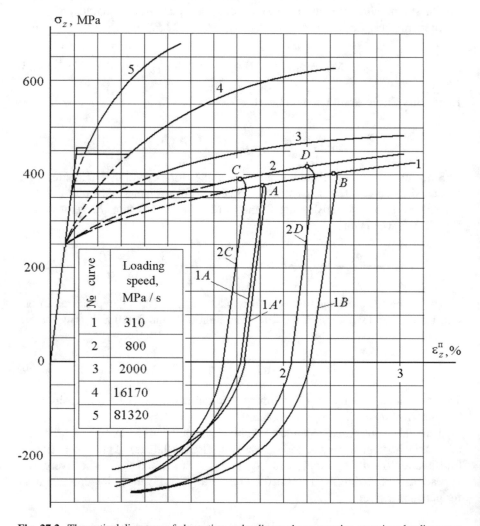

№ curve	Loading speed, MPa / s
1	310
2	800
3	2000
4	16170
5	81320

Fig. 27.2 Theoretical diagrams of elongation, unloading and compression at various loading rates

Fig. 27.3 Effects of the differential part of the shear resistance operator on the nature of the hardening curve

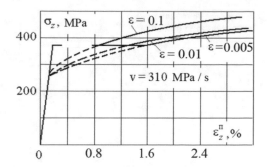

Hardening curves in elongation are built with the following constant values: $E = 2 \cdot 10^5$ MPa, $a = 0.334$, $b = 0.668$, $c = 5.8$, $A = 0.42$, $B = 0.81$, $\varepsilon = 0.01$, $m = 0$, and $k = 0.36$. Sections of hardening curves that are not visible are shown by dash lines in Fig. 27.2. The yield plateau for each loading rate is depicted by a horizontal section at the level of the yield stress from Hooke's straight line until crossing with the hardening curve. As Fig. 27.2 shows, the diagrams qualitatively correctly reflect the nature of dependency between elongation diagrams of real plastic materials on the loading rate: as the loading rate grows, the yield stress goes up, the yield plateau is decreased, and the hardening curve becomes steeper. For curves 1 and 2, loading diagrams are calculated for loading from the points A, B (curve 1) and C, D (curve 2), as well as compression diagrams after unloading with the full Bauschinger effect, so that the loading rate in compression for curves $1A$, $1B$ and $2C$, $2D$ is adopted the same as in elongation for curves 1 and 2, respectively. The curve $1A'$ is built at the loading and compression rate equal to $1/6$ of the loading rate in elongation. Compression diagrams were built upon the dependencies (27.45)–(27.47).

Figure 27.3 demonstrates sensitivity to the differential part of the shear resistance operator. As the parameter ε grows, the curvature of the hardening curve rises, and the yield plateau length decreases. As in Fig. 27.2, dash lines here show the sections of hardening curves invisible in experiments.

Here all three hardening curves are built for the loading rate of 310 MPa/s. Other constant values (except for the varied parameter ε) are left the same as given in p. 365.

Figure 27.4 shows a series of unloading curves from the point A of curve 1 in Fig. 27.2 for various unloading rates v. The diagrams are built based on formulas (27.34) and (27.39). Each unloading curve is built to the point where the plastic strain rate turns to zero. The coordinates of this point were defined using formula (27.40). As we already said (p. 315), an increase in plastic strain in unloading similar to that given in Fig. 27.4 is observed in experiments with plastic bodies.

Diagrams shown in Fig. 27.5 reflect the strain creep process for two various stress levels corresponding to the points A and B in curve 1 of Fig. 27.2. The charts are built based on formula (27.50), and their asymptotes shown by dashed lines are calculated under formula (27.51). The figure shows that for the selected constant

Fig. 27.4 Increment of
plastic strain in unloading

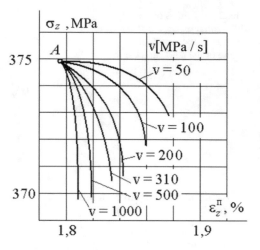

Fig. 27.5 Strain creep curves
in constant loading

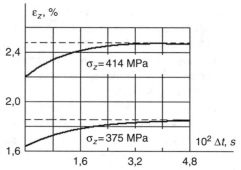

values of the material, the creep strain rapidly approaches its asymptotic value,
which accords with experimental results [8].

Figure 27.6 gives a stress relaxation curve calculated upon Eq. (27.55) taking into
account formulas (27.34)–(27.34), whereas the initial stress corresponds to the point
B in curve 1 of Fig. 27.2. For the selected constant values, the relaxation process also
goes rather rapidly.

Figure 27.7 gives comparison for torsion and elongation in the coordinates
"octahedral plastic shear–octahedral tangential stress." The diagrams are calculated
for $A = B = \varepsilon = 0$. As the figure shows, the diagrams do not coincide. This means
that the assumption on the unified universal curve "stress intensity–strain intensity"
in the developed model of plastic medium adopted in the strain theory of plasticity
is not fulfilled. However, it should be noted that the discrepancy between intensity
diagrams of elongation and torsion does not exceed 10%, which correlates with
experimental data [7].

The considered examples of diagrams built upon model dependencies in
Figs. 27.2–27.6 show that by setting the shear resistance operator as (25.2)–(25.3),

Fig. 27.6 Stress relaxation at fixed strain

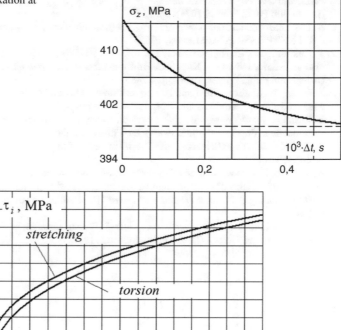

Fig. 27.7 Diagrams "octahedral stress—octahedral plastic shear" in elongation and pure shear

we can describe a wide range of phenomena with high quality, which are observed in tests of plastic materials.

References

1. E. Blinov, Plasticheskaya deformatsiya pri proizvol'nom nagruzhenii uprochnyayushchegosya tela [Plastic deformation under arbitrary loading hardening of the body], Dr. philos. sci. diss. Frunze, "Ilim" Publication, PhD Thesis, In-t Phys. and Math. An Kirg. SSR, 1972
2. J. Hendrickson, D. Wood, The effect of rate of stress application and temperature on the upper yield stress of annealed mild steel. Trans. ASM. V. **50**, 498–516 (1958)
3. M. Leonov, A. Baiterekov, *Deformatsiya pri plasticheskom sdvige* (Deformation during plastic shear). Izv. Academy of Sciences of Kyrghyz. SSR, no. 3 (1972), pp. 6–13
4. M. Leonov, Y. Klyshevich, Z. Sulaimanov, *Prosteishaya zadacha plasticheskogo techeniya* (The simplest task plastic flow). Izv. AN Kirg. SSR, no. 6 (1971), pp. 4–13
5. M. Leonov, V. Molotnikov, B. Rychkov, Nekotorye obobshcheniya kontseptsii skol'zheniya v teorii plastichnosti (Some generalizations the concept of slip theory of plasticity), in

Polzuchest' tverdogo tela : sb. nauchn. tr. (Creep solid body: Collection of scientific. work) (Izd-vo Ilim Publ., Frunze, 1974), pp. 12–35

6. A. Lyav, *Matematicheskaya teoriya uprugosti* (Mathematical theory of elasticity) (Leningrad, ONTI NKTP SSSR Publ., Moscow, 1935)

7. Y. Parkhomenko, *Znakoperemennaya plasticheskaya deformatsiya* (Alternating plastic deformation. Frunze, Izd-vo AN Kirg. SSR Publ.). *Poluchest' tverdogo tela* (Creep of a solid body) (Izd-vo Ilim Publ., Frunze, 1974), pp. 77–83

8. J. Rogan, A. Shelton, Yield and subsequent flow behaviour of some annealed steels under combined stress. J. Strain Analy. 4(2), 127–137 (1969)

9. K. Rusinko, *Nekotorye voprosy deformatsii plastichnykh i khrupkikh tel* (Some of the issues of deformation in ductile and brittle solids), Dr. Philos. Sci. Diss. Novosibirsk, Siberian branch of the USSR Academy of Sciences, PhD Thesis, Institute of Mechanics of the USSR Academy of Sciences, 1971

10. K. Rusinko, E. Blinov, Analiticheskoe issledovanie sootnosheniya napryazhenie–deformatsiya pri proizvol'noi traektorii nagruzheniya (Analytical study of the ratio stress-strain at an arbitrary loading path), in *Mekhanika polimerov* [Mechanics of polymers], no. 6 (1971), pp. 981–986

11. E. Yanke, F. Ehmde, F. Lesh, *Spetsial'nye funktsii* (Formuly, grafiki, tablitsy) [Special functions (formulas, graphs, tables)] (Nauka Publ., Moscow, 1977)

Chapter 28
Module of Additional Orthogonal Load

28.1 Problem Statement

In the studies of material behavior in the case of complex loading, a special place is taken [6] by the problem of finding a link between the increments of stresses and strain in the vicinity of the loading trajectory break (angular point). We have already said (p. 286) that this type of problem based on the Batdorf and Budiansky model [1] was first stated by Cicala [3]. This link is a kind of an applicability test for the ratios of various variants of plasticity theory in the case of non-proportional loading. Moreover, the knowledge of this link is necessary in the study of the flexure–torsion form of stability loss of structural elements beyond the yield stress.

The simplest form of an angular point represents a so-called orthogonal loading whose partial case is torsion with constant pre-applied elongating stress of a higher stress yield. We are interested in the quantity of the instantaneous shear modulus G_i equal to the ratio of the tangential stress increment $\Delta\tau$ to the shear strain increment $\Delta\gamma$ at the moment after the loading trajectory break: $G_i = \Delta\tau/\Delta\gamma$.

28.2 Determining the Intensity of Additional Slips

Assume that an infinitely low tangential stress $\Delta\tau_{xz}$ is applied to a plastic material element elongated beyond the yield stress with the constant elongating stress σ_z ($\Delta\sigma_z = 0$). Let us find the component of the plastic strain $\Delta\gamma_{xz}$ caused by the action of the component $\Delta\tau_{xz}$. To simplify the research, assume as follows in formulas (25.2)–(25.3): $\varepsilon = b = g = 0$, i. e. let us represent the shear resistance as

$$S_{\nu\lambda}\varphi_{nl} = \psi + \Psi(a\varphi_{\nu\lambda} + c\gamma_{\nu\lambda}) - A\Sigma_{\nu\lambda} - B\Sigma. \tag{28.1}$$

V. Molotnikov, A. Molotnikova, *Theory of Elasticity and Plasticity*,
https://doi.org/10.1007/978-3-030-66622-4_28

The application of the stress $\Delta\tau_{xz}$ for constant σ_z will cause the increment of tangential stress ($\tau_{\nu\lambda}$) relative to the axes ν and λ that we will find based on the formula (20.8):

$$\Delta\tau_{\nu\lambda}\varphi_{nl} = (\lambda_x\nu_z + \lambda_z\nu_x)\Delta\tau_{xz}. \tag{28.2}$$

Octahedral (τ_i) and maximum tangential (τ_m) stresses will not obtain increments (with the accuracy up to infinitely low values of the second order): $\Delta\tau_i = \Delta\tau_m = 0$. Taking this into account, the shear resistance increment (28.1) at the moment of the loading trajectory break will be

$$\Delta S_{\nu\lambda}\varphi_{nl} = \Psi(a\Delta\varphi_{\nu\lambda} + c\gamma_{\nu\lambda}) - A\Sigma\left(\frac{\Delta\gamma_{\nu\lambda}}{\gamma_m} - \frac{\gamma_{\nu\lambda}\Delta\varepsilon_z}{\gamma_m^2}\right), \tag{28.3}$$

where $\Delta\gamma_{\nu\lambda}$ is the shear occurring due to additional slips caused by the application of the stress $\Delta\tau_{xz}$. Let us use $\Delta\varepsilon_x$, $\Delta\varepsilon_y$, $\Delta\varepsilon_z$, and $\Delta\gamma_{xz}$ ($\Delta\gamma_{xy} = \Delta\gamma_{yz} = 0$) to designate the increment of plastic strain components: let us find

$$\Delta\gamma_{\nu\lambda} = 2(\lambda_x\nu_x\Delta\varepsilon_x + \lambda_y\nu_y\Delta\varepsilon_y + \lambda_z\nu_z\Sigma\varepsilon_z) + (\lambda_x\nu_z + \lambda_z\nu_x)\Delta\gamma_{xz}. \tag{28.4}$$

At the moment preceding the loading trajectory, slips take place in the planes set by the angle ν ($\nu = \pi/4 - \alpha_0$) and in each of these planes in the directions defined by the angle ω_0, whereas, according to the solution of the problem of elongation (27.24), (27.29), and (27.22), these angles satisfy the inequations

$$\begin{aligned} -\nu^* \leqslant \nu \leqslant \nu^*, \quad \cos 2\nu^* = \overline{\lambda}_0/\overline{\lambda}_1; \\ -\Omega \leqslant \omega_0 \leqslant \Omega, \quad \cos 2\nu \cos \Omega = \overline{\lambda}_0/\overline{\lambda}_1. \end{aligned} \tag{28.5}$$

After breaking the loading trajectory, additional slips cannot occur beyond those planes and directions where plastic shears took place at the moment preceding the break of the loading trajectory. This means that beyond these planes and directions, the shear resistance remains higher than the respective components of tangential stress. Additional plastic strains will occur in that part of the area (28.5) where

$$\Delta S_{\nu\lambda}\varphi_{nl} = \Delta\tau_{\nu\lambda}. \tag{28.6}$$

In the part of the area (28.5) where $\Delta\tau_{\nu\lambda} < \Delta S_{\nu\lambda}\varphi_{nl}$, no additional slips occur. From equation (28.6), let us find the intensity of additional slips

$$\Delta\varphi_{\nu\lambda} = \frac{\Delta\tau_{\nu\lambda}}{a\Psi} - \left(\frac{c}{a} - \frac{A\Sigma}{a\Psi\gamma_m}\right)\Delta\gamma_{\nu\lambda} - \frac{2A\Sigma}{a\Psi\gamma_m}\nu_z\lambda_z\Delta\varepsilon_z. \tag{28.7}$$

The boundary of the area of additional slips is found from the condition that the function (28.7) equals zero at this boundary:

$$\Delta\varphi_{\nu\lambda} = 0. \qquad (28.8)$$

28.3 Calculation of the Strain Increments and Additional Loading Modulus

The sought increments of plastic strain components are found using the formulas arising from (20.5):

$$\Delta\varphi_{ij} = \frac{1}{2} \int_{\Omega'} d\Omega \int_{\Omega_1(\alpha_0,\beta_0)}^{\Omega_2(\alpha_0,\beta_0)} \Delta\varphi_{nl}(n_i l_j + n_j l_i) d\omega_0, \quad (i, j = x, y, z),$$

$$\Delta\varepsilon_x = \frac{1}{2}\Delta\gamma_{xx},$$

(28.9)

where Ω', Ω_1, and Ω_2 are the boundaries of the area of additional slips defined from the condition (28.8). The solution of the obtained problem to determine the components $\Delta\gamma_{ij}$ can be obtained in a closed form if we deem that $c\gamma_m \ll \tau_m$, $A\Sigma \leqslant \tau_m$ and neglect the square of ratios $c\gamma_m/\tau_m$, $A\Sigma/\tau_m$. Indeed, since, in the expression (28.7), the components $\Delta\gamma_{\nu\lambda}$ and $\Delta\varepsilon_z$ have multipliers c and $A\Sigma$, $\Delta\gamma_{\nu\lambda}$ and $\Delta\varepsilon_z$ can be replaced by the values at $c = A = 0$. These values are calculated in the paper [9]. By substituting them into the expressions (28.4) and (28.7), we obtain a new value $\Delta\varphi_{\nu\lambda}$. Then the condition (28.8) will be used to find the boundaries of the area of additional slips, and then formulas (28.9) are used to find a new value $\Delta\gamma_{xz}$. By omitting intermediate calculations, we will give only the final result:

$$\Delta\gamma_{xz} = \frac{1}{a\Psi}\left[\eta_{xz} + \left(\frac{2c}{a} - \frac{A\Sigma}{a\Psi\gamma_m}\right)\left(\eta_x^2 + \eta_y^2 + \frac{\eta_{xz}^2}{2}\right)\right]\Delta\tau_{xz}, \qquad (28.10)$$

where

$$\eta_{xz} = \frac{2\pi\delta\sqrt{2}}{5\cos v^*}\left[\Pi(e_1, e_2) - (1 - 2\cos^4 v^*)K(e_2) - 2\cos^2 v^* E(e_2)\right];$$

$$\eta_x = \frac{1}{15}\left[\frac{(5 - 3\delta - 10\delta^2 - 4\delta^3)\sqrt{2\delta}}{(1 + \delta)\,\mathrm{tg}\,v^*}K + \frac{2\,\mathrm{tg}\,v^*}{\sqrt{2\delta}}\Pi_1 + \right.$$

$$\left. + \frac{15\delta^2 - 3}{\sqrt{2\delta}\,\mathrm{tg}\,v^*}\Pi_2 - -5\,\mathrm{tg}\,v^*\sqrt{2\delta}(1 + \delta)E\right];$$

$$\eta_y = \frac{1}{15}\left[\frac{2\delta\sqrt{2\delta}(3 - \delta^2)}{(1 + \delta)\,\mathrm{tg}\,v^*}K - \frac{4\,\mathrm{tg}\,v^*}{\sqrt{2\delta}}\Pi_1 - \frac{4}{\sqrt{2\delta}\,\mathrm{tg}\,v^*}\Pi_2\right];$$

$$K(e_2) = \frac{1}{15} \int_0^{\pi/2} \frac{dx}{\sqrt{2 - e_2^2 \sin^2 x}}; \quad E(e_2) = \int_0^{\pi/2} \sqrt{1 - e_2^2 \sin^2 x}\, dx;$$

$$\Pi(e_1, e_2) = \int_0^{\pi/2} \frac{dx}{(1 + e_1 \sin^2 x)\sqrt{1 - e_2^2 \sin^2 x}}; \quad \delta = \cos 2v^*;$$

$$e_1 = -2 \sin^2 v^*; \quad e_2 = \mathrm{tg}\, v^*, \tag{28.11}$$

K and E are full elliptical intervals of the first and second type with the modulus $1/\sqrt{2}$, and Π_1 and Π_2 are elliptical integrals of the third type defined by Gradshtein and Ryzhik [4] with the formulas

$$\Pi_1 = \int_0^1 \left(1 - \frac{\cos^2 v^*}{\delta} y^2\right)^{-1} \left(1 - y^2\right)^{-1/2} \left(1 - \frac{y^2}{2}\right)^{-1/2} dy,$$

$$\Pi_2 = \int_0^1 \left(1 + \frac{\sin^2 v^*}{\delta} x^2\right)^{-1} \left(1 - x^2\right)^{-1/2} \left(1 - \frac{x^2}{2}\right)^{-1/2} dx.$$

Using the results of the papers [8, 9], the link between stress and plastic strain (with the shear resistance operator selected here) at the moment preceding the loading trajectory break can be represented as follows:

$$\frac{\sigma_z}{\varepsilon_z \Psi} = \frac{3}{2} \cdot \frac{(a + 2c\eta_{xz})}{1 + 2A\Sigma/\sigma_z)\eta_{xz}}.$$

From this result and formula (28.10), it follows

$$\Delta\gamma_{xz} = \frac{3}{2} \cdot \frac{\varepsilon_z}{\sigma_z} \cdot \frac{\left[1 - D(v^*)\left(\frac{c}{a} - \frac{A\Sigma}{a\gamma_m \Psi}\right)\right] \Delta\tau_{xz}}{1 + \frac{2A\Sigma}{\sigma_z}}, \tag{28.12}$$

where

$$D(v^*) = 3\eta_{xz} + 2\frac{\eta_x^2 + \eta_y^2}{\eta_{xz}}.$$

Let us write the components of plastic strain $\Delta\gamma_{xz}$ and ε_z as the difference of the full and elastic components. Formula (28.12) will look as follows:

$$\frac{\Delta\gamma_{xz}^p}{\Delta\tau_{xz}} - \frac{\Delta\gamma_{xz}^e}{\Delta\tau_{xz}} = \frac{3}{2}\left(\frac{\varepsilon_z^p}{\sigma_z} - \frac{\varepsilon_z^e}{\sigma_z}\right) \cdot \frac{\left[1 - D(v^*)\left(\frac{c}{a} - \frac{A\Sigma}{a\gamma_m \Psi}\right)\right]}{1 + 2A\Sigma/\sigma_z},$$

or

$$\frac{1}{G_i} - \frac{1}{G_0} = \frac{3}{2}\left(\frac{1}{E_s} - \frac{1}{E}\right)\left[1 - D(v^*)\left(\frac{c}{a} - \frac{A\Sigma}{a\gamma_m\Psi}\right)\right]\Big/\left[1 + \frac{2A\Sigma}{\sigma_z}\right],$$

where G_i is the additional loading modulus, G_0 is the shear modulus, E is the Young modulus, and E_s is the secant modulus in elongation.

From the latter formula, we find [6]

$$G_i = \frac{G_o}{1 + \frac{3}{2}G_0\left(\frac{1}{E_s} - \frac{1}{E}\right)\left[1 - D(v^*)\left(\frac{c}{a} - \frac{A\Sigma}{a\gamma_m\Psi}\right)\right]\Big/\left(1 + \frac{2A\Sigma}{\sigma_z}\right)}. \qquad (28.13)$$

Formula (28.13) expresses the sought shear modulus in orthogonal additional loading. For $c = A = 0$, this result coincides with the formula (see p. 287) of Cicala [3] obtained based on the Batdorf and Budiansky model [2].

28.4 Analysis of Results and Conclusions

Figure 28.1 represents [7] experimental charts of dependency between $\Delta\gamma_{xz}$ and $\Delta\tau_{xz}$ for various values of preliminary plastic strain (low circles and continuous lines approximated to them). According to strain theory, the additional loading modulus G_i must coincide with the secant modulus $G_s(\gamma_0)$. If we assume that the neutral additional loading is accompanied by elastic strain only, the additional loading modulus must coincide with the shear modulus $G_i = G$. Figure 28.1 shows straight lines corresponding to the cases $G_i = G_s(\gamma_0)$ (thin solid lines) and $G_i = G$ (dashed lines), whereas γ_0 is the octahedral shear at the moment preceding the loading trajectory break.

In the case of low plastic strain of elongation, experimental values of additional loading are closer to the shear modulus than the secant modulus. However, for a significant initial strain, $G_i \ll G$, but still significantly higher than the value predicted by the strain theory of plasticity.

It is known [5, 7, 10] that the Cicala formula also gives a significantly lower modulus of orthogonal additional loading than that observed in experiments. This was one of the reasons for relieving hopes related to accounting for the physical mechanism of the plasticity phenomena in the slip concept and for disappointments and "pessimistic conclusions relative to the possible progress of plasticity theory" [7, p. 104].

However, the analysis of formula (28.13) shows that introducing the addend $c\gamma_{\nu\lambda}$ and the components of structural softening into the shear resistance increases the modulus of orthogonal additional loading only if the condition is fulfilled

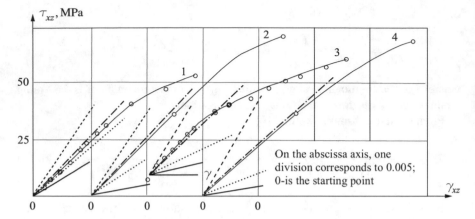

Fig. 28.1 Comparison of experimental and theoretical results in orthogonal additional loading

$$D(v^*)\left[\frac{c}{a} + 1 \middle/ \left(3a\Psi\left(\frac{1}{E_s} - \frac{1}{E}\right)\right)\right] < 1. \tag{28.14}$$

By the respective selection of model parameters, we manage to achieve matching of experimental and model values of the additional modulus values for both low and developed initial elongation strains. The results are shown in Fig. 28.1 with dash lines. Point lines indicate the results obtained under the Cicala formula (see, for example, [3]).

In this manner, setting the coefficients c, A and B may adjust the additional modulus value when required, which eliminates the described disadvantage of the Batdorf and Budiansky theory and rehabilitates the slip concept in plasticity theory (at least for materials with structural softening).

We note that if we know the modulus of orthogonal additional loading from experiments, the results (28.13) and (28.14) can be used to determine the model parameters.

References

1. S. Batdorf, B. Budiansky, Splitting tests: an alternative to determine the dynamic tensile strength of ceramic materials, in *NACA TC1871* (1949), pp. 1–31
2. S. Batdorf, B. Budyanskii, *Mekhanika: sb. perev., no 5(33)* (Mechanics: a collection of translations, no 5(33)). Moscow, Inostrannaya literatura Publ., chapter *Zavisimost' mezhdu napryazheniyami i deformatsiyami dlya uprochnyayushchegosya metalla pri slozhnom napryazhennom sostoyanii* (The relationship between stress and deformations for the hardening of metals under complex stress state), Mekhanika: sb. perev., no 5(33) (Mechanics: a collection of translations) (1955), pp. 120–127
3. P. Cicala, On the plastic deformation, in *Atti Accad. naz. Lincei. Rend. Cl. Sci. fis. e. nature. V. 8* (1950), pp. 583–586

4. I. Gradshtein, I. Ryzhik, *Tablitsy integralov, summ, ryadov i proizvedenii* (Tables of integrals, sums, series, and works). (Nauka Publ., Moscow, 1971)
5. I. Iosimura, *Mekhanika. Periodich. sb. perev. inostr. statei.* (Mechanics. Periodicity. collection of pens. foreign. articles), no. 2 (60), Moscow, Mir Publ., chapter Zamechaniya k teorii skol'zheniya Batdorfa i Budyanskogo (Remarks on the theory of sliding by Batdorf and Budiansky) (1960), pp. 109–116
6. V. Molotnikov, A. Molotnikova, Zadacha chikala dlya materialov so strukturnym razuprochne-niem (Chikala task for materials with structural loosening). Izv. vuzov. Sev.–Kavkaz. region. Estestv. nauki (Izvestiya vuzov. North.–The Caucasus. region. Nature. science) **2**, 27–30 (2010)
7. Y. Rabotnov, *Polzuchest' ehlementov konstruktsii* (Creep of structural elements). (Moscow, Nauka Publ. 1966)
8. K. Rusinko, *Deformatsiya neuprugogo tela (Deformation of an inelastic body)*, Frunze, Izd-vo Ilim Publ., chapter K teorii plastichnosti, osnovannoi na kontseptsii skol'zheniya (To the theory of plasticity based on the concept of sliding) (1971), pp. 3–14
9. K. Rusinko, Obobshchenie formuly chikala (Generalization of the Chikala formula). Izv. AN SSSR, Mekh-ka tv. tela (Izvestia of the USSR Academy of Sciences, solid state Mechanics)**6**, 37–44 (1971)
10. V. Sveshnikova, O plasticheskom deformirovanii plasticheskikh metallov (On plastic deformation of plastic metals). Izv. AN SSSR. OTN. (News of the USSR Academy of Sciences. Department of Technical Sciences) **1**, 155–162 (1956)

Chapter 29
Plane-Plastic Strain

29.1 Theorem of Strain in Pure Shear

Assume that all stress tensor components except for τ_{xy} equal zero. Based on the previously proved lemma (p. 312), the plastic strain component (γ_{xy}) is expressed via the components $(r_{x'x'}, r_{x'y'})$ of slip tensor intensity using formula (20.33):

$$\gamma_{xy} = \frac{1}{2} \int_0^\pi (r_{x'x'} \sin 2\beta_0 + r_{x'y'} \cos 2\beta_0) d\beta_0.$$

By using this formula, we will prove the following conclusion.

Theorem *If the opening of the slip plane fan in the angle α_0 is small, the following formula takes place for the plastic strain component (γ_{xy}) in pure shift:*

$$\gamma_{xy} \approx \frac{1}{2} \int_0^\pi r_{x'y'} \cos 2\beta_0 d\beta_0. \tag{29.1}$$

Proving From the definitions (20.16)–(20.17) and formulas (20.31) in the considered case, it follows:

$$r_{x'y'} = 2 \int_{\alpha_1}^{\alpha_2} \sin^2 \alpha_0 d\alpha_0 \int_{\Omega_1}^{\Omega_2} \varphi(\alpha_0, \beta_0, \omega_0) \sin \omega_0 d\omega_0,$$

$$r_{x'x'} = -2 \int_{\alpha_1}^{\alpha_2} \cos \alpha_0 \sin^2 \alpha_0 d\alpha_0 \int_{\Omega_1}^{\Omega_2} \varphi(\alpha_0, \beta_0, \omega_0) \cos \omega_0 d\omega_0, \quad (29.2)$$

$$\left(\alpha_1 = \alpha_1(\beta_0), \ \alpha_2 = \alpha_2(\beta_0), \ \Omega_1 = \Omega_1(\alpha_0, \beta_0), \ \Omega_2 = \Omega_2(\alpha_0, \beta_0) \right),$$

where $\Omega_{1,2}$ are the boundary values of the angle ω_0 defining the opening of the slip direction fan in the plane (α_0, β_0) and $\alpha_{1,2}$ are the boundaries of the slip plane fan in the angle α_0.

In the latter formulas, let us replace the variables ω_0 with ξ and α_0 with v_0, assuming that

$$\omega_0 = \xi + \Omega_c, \ \ \Omega_c = \frac{1}{2}(\Omega_1 + \Omega_2), \ \ \Omega = \frac{1}{2}(\Omega_2 - \Omega_1), \ \ \alpha_0 = \frac{\pi}{2} - v_0,$$

$$v_1(\beta_0) = \frac{\pi}{2} - \alpha_1(\beta_0), \ \ v_2(\beta_0) = \frac{\pi}{2} - \alpha_2(\beta_0),$$

$$\varphi[v_0, \beta_0, \xi] = \varphi\left(\frac{\pi}{2} - v_0, \beta_0, \xi + \Omega_c\right).$$

$$(29.3)$$

Then we obtain

$$r_{x'y'} = \int_{v_1}^{v_2} F[v_0, \beta_0] dv_0,$$

$$r_{x'x'} = -\int_{v_1}^{v_2} \operatorname{ctg} \Omega_c \sin v_0 F[v_0, \beta_0] dv_0,$$

$$(29.4)$$

where

$$F[v_0, \beta_0] = 2 \cos^2 v_0 \sin \Omega_c \int_{-\Omega}^{\Omega} \varphi[v_0, \beta_0, \xi] \cos \xi d\xi.$$

It is taken into account that in the considered case of pure shear, due to the symmetry of the stress–strain state relative to the plane xOy, the function $\varphi[v_0, \beta_0, \xi]$ is even upon the argument ξ, so

$$\int\limits_{-\Omega}^{\Omega} \varphi[v_0, \beta_0, \xi] \sin \xi \, d\xi = 0.$$

Due to the symmetry in pure shear, the resultant of all slips in the slip plane coincides in direction with the maximum tangential stress in this plane, i.e.

$$\operatorname{ctg} \Omega_c = -\sin v_0 \operatorname{tg} 2\beta_0.$$

Taking into account this dependency, the second of formulas (29.4) looks as follows:

$$r_{x'x'} = \operatorname{tg} 2\beta_0 \int\limits_{v_1}^{v_2} F[v_0, \beta_0] \sin^2 v_0 dv_0. \tag{29.5}$$

By applying the mean value theorem to the integrals (29.3) and (29.4), we can represent as follows:

$$r_{x'x'} = \operatorname{tg} 2\beta_0 F[\bar{v}, \beta_0](v_2 - v_1) \sin^2 \bar{v}, \quad v_1 \leqslant \bar{v} \leqslant v_2,$$

$$r_{x'y'} = F[\tilde{v}, \beta_0](v_2 - v_1), \quad v_1 \leqslant \tilde{v} \leqslant v_2.$$

The latter dependencies follow that if the segment $[v_1, v_2]$ is small, the component $r_{x'x'}$ is small and of the second order as compared to $r_{x'y'}$ and can be neglected, which proves the formulated theorem.

29.2   General Dependencies in Pure Shear

Let us assume $a = b = 0$ in formula (25.3) and represent [3] the operator of shear resistance $S_{v\lambda}\varphi_{nl}$ as follows:

$$S_{v\lambda}\varphi_{nl} = \psi + \Psi \left[g \left((r_{v\lambda} + \varepsilon \dot{r}_{v\lambda}) + c \left(\gamma_{v\lambda} + \varepsilon \dot{\gamma}_{v\lambda} \right) \right] - A\Sigma_{v\lambda} - B\Sigma. \tag{29.6}$$

In the case of pure shear ($\tau_{xy} \neq 0$),

$$\tau_{v\lambda} = \tau_{xy}(\sin \omega \cos 2\beta - \cos \omega \cos \alpha\alpha \sin 2\beta),$$
$$\gamma_{v\lambda} = \gamma_{xy}(\sin \omega \cos 2\beta - \cos \omega \cos \alpha\alpha \sin 2\beta). \tag{29.7}$$

Shear resistance in the slip area equals the tangential stress: $S_{v\lambda}\varphi_{nl} = \tau_{v\lambda}$. When substituting the expressions (29.6)–(29.7) into this equation at $\alpha = \omega = \pi/2$, we obtain the following differential ratio $r_{x'y'} \equiv r(\beta_0, t)$ for the tensor intensity component in the planes parallel to the axis Oz :

$$r(\beta_0, t) + \varepsilon \dot{r}(\beta_0, t) = \frac{1}{g} \left[\bar{\lambda}(t) \cos 2\beta_0 - \bar{\lambda}_0(t) \right], \qquad (29.8)$$

where

$$\bar{\lambda}(t) = \frac{\tau_{xy}(t) + A\Sigma}{\Psi[t]} - c[\gamma_{xy}(t) + \varepsilon \dot{\gamma}_{xy}(t)],$$

$$\bar{\lambda}_0(t) = \frac{\psi[t] - B\Sigma}{\Psi[t]}. \qquad (29.9)$$

With the function $r(\beta_0(t))$ known, the shear resistance in an arbitrary plane inclined to the axis Oz in the direction $\omega = \pi/2$ based on formulas (29.6)–(29.9) will be

$$S_{v\lambda}\varphi_{nl} = \psi + \Psi[\bar{\lambda}\cos 2\beta - \bar{\lambda}_0 - c(\gamma_{xy} + \varepsilon \dot{\gamma}_{xy}) \sin \alpha \cos 2\beta]. \qquad (29.10)$$

The first of formulas (29.7) at $\omega = \pi/2$ gives

$$\tau_{v\lambda} = \tau_{xy} \sin \alpha \cos 2\beta. \qquad (29.11)$$

By comparing formulas (29.10) and (29.11), we can write that the shear resistance (29.10) equals the tangential stress (29.11) only if $\alpha = \pi/2$; if $\alpha \neq \pi/2$, then $S_{v\lambda}\varphi_{nl} > \tau_{v\lambda}$, i.e. slips take place only in the planes parallel to the axis Oz [1, 3]. It means that the opening of the slip plane fan in the angle α equals zero. Based on the proved theorem, we have

$$\gamma_{xy} = \frac{1}{2} \int_{\beta_1}^{\beta_2} r_{x'y'} \cos 2\beta_0 d\beta_0, \qquad (29.12)$$

where $\beta_{1,2}$ are the boundaries of the slip plane fan. Hence, we can calculate

$$\gamma_{xy}(t) + \varepsilon \dot{\gamma}_{xy}(t) = \frac{1}{2} \int_{\beta_1}^{\beta_2} [r(\beta_0, t) + \varepsilon \dot{r}(\beta_0, t)] \cos 2\beta_0 d\beta_0. \qquad (29.13)$$

By substituting the expression for the tensor intensity of slips (29.8) into formula (29.13), after calculating the integral:

$$\gamma_{xy} + \varepsilon \dot{\gamma}_{xy} = \frac{\bar{\lambda}(t)}{2g} \left[\Theta(t) - \frac{1}{4} \sin 4\Theta(t) \right], \qquad (29.14)$$

where

$$\Theta(t) = \frac{1}{2} \arccos \left[\frac{\overline{\lambda}_0(t)}{\overline{\lambda}(t)} \right]. \tag{29.15}$$

From the latter formulas and expressions (29.9) for the function $\overline{\lambda}_0(t)$, we obtain as follows:

$$\frac{\tau_{xy}(t)}{\psi[t] - B\Sigma} = \frac{1}{2\cos\Theta} \left[1 + \frac{c}{2g} \left(\Theta - \frac{1}{4} \sin 4\Theta \right) \right]. \tag{29.16}$$

Formulas (29.14)–(29.16) set a link between stress, strain, and strain rate in pure shift. If the loading law $\tau_{xy}(t)$ is stated, the latter equation is used to find the dependency $\Theta(t)$. Then we find as follows from the differential equation (29.14) at the initial conditions $t = 0$: $\gamma_{xy}(0) = 0$:

$$\gamma_{xy} = \frac{1}{2g\varepsilon} \exp\left(-\frac{t}{\varepsilon}\right) \int_0^t \frac{\psi[\zeta] - B\Sigma}{\Psi[\zeta]} \cdot \frac{\Theta[\zeta] - \frac{1}{4}\sin 4\Theta[\zeta]}{\cos 2\Theta[\zeta]} \exp\left(\frac{\zeta}{\varepsilon}\right) d\zeta. \tag{29.17}$$

From formulas (29.11)–(29.17) at $A = B = \varepsilon = 0$ as a partial case, dependencies are obtained between stress and strain found in the paper [2] based on the model of the so-called plane-plastic medium. As shown above, when setting the shear resistance as (29.6)–(29.7), slips in pure shear occur only in the planes parallel to the axis Oz and in the directions perpendicular to that axis. This slip mechanism first proposed in the paper [2] rather conditionally reflects the true situation of slips.

29.3 Monotonous Plane-Plastic Strain

29.3.1 Preparation of Initial Dependencies

For plane-plastic strain ($\varepsilon_z = 0$) from formulas (20.8) and (27.1), we have

$$\tau_{\nu\lambda} = -\frac{\sigma_x - \sigma_y}{2} \sin\alpha \sin\omega \sin 2\beta -$$
$$- \sin\alpha \cos\omega \cos\alpha [\sigma_x \cos^2\beta + \sigma_y \sin^2\beta] +$$
$$+ \sigma_z \cos\alpha \sin\alpha \cos\omega + \tau_{xy} \sin\alpha (\sin\omega \cos\cos 2\beta - \cos\alpha \cos\omega \sin 2\beta),$$
$$\gamma_{\nu\lambda} = -2(\varepsilon_x - \varepsilon_y)(\sin\omega \sin 2\beta -$$
$$- \cos\omega \cos\alpha \cos 2\beta) \sin\alpha +$$
$$+ \gamma_{xy} \sin\alpha (\sin\omega \cos 2\beta - \cos\alpha \cos\omega \sin 2\beta). \tag{29.18}$$

As for pure shift, assume in the latter formulas $\alpha = \omega = \pi/2$, i.e. we will consider slips only in the plane xOy. We obtain

$$\tau_{\nu\lambda} = -\frac{1}{2}(\sigma_x - \sigma_y)\sin\sin 2\beta + \tau_{xy}\cos 2\beta,$$
$$\gamma_{\nu\lambda} = -(\varepsilon_x - \varepsilon_y)\sin 2\beta + \gamma_{xy}\cos 2\beta. \tag{29.19}$$

If we assume stress tensor components as known time functions, formulas (29.19) can be represented as

$$\tau_{\nu\lambda}(t) = \tau_m(t)\cos 2[\Theta_0 - \chi(t)],$$
$$\gamma_{\nu\lambda}(t) = \gamma_m(t)\cos 2[\Theta_0 - \chi(t)], \tag{29.20}$$

where

$$\tau_m(t) = \frac{1}{2}\sqrt{[\sigma_x(t) - \sigma_y(t)]^2 + 4\tau_{xy}^2(t)},$$
$$\gamma_m(t) = \sqrt{[\varepsilon_x(t) - \varepsilon_y(t)]^2 + \gamma_{xy}^2(t)},$$
$$\chi(t) = \chi_1(t) - \chi_0(t_0), \tag{29.21}$$

whereas

$$\chi_1(t) = \frac{1}{2}\operatorname{arctg}\frac{2\tau_{xy}(t)}{\sigma_x(t) - \sigma_y(t)},$$
$$\chi_0(t_0) = \frac{1}{2}\operatorname{arctg}\frac{2\tau_{xy}(t_0)}{\sigma_x(t_0) - \sigma_y(t_0)}, \tag{29.22}$$

and t_0 is the time moment corresponding to the occurrence of plastic strain when the condition (20.10), $(\varphi(t_0) = 0)$ indicating no slips is still fulfilled. The angle Θ_0 in formulas (29.20) is counted in the counterclockwise direction starting from the direction of the maximum tangential stress $\tau_m(t_0)$ at the moment t_0 (Fig. 29.1).

29.3.2 Determinant Ratios

Let us call plastic strain *monotonous,* if the tensor intensity of slips grows with time. A formal expression of this definition is the fulfillment of the condition

$$\frac{\partial r(\Theta_0, t)}{\partial t} > 0, \quad (\Theta_0 \in [-\Theta_1(t), \Theta_2(t)]), \tag{29.23}$$

where $\Theta_1(t)$, $\Theta_2(t)$ are the boundaries of the fan of slip direction (Fig. 29.1).

If the condition (29.23) of the function $\Theta_{1,2}(t)$ is fulfilled, the determinant boundaries of the slip plane fan in the coordinate Θ_0 are monotonously increasing

Fig. 29.1 Fan of slip directions in monotonous plane-plastic strain

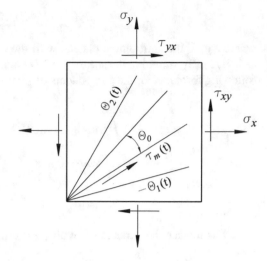

or at least decreasing functions so that

$$\dot{\Theta}_{1,2}(t) > 0, \quad \left(t \leqslant t_0, \quad \dot{\Theta} = \frac{d\Theta}{dt}\right).$$

By equaling the shear resistance (29.23) to the tangential stress (29.20), let us find as follows in the slip direction:

$$g\left[r(\Theta_0, t) + \varepsilon \dot{r}(\Theta_0, t)\right] = \left[\frac{\tau_n(t) + A\Sigma}{\Psi[t]} - c\gamma_m(t) - c\varepsilon\dot{\gamma}_m(t)\right] \times$$
$$\times \cos 2[\Theta_0 - \chi(t)] - 2c\varepsilon\gamma_m(t)\dot{\chi}(t)\sin 2[\Theta_0 - \chi(t)] - \frac{\psi[t] - B\Sigma}{\Psi[t]}. \tag{29.24}$$

Let us introduce designations:

$$\lambda_4(t) = \frac{\tau_m(t) + A\Sigma}{\Psi[t]} - c\left[\gamma_m(t) + \varepsilon\dot{\gamma}_m(t)\right], \tag{29.25}$$
$$\lambda_5(t) = 2c\varepsilon\gamma_m(t)\dot{\chi}(t).$$

Then Eq. (29.24) will be written as follows:

$$g\left[r(\Theta_0, t) + \varepsilon\dot{r}(\Theta_0, t)\right] = \lambda_4(t)\cos 2[\Theta_0 - \chi(t)] -$$
$$-\lambda_5(t)\sin 2[\Theta_0 - \chi(t)] - \bar{\lambda}_0(t), \quad (\Theta_0 \in [-\Theta_1(t), \Theta_2(t)]), \tag{29.26}$$

whereas $\bar{\lambda}_0(t)$ is again defined using formula (27.6).

In the considered case of plane-plastic strain, the formula (20.20) looks as follows:

$$dy_v = r(\Theta_0, t)d\Theta_0, \tag{29.27}$$

where dy_v is the shift on the plateau with the normal line v, $(dy_v = dy_{v\lambda}$ at $\alpha = \omega = \pi/2)$. By summing the shears (29.27) under the tensor rule, we will obtain the following formulas for the components of plane-plastic strain:

$$\gamma_{xy}(t) = \int_{-\Theta_1(t)}^{\Theta_2(t)} r(\Theta_0, t) \sin 2(\Theta_0 + \chi_0)d\Theta_0,$$

$$\varepsilon_x(t) = -\varepsilon_y(t) = \frac{1}{2} \int_{-\Theta_1(t)}^{\Theta_2(t)} r(\Theta_0, t) \cos 2(\Theta_0 + \chi_0)d\Theta_0. \tag{29.28}$$

Let us replace the variable Θ_0 with ζ in formulas (29.26) and (29.28) assuming that

$$\zeta = \Theta_0 - \delta_0, \quad \delta_0 = \frac{1}{2}(\Theta_2 - \Theta_1),$$

$$\Theta = \frac{1}{2}(\Theta_1 + \Theta_2), \quad r[\zeta, t] = r(\zeta + \delta_0, t). \tag{29.29}$$

These formulas will be converted into

$$g\{r[\zeta, t] + \varepsilon \dot{r}[\zeta, t]\} = \lambda_4(t) \cos 2[\zeta + \delta_0 - \chi(t)] - \\ -\lambda_5(t) \sin 2[\zeta + \delta_0 - \chi(t)] - \bar{\lambda}_0(t), \tag{29.30}$$

$$\gamma_{xy}(t) = \int_{-\Theta(t)}^{\Theta(t)} r[\zeta, t] \sin 2[\zeta + \delta_0 + \chi_0]d\zeta,$$

$$\varepsilon_x(t) = -\varepsilon_y(t) = \frac{1}{2} \int_{-\Theta(t)}^{\Theta(t)} r[\zeta, t] \cos 2[\zeta + \delta_0 + \chi_0]d\zeta. \tag{29.31}$$

29.3.3   Continuity Condition

From the condition (25.6) of slip rate continuity at the boundary of the slip area, it follows that at this boundary ($\zeta = \pm\Theta$), the tensor intensity of slips and its rate turn to zero:

$$r[\pm\Theta, t] + \varepsilon \dot{r}[\pm\Theta, t] = 0.$$

Hence we find

$$\lambda_4[\cos 2\Theta \cos 2(\delta_0 - \chi) - \sin 2\Theta \sin 2(\delta_0 - \chi)] -$$
$$-\lambda_5[\sin 2\Theta \cos 2(\delta_0 - \chi) + \cos 2\Theta \sin 2(\delta_0 - \chi)] = \bar{\lambda}_0,$$

$$\lambda_4[\cos 2\Theta \cos 2(\delta_0 - \chi) + \sin 2\Theta \sin 2(\delta_0 - \chi)] -$$
$$-\lambda_5[-\sin 2\Theta \cos 2(\delta_0 - \chi) + \cos 2\Theta \sin 2(\delta_0 - \chi)] = \bar{\lambda}_0.$$

These equations give

$$\cos 2\Theta(t) = \frac{\bar{\lambda}_0(t)}{\sqrt{\lambda_4^2(t) + \lambda_5^2(t)}},$$

$$\operatorname{tg} 2[\delta_0 - \chi(t)] = -\frac{\lambda_5(t)}{\lambda_4(t)}. \tag{29.32}$$

According to formulas (29.29) and (29.22), the angle δ_0 defines the direction of the symmetry axis of the slip fan in the plane xOy at the moment t, and the angle $\chi(t)$ defines the direction of maximum tangential stress at this moment. The latter result shows that these directions coincide only in two cases: either when there is proportional loading ($\dot{\chi} = 0$, $\lambda_5 = 0$) or in case $\varepsilon = 0$. In the general case, the directions of the maximum tangential stress and the symmetry axis of the slip fan make an angle between each other defined from formulas (29.32).

Using the results (29.32), we can write formula (29.30) as follows:

$$g[r(\zeta, t) + \varepsilon \dot{r}(\zeta, t)] = \sqrt{\lambda_4^2(t) + \lambda_5^2(t)} \, \cos 2\zeta - \bar{\lambda}_0(t). \tag{29.33}$$

To get ratios linking stresses with strains and their rates, let us make the following transformations. By differentiating formulas (29.31), we can easily find

$$\dot{\gamma}_{xy}(t) = \int_{-\Theta(t)}^{\Theta(t)} \{\dot{r}[\zeta, t] \sin 2[\zeta + \delta_0(t) + \chi_0] +$$
$$+2r[\zeta, t]\dot{\delta}_0(t) \cos 2[\zeta + \delta_0(t) + \chi_0]\} \, d\zeta,$$

$$\dot{\varepsilon}_x(t) = -\dot{\varepsilon}_y(t) = \frac{1}{2} \int_{-\Theta(t)}^{\Theta(t)} \{\dot{r}[\zeta, t] \cos 2[\zeta + \delta_0(t) + \chi_0] -$$
$$- 2r[\zeta, t]\dot{\delta}_0(t) \sin 2[\zeta + \delta_0(t) + \chi_0]\} \, d\zeta.$$

Hence, using identical transformations, we can obtain the following equations:

$$\gamma_{xy}(t) + \varepsilon\dot{\gamma}_{xy}(t) - 4\varepsilon\varepsilon_x(t)\dot{\delta}_0(t) =$$

$$= \int_{-\Theta(t)}^{\Theta(t)} \{r[\zeta, t] + \varepsilon\dot{r}[\zeta, t]\} \sin 2[\zeta + \delta_0(t) + \chi_0]d\zeta, \qquad (29.34)$$

$$\varepsilon_x(t) + \varepsilon\dot{\varepsilon}_x(t) + \varepsilon\gamma_{xy}(t)\dot{\delta}_0(t) =$$

$$= \int_{-\Theta(t)}^{\Theta(t)} \{r[\zeta, t] + \varepsilon\dot{r}[\zeta, t]\} \cos 2[\zeta + \delta_0(t) + \chi_0]d\zeta.$$

Let us calculate integrals in the right parts of formulas 29.35). To do it, we use the expressions $(r + \varepsilon\dot{r})$ for the sum (29.33). We obtain:

$$\int_{-\Theta(t)}^{\Theta(t)} \{r[\zeta, t] + \varepsilon\dot{r}[\zeta, t]\} \sin 2[\zeta + \delta_0(t) + \chi_0]d\zeta =$$

$$= \frac{\overline{\lambda}_0(t)}{g} N(\Theta) \sin 2[\delta_0(t) + \chi_0],$$

$$\int_{-\Theta(t)}^{\Theta(t)} \{r[\zeta, t] + \varepsilon\dot{r}[\zeta, t]\} \cos 2[\zeta + \delta_0(t) + \chi_0]d\zeta =$$

$$= \frac{\overline{\lambda}_0(t)}{g} N(\Theta) \cos 2[\delta_0(t) + \chi_0],$$

where

$$N(\Theta) = \frac{1}{\cos 2\Theta}\left[\Theta - \frac{1}{4}\sin 4\Theta\right].$$

Taking into account these expressions, formulas (29.35) give as follows:

$$\gamma_{xy}^2(t) + \varepsilon\dot{\gamma}_{xy}(t)\gamma_{xy}(t) + [\varepsilon_x(t) - \varepsilon_y(t)]^2 +$$

$$+\varepsilon[\dot{\varepsilon}_x(t) - \dot{\varepsilon}_y(t)][\varepsilon_x(t) - \varepsilon_y(t)] =$$

$$= \frac{\overline{\lambda}_0(t)}{g} N(\Theta) \{\gamma_{xy}(t) \sin 2[\delta_0(t) + \chi_0] +$$

$$+[\varepsilon_x(t) - \varepsilon_y(t)] \cos 2[\delta_0(t) + \chi_0]\}.$$

Taking into account equations (29.32) and (29.21), we can make sure that the latter equation is identical to the following:

$$\gamma_m(t) + \varepsilon \dot{\gamma}_m(t) = \frac{\lambda_4(t)\tau_m(t)}{g\tau_{xy}(t)} N(\Theta) \cos 2\Theta \sin 2\chi_1(t). \tag{29.35}$$

Having in mind the designations (29.25), we will find

$$\gamma_m(t) + \varepsilon \dot{\gamma}_m(t) = \frac{\tau_m(t) + A\Sigma}{g\Psi[t]} \cdot \frac{\Theta - \frac{1}{4}\sin 4\Theta}{1 + \frac{c}{g}\left(\Theta - \frac{1}{4}\sin 4\Theta\right)}. \tag{29.36}$$

Finally, from Eq. (29.36) and the first of formulas (29.16), we obtain

$$\cos 2\Theta = \frac{\bar{\lambda}_0(t)}{\sqrt{\lambda_5^2(t) + \left\{\dfrac{\tau_m(t) + A\Sigma}{\Psi[t][1 + (\Theta - 0.25\sin 4\Theta)c/g]}\right\}^2}}. \tag{29.37}$$

Formulas (29.36) and (29.37) set a link between stresses, strains, and strain rates in monotonous plane-plastic strain. Calculations under these formulas can be done by numerical methods using a computer. In partial cases, these calculations are simplified. For example, for proportional loading, Eq. (29.27) sets the dependency of the parameter Θ on the time t. Then Eq. (29.36) can be integrated, and the solution does not differ in essence from the above solutions for uniaxial elongation and pure shift.

Based on the obtained results (29.36)–(29.37), we can make a conclusion that the ratios of the link between stresses and strains depend on the loading history. Therefore, the loading duration plays an important role and monotonous loading to some stress τ_m cannot be replaced with a proportional one as was the case for monotonous strain of metals considered by the authors of [2] at $A = B = \varepsilon = 0$.

29.3.4 Monotony Conditions

From the definition (p. 382) of monotonous plastic strain, it follows that the functions $\Theta_{1,2}(t)$ limiting the slip plane fan must be non-decreasing:

$$\dot{\Theta}_{1,2}(t) \geqslant 0, \quad (t \geqslant t_0), \tag{29.38}$$

where t_0 is the moment when plastic strain occurs. Referring to formulas (29.29) of the last paragraph, the condition (29.38) can be written otherwise

$$\dot{\Theta}(t) - \dot{\delta}_0(t) \geqslant 0, \quad \dot{\Theta}(t) + \dot{\delta}_0(t) \geqslant 0,$$

or

$$\dot{\Theta}(t) \geqslant |\dot{\delta}_0(t)|, \quad (t \geqslant t_0). \tag{29.39}$$

Using the results of the previous paragraph, let us represent the condition (29.39) otherwise. From formulas (29.32), we can express

$$\lambda_5 = \pm \frac{1}{\cos 2\Theta} \sqrt{\bar{\lambda}_0^2 - \lambda_4^2 \cos^2 2\Theta}, \qquad (29.40)$$

whereas the upper sign is used in the latter formula if the rotation of principal stresses during loading occurs counterclockwise (Fig. 29.1) and the lower sign is used in the contrary case. The differentiation of formulas (29.32) results in

$$\dot{\Theta} = \frac{\bar{\lambda}_0}{\sqrt{\lambda_4^2 + \lambda_5^2 - \bar{\lambda}_0^2}} \left[\frac{\lambda_4 \dot{\lambda}_4 + \lambda_5 \dot{\lambda}_5}{\lambda_4^2 + \lambda_5^2} - \frac{\dot{\bar{\lambda}}_0}{\bar{\lambda}_0} \right],$$

$$\dot{\delta}_0 = \dot{\chi} - \frac{1}{2} \frac{\lambda_4 \dot{\lambda}_5 - \lambda_5 \dot{\lambda}_4}{\lambda_4^2 + \lambda_5^2}. \qquad (29.41)$$

When substituting the expressions (29.41) into the condition (29.39), we obtain

$$\frac{1}{2} \frac{\lambda_4 \dot{\lambda}_5 - \lambda_5 \dot{\lambda}_4}{\lambda_4^2 + \lambda_5^2} - \frac{\bar{\lambda}_0}{\sqrt{\lambda_4^2 + \lambda_5^2 - \bar{\lambda}_0^2}} \left[\frac{\lambda_4 \dot{\lambda}_4 + \lambda_5 \dot{\lambda}_5}{\lambda_4^2 + \lambda_5^2} - \frac{\dot{\bar{\lambda}}_0}{\bar{\lambda}_0} \right] \leqslant$$

$$\dot{\chi} \leqslant \frac{1}{2} \frac{\lambda_4 \dot{\lambda}_5 - \lambda_5 \dot{\lambda}_4}{\lambda_4^2 + \lambda_5^2} - \frac{\bar{\lambda}_0}{\sqrt{\lambda_4^2 + \lambda_5^2 - \bar{\lambda}_0^2}} \left[\frac{\lambda_4 \dot{\lambda}_4 + \lambda_5 \dot{\lambda}_5}{\lambda_4^2 + \lambda_5^2} - \frac{\dot{\bar{\lambda}}_0}{\bar{\lambda}_0} \right]. \qquad (29.42)$$

Let us consider at first the case when while loading principal stresses rotate counterclockwise. Substituting formula (29.20) into the condition (29.42) gives

$$\frac{1}{2\bar{\lambda}_0^2 \sqrt{\bar{\lambda}_0^2 - \lambda_4^2 \cos^2 2\Theta}} \left[\bar{\lambda}_0 (\dot{\bar{\lambda}}_0 \cos 2\Theta + 2\bar{\lambda}_0 \dot{\Theta} \sin 2\Theta) (\lambda_4 - \right.$$

$$\left. - 2\sqrt{\bar{\lambda}_0^2 - \lambda_4^2 \cos^2 2\Theta} \right) + (\dot{\lambda}_4 \bar{\lambda}_0^2 - 3\lambda_4^2 \dot{\lambda}_4 \cos^2 2\Theta) \cos 2\Theta \bigg] +$$

$$+ \frac{\dot{\bar{\lambda}}_0}{\bar{\lambda}_0} \cos 2\Theta \leqslant \dot{\chi}(t) \leqslant \frac{1}{2\bar{\lambda}_0^2 \sqrt{\bar{\lambda}_0^2 - \lambda_4^2 \cos^2 2\Theta}} \times \qquad (29.43)$$

$$\times \left[\bar{\lambda}_0 (\dot{\bar{\lambda}}_0 \cos 2\Theta + 2\bar{\lambda}_0 \dot{\Theta} \sin 2\Theta)(\lambda_4 + 2\sqrt{\bar{\lambda}_0^2 - \lambda_4^2 \cos^2 2\Theta}) + \right.$$

$$\left. + (\dot{\lambda}_4 \bar{\lambda}_0^2 - 3\lambda_4^2 \dot{\lambda}_4 \cos^2 2\Theta) \cos 2\Theta \right] - \frac{\dot{\bar{\lambda}}_0}{\bar{\lambda}} \cos 2\Theta.$$

If principal stresses rotate clockwise, we will find as follows from the condition (29.42) and formula (29.40):

$$\frac{1}{2\bar{\lambda}_0^2\sqrt{\bar{\lambda}_0^2 - \lambda_4^2\cos^2 2\Theta}}\left[\bar{\lambda}_0(\dot{\bar{\lambda}}_0\cos 2\Theta + 2\bar{\lambda}_0\dot{\Theta}\sin 2\Theta)(\lambda_4 - \right.$$

$$\left. - 2\sqrt{\bar{\lambda}_0^2 - \lambda_4^2\cos^2 2\Theta}\right) + (\dot{\lambda}_4\bar{\lambda}_0^2 + \lambda_4^2\dot{\lambda}_4\cos^2 2\Theta)\cos 2\Theta\right] +$$

$$+\frac{\dot{\bar{\lambda}}_0}{\bar{\lambda}_0}\cos 2\Theta \leqslant \dot{\chi}(t) \leqslant \frac{1}{2\bar{\lambda}_0^2\sqrt{\bar{\lambda}_0^2 - \lambda_4^2\cos^2 2\Theta}}\times \tag{29.44}$$

$$\times\left[\bar{\lambda}_0(\dot{\bar{\lambda}}_0\cos 2\Theta + 2\bar{\lambda}_0\dot{\Theta}\sin 2\Theta)(\lambda_4 + 2\sqrt{\bar{\lambda}_0^2 - \lambda_4^2\cos^2 2\Theta}) + \right.$$

$$\left. +(\dot{\lambda}_4\bar{\lambda}_0^2 + \lambda_4^2\dot{\lambda}_4\cos^2 2\Theta)\cos 2\Theta\right] - \frac{\dot{\bar{\lambda}}_0}{\bar{\lambda}}\cos 2\Theta.$$

Thus the conditions (29.43)–(29.44) limit the rotation velocity of principal stresses ($\dot{\chi}(t)$) when monotonous plastic strain takes place.

Let us note that if in the shear resistance operator (29.7) we bring the constant value g to zero, we will obtain based on formulas (29.32)–(29.33) and (29.29) that $\zeta \equiv \pm\Theta, = 0$, e.g. slips will take place at any moment in time in a single direction only. In this case, formulas (29.39) show that $\dot{\chi}(t) = 0$. This means that for $g = 0$ monotonous strain occurs only in the case of proportional loading. Using the operator (29.7) for $g = 0$ is rather efficient to solve problems at sufficiently non-monotonous strain due to the extreme simplicity of such operator.

References

1. M. Leonov, V. Molotnikov, *K teorii deformatsii metallov s yarko vyrazhennym predelom tekuchesti* (On the theory of deformations of metals with bright expressed yield strength) Izv. AN Kirg. SSR [Izv. Academy of Sciences of Kyrghyz. SSR] **6**, 3–10 (1974)
2. M. Leonov, N. Shvaiko, *Slozhnaya ploskaya deformatsiya* (Complex plane deformation), dan sssr (reports of the USSR Academy of Sciences) DAN SSSR [Reports of the USSR Academy of Sciences] (5) **159**, 1007–1010 (1964)
3. V. Molotnikov, A. Molotnikova, Plosko-plasticheskaya deformatsiya (flat-plastic deformation), in *Sostoyanie i perspektivy razvitiya sel'skokhozyaistvennogo mashinostroeniya: materialy mezhdunar. nauchn.-pr. konf. "Interagromash"* (State and prospects of agricultural engineering development: materials of the international conference. nauchn. - Ave. Conf. "Interagromash") (2009), pp. 186–189

Part IV
Non-elastic Strain of Geomaterials

Chapter 30
Complex Strain of Soils

30.1 Real State of the Mechanics of Non-elastic Strains

An overview of plasticity theory and the state of the primary variants of its development described in the previous two parts of the book makes us agree that the theory of non-elastic solid body strain remains one of the primary problems in modern mechanics. Today, we have many models, concepts, postulates, and principles, the most significant of which were included in these two sections. However, we should admit that there is still no theory that would be, in terms of correspondence to experience, comparable with the Cauchy–Navier elasticity theory or the theory of motion of viscous liquids of Navier–Stokes [4, 6].

As M. Ya. Leonov believes [6], such prolonged stagnation in the development of mathematical plasticity theory is caused not only by an extreme complexity of the problem but also by a tradition not to consider the physical mechanism of the respective processes. Non-elastic strains result from structural changes in some volumes, and in essence, they are defined by the discrete structure of solid bodies. For this reason, when building determinant ratios recently, more attention has been paid to various ways of accounting for the physical mechanism of the plasticity phenomenon. These approaches often undergo stages of hopes and disappointments. The most shining example here is the Batdorf–Budiansky plasticity theory [2] that we discussed in the previous section of the book. Experiments to check the ratios of the Batdorf–Budiansky theory showed that its dependencies agreed with experimental data for some loading paths (close to proportional) and were not confirmed by other experiments [11, 13–15].

We have shown that the update of slip concepts undertaken in the previous section removes the disadvantages of the Batdorf–Budiansky theory; however, it significantly complicates the theory. A significant advantage of the proposed slip model is its two-dimensional variant and a revealed opportunity to expand the obtained results to the spatial case using the Ilyushin isotropy postulate.

V. Molotnikov, A. Molotnikova, *Theory of Elasticity and Plasticity*,
https://doi.org/10.1007/978-3-030-66622-4_30

As we said in Sect. 12.7, due to the complexity of building the general mathematical plasticity theory, currently it can be universally deemed that it is impossible to create a single, universal, and sufficiently full plasticity theory by using either classical methods of solid body mechanics or philosophical synergetic methods. Therefore, we believe (see also p. 158) that much attention in the upcoming decades will be paid to simplified settings and solutions of problems of non-elastic strains reproducing the primary and most important properties of real bodies.

30.2   Simple Strain Model of Hardening Dense Soils

The studies show [1, 5] that low-moistened clay and loamy soils have a relatively high strength. Ultimate compressive strength when compressing loamy gray soil at a moisture of 0.61% reaches the strength of a grade 100 brick or even exceeds it ([5], p. 276). Permanent strains (especially during elongation tests) are almost as elastic strains. Such bodies are often called [12] semi-brittle.

During the non-elastic strain of a semi-brittle body [8], loosening or compaction of the material structure occurs that is caused by the formation and growth or closure of micro- and macro-fractures. This process manifests itself in a residual change in volume.

Let us represent the full strain of such body as a sum of elastic, purely plastic strain, as well as the strain of compaction (loosening). Let us write elastic strain components in the principal axes (1, 2, 3) as

$$\varepsilon_j^y = \frac{p}{K} + \frac{\sigma_j - p}{2G_o}, \quad (j = 1, 2, 3), \tag{30.1}$$

where $K/3$ is an elastic modulus of volumetric expansion, G_o is the elastic shear modulus, and the average stress p is defined by the formula as before:

$$p = \frac{1}{3}(\sigma_1 + \sigma_2 + \sigma_3). \tag{30.2}$$

Let us represent purely plastic strain as

$$\varepsilon_j^p = (\sigma_j - p)\left(\frac{1}{2G} - \frac{1}{2G_o}\right), \quad (j = 1, 2, 3), \tag{30.3}$$

whereas G is a still unknown function.

Let us represent the components of compaction (loosening) strain as

$$\varepsilon_j^p = f(\sigma_n, \sigma_m, p)|\sigma_j - p|\left(\frac{1}{G} - \frac{1}{G_o}\right), \tag{30.4}$$

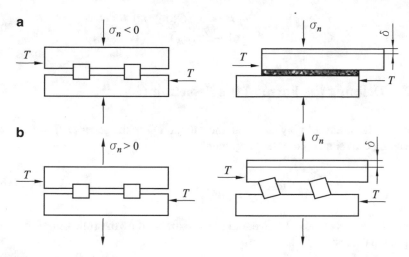

Fig. 30.1 Scheme of the compaction–loosening mechanism

where σ_m is the maximum normal stress (in terms of the absolute value) at this moment of stress, σ_n is the normal stress on the plateau with maximum tangential stress, and f is some function of these arguments.

Let us consider the mechanism of compaction (loosening). Observations show that non-elastic strain occurs by sliding soil blocks relative to each other, and initially, it has local nature as in the case of metals. These slips may face obstacles in the form of grains of mother rock, inclusions of stronger particles, etc. Two cases can be represented.

1. Assume that the normal stress on the slip plateau is negative (Fig. 30.1a). In this case, an obstacle may re-orient or structural changes may occur in the case of slip in the block material in the vicinity of the slip area. As a result, slipping blocks will come together by the quantity δ as shown in the right figure in position $a)$. This approach will cause the compaction effect.
2. If $\sigma_n \geqslant 0$, the contacting blocks are distanced from each other during the slip (Fig. 30.1b) causing loosening of the material.

We will account for these effects by assuming that

$$f(\sigma_n, \sigma_m, p) = \frac{\eta}{1 - p/\sigma_m} SG(\sigma_n),$$
$$SG(\sigma_n) = \begin{cases} 1 & \text{at } \sigma_n \geqslant 0, \\ -1 & \text{at } \sigma_n < 0, \end{cases}$$
$$\eta = const.$$

(30.5)

Then the full strain in the principal directions will be

$$\varepsilon_j = \frac{p}{K} + \frac{\sigma_j - p}{2G} + \frac{\eta|\sigma_j - p|SG(\sigma_n)}{1 - (p/\sigma_m)} \left(\frac{1}{G} - \frac{1}{G_o}\right). \tag{30.6}$$

30.3   Defining the Form of the Function G

To model the strains of gray cast iron, the function G in the paper [12] was deemed depending on a single (complex) argument:

$$u = \frac{\tau_o^2}{\tau_o - cp}, \quad (c = const), \tag{30.7}$$

where τ_o means octahedral tangential stress expressed by the following formula in the principal axes:

$$\tau_o = \frac{1}{3}\left[(\sigma_1 - \sigma_2)^2 + (\sigma_2 - \sigma_3)^2 + (\sigma_3 - \sigma_1^2)\right]^{1/2}. \tag{30.8}$$

Experiments show [9] that in the case of strain of specimens of non-disturbed low-moistened clay loams with moderate rates, stress diagrams are qualitatively similar to the respective diagrams for cast irons. Along with that, strains of soils substantially depend on time effects. This circumstance makes us treat the function G as dependent not only on stresses but on their rate as well. We will account for these effects by assuming the following dependency for the function G:

$$G + \tau\dot{G} = \Bbbk(u), \quad \left(\dot{G} = \frac{\partial G}{\partial t}; \ \tau - const\right), \tag{30.9}$$

where t is the time and $\Bbbk$ is the function of the specified argument that we will write as

$$u = \varrho\tau_o, \quad \varrho = \frac{1 + p/\sigma_m}{1 - cp/\sigma_m}. \tag{30.10}$$

Representing the function G as a dependency (30.9) describes a wide range of events observed during the strain of considered materials. The problem lies in determining the form of the function $\Bbbk$ and the parameters of the formulated model from a limited range of experiments. For this purpose, we will consider a case of loading with a constant change rate of octahedral tangential stress.

30.3.1   Building the G Function for a Material with High Hardening

The calculations show [10] that in this case, we can assume as follows:

$$\Bbbk(u) = G_o - bu_2, \quad (b = const). \tag{30.11}$$

By substituting the expression (30.11) into formula (30.9), we come to the equation:

$$\dot{G} + \frac{1}{\tau}G = \frac{1}{\tau}(G_o - b\varrho^2 v^2 t^2), \tag{30.12}$$

where $v = const$ is the change rate of octahedral tangential stress. By integrating the linear Eq. (30.12) at the initial condition $G(0) = G_o$, we find

$$G = G_o - b\varrho^2 \left[\tau_o^2 - 2\tau v\tau_o + 2v^2\tau^2 \left(1 - \exp(-\tau_o/v\tau)\right) \right]. \tag{30.13}$$

30.3.2   Universal G Function for Hardening Soils

Below we show that the function G defined by formula (30.13) well describes the strains of soils with high compaction. However, it decreases faster and, starting with some values of the argument, it becomes negative so that its physical sense is lost. The analysis of formulas (30.6) leads to a conclusion that the G function must have a form of a curve of normal distribution law [7]. We will satisfy this condition by assuming that

$$\Bbbk(u) = G_o \exp(-(au)^2), \tag{30.14}$$

where a is the parameter that we will deem unchanged over the loading time.

Assume that stress at a specified rate is applied to a non-loaded specimen at the point of time $t = t_o$. Let us designate the change rate of octahedral tangential stress as $v(t)$. Acting stress at an arbitrary point of time t will be

$$\tau_o(t) = \int_{t_o}^{t} v(\xi)d\xi. \tag{30.15}$$

By substituting the expression (30.15) into formulas (30.10)–(30.11), we will obtain the following linear differential equation relative to the sought G function from the ratio (30.9):

$$\dot{G} + \frac{1}{\tau} G = \frac{1}{\tau} \exp \left[-a\varrho \int\limits_{t_o}^{t} v(\xi) d\xi \right]^2. \tag{30.16}$$

The solution to Eq. (30.16) at $v = const$, $t_o = 0$ and the initial condition $G(0) = G_o$ is the function [3]

$$G(u) = G_o \exp \left(-\frac{u}{\varrho v \tau} \right) \left[Q\sqrt{\pi} \exp(Q^2) \left(\Phi(au - Q) - \Phi(-Q) \right) + 1 \right], \tag{30.17}$$

where the constant value Q is defined by the formula

$$Q = \frac{1}{2a\tau \varrho v}, \tag{30.18}$$

and $\Phi(x)$ is the probability integral [3]:

$$\Phi(x) = \frac{2}{\sqrt{\pi}} \int\limits_{0}^{x} \exp(-t^2) dt. \tag{30.19}$$

The found function (30.17) equally well describes strains in any monotonous hardening of material and is universal in this sense. Now let us go over to considering partial types of loadings at constant rates.

References

1. P. Bakhtin, *Trudy Stavropol'skogo NII s.-kh.* (Proceedings of the Stavropol research Institute of agriculture), Frunze, Izd-vo Ilim Publ., chapter Znachenie fiziko-mekhanicheskikh i tekhnologicheskikh svoistv osnovnykh tipov pochv (Significance of physical, mechanical and technological properties of the main types of soils) (1971), pp. 333–342
2. S. Batdorf, B. Budyanskii, *Mekhanika: sb. perev., no 5(33)* (Mechanics: a collection of translations, no 5(33)). (Moscow, Inostrannaya literatura Publ., chapter *Zavisimost' mezhdu napryazheniyami i deformatsiyami dlya uprochnyayushchegosya metalla pri slozhnom napryazhennom sostoyanii* (The relationship between stress and deformations for the hardening of metals under complex stress state). Mekhanika: sb. perev., no 5(33) [Mechanics: a collection of translations] (1955) pp. 120–127
3. I. Gradshtein, I. Ryzhik, *Tablitsy integralov, summ, ryadov i proizvedenii* (Tables of integrals, sums, series, and works). (Nauka Publ., Moscow, 1971)
4. A. Il'yushin, *Plastichnost'. (Osnovy obshchei matematicheskoi teorii)* (Plasticity. Fundamentals of General mathematical theory). (Izd-vo AS USSR Publ., Moscow, 1963)
5. A. Ioffe, I. Revut, *Osnovy agrofiziki / red. Ioffe A.F., Revut I.B.* (Fundamentals of agrophysics). ed. by A.F. Ioffe, I.B. Revut (Fizmatlit Publ., Moscow, 1959)
6. M. Leonov, *Mekhanika deformatsii i razrusheniya* (Deformation and fracture mechanics) (Izd-vo Ilim Publ., Frunze, 1981)

7. E. L'vovskii, *Statisticheskie metody postroeniya ehmpiricheskikh formul* (Statistical methods for constructing empirical formulas). (Vyssh. shk. Publ., Moscow, 1982)

8. V. Molotnikov, *Plastichnost' i prochnost' materialov i konstruktsii* (Plasticity and strength of materials and structures). Frunze, Izd-vo Frunz. politekhn. in-ta Publ., chapter O kraevykh zadachakh mekhaniki neuprugogo tv"erdogo tela (On boundary value problems in inelastic solid mechanics) (1981), pp. 71–79

9. V. Molotnikov, *Metody modelirovaniya v zemledel'cheskoi mekhanike* (Modeling methods in agricultural mechanics). PhD thesis, Sankt-Peterburg, ASFI., 1994

10. V. Molotnikov, *Min-vo s.-kh. RF, DoNGAU. Dep. v NIIITEHISKh 21.01.94, №6 VS–94 Dep.* (Ministry of agricultural of the Russian Federation, don state agrarian University.), Moscow, Min. of agr. of the RF Publ., chapter O kraevykh zadachakh mekhaniki neuprugogo tv"erdogo tela (On boundary value problems in inelastic solid mechanics) (1994), pp. 1–13

11. Nakhdi, Rouli, *Mekhanika: sb. perev* (Mechanics: collected transfers) no. 3 (31), pp. 138–147., Moscow, IL Publ., chapter Ehksperimental'noe izuchenie zavisimostei mezhdu napryazheniyami i deformatsiyami v plasticheskoi oblasti pri dvukhosnom napryazhennom sostoyanii (Experimental study of the relationship between stresses and deformations in the plastic region under biaxial stress state) (1955), pp. 138–147

12. V. Panyaev, K. Rusinko, *O deformatsiyakh i razrushenii polukhrupkikh tel* (On deformations and destruction of semi-brittle bodies). *Deformatsiya neuprugogo tela: sb. nauchn. statei* (Deformation of an inelastic body: sci. articles) (1970), pp. 98–100

13. V. Sveshnikova, *O plasticheskom deformirovanii plasticheskikh metallov* (On plastic deformation of plastic metals). Izv. AN SSSR. OTN. (News of the USSR Academy of Sciences. Department of Technical Sciences) **1**, 155–162 (1956)

14. A. Zhukov, *Plasticheskie deformatsii stali pri slozhnom nagruzhenii* (Plastic deformations of steel under complex loading). Izv. AN SSSR, OTN (Izvestia of the USSR Academy of Sciences, Department of technical Sciences) **11**, 53–62 (1954)

15. A. Zhukov, Y. Rabotnov, *Issledavanie plasticheskoi deformatsii stali pri slozhnom nagruzhenii* (Investigation of plastic deformation of steel under complex loading). Inzh. sb. (Engineering collection) **18**, 105–112 (1954)

Chapter 31
Simple Loadings of Geomaterials

31.1 Uniaxial Compression

In this case, $\sigma_1 = \sigma_2 = 0$, $\sigma_3 < 0$; $\sigma_n < 0$, $SG(\sigma_n) = -1$. Let us designate $\sigma = |\sigma_3|$ and $\varepsilon = |\varepsilon_3|$. Then $\sigma_m = \sigma$ and

$$p = -\frac{1}{3}\sigma; \quad \tau_o = \frac{\sqrt{2}}{3}\sigma; \quad \varrho = \varrho_c = \frac{2}{3+c}; \quad u = \varrho_c\tau_o. \tag{31.1}$$

Taking into account formulas (31.1) from (30.6) in the considered case, it follows that:

$$
\begin{aligned}
\varepsilon_1 = \varepsilon_2 &= \frac{\sigma}{3}\left[-\frac{1}{K} + \frac{1}{2G} - \frac{3}{4}\eta\left(\frac{1}{G} + \frac{1}{G_o}\right)\right]; \\
\varepsilon = |\varepsilon_3| &= \frac{\sigma}{3}\left[\frac{1}{K} + \frac{1}{2G} + \frac{3}{2}\eta\left(\frac{1}{G} + \frac{1}{G_o}\right)\right].
\end{aligned}
\tag{31.2}
$$

In this manner, a link between stresses and strains is given by formulas (31.2), where G is defined according to (30.13) or (30.17).

31.2 Creep in Uniaxial Compression

Assume that the stress $\sigma(t^*) = \sigma^*$ is recorded at some point in time t^* during uniaxial compression. By designating

$$u^* = u(\tau_o^*) = \varrho_c\frac{\sigma^*}{3}\sqrt{2},$$

© The Author(s), under exclusive license to Springer Nature Switzerland AG 2021
V. Molotnikov, A. Molotnikova, *Theory of Elasticity and Plasticity*,
https://doi.org/10.1007/978-3-030-66622-4_31

and using the representation (30.11) for $\Bbbk(u)$, we have as follows based on (30.9):

$$G(t) + \tau \dot{G} = G_o - bu^{*2}. \tag{31.3}$$

By integrating this equation within t^* to t with the initial condition $G(t^*) = G^*$, whereas G^* is the value G calculated upon formula (30.13) at $t = t^*$, $\varrho = \varrho_c$, we obtain

$$G(t) = (G_o - bu^{*2})\left[1 - \exp\left(-\frac{t - t^*}{\tau}\right)\right] + G^* \exp\left(-\frac{t - t^*}{\tau}\right). \tag{31.4}$$

Based on the second of formulas (31.2), the creep law will be

$$\varepsilon = \frac{\sigma^*}{3}\left[\frac{1}{K} + \frac{1}{G} + \frac{3}{2}\eta\left(\frac{1}{G} - \frac{1}{G_o}\right)\right], \quad (t \geqslant t^*), \tag{31.5}$$

where G is defined by formula (31.4).

To get the creep law for soil with arbitrary hardening, we use formula (30.14) for $\Bbbk(u)$. Then the expression for the G function in creep will be obtained by replacing the expression $(G_o - bu^{*2})$ with $G_o \exp(-au^*)^2$ in formula (31.4), and the value of the function (30.17) at $u = u^*$ must be placed in (31.4) instead of G^*. After such replacement, formula (31.5) will define the sought creep law of the considered material.

31.3   Uniaxial Elongation

In the case of elongation, we have: $\sigma_1 = \sigma$, $\sigma_2 = \sigma_3 = 0$, $\sigma_m = \sigma$, $p = \sigma/3$, $\tau_o = \frac{\sigma}{3}\sqrt{2}$, $\varrho = \varrho_p = \frac{4}{3(1 - c/3)}$, $SG(\sigma_n) = 1$. From formula (30.6), we obtain as follows:

$$\begin{aligned} \varepsilon = \varepsilon_1 &= \frac{\sigma}{3}\left[\frac{1}{K} + \frac{1}{G} + 3\eta\left(\frac{1}{G} - \frac{1}{G_o}\right)\right], \\ \varepsilon_2 = \varepsilon_3 &= -\frac{\sigma}{3}\left[-\frac{1}{K} + \frac{1}{G} - 3\eta\left(\frac{1}{G} - \frac{1}{G_o}\right)\right], \end{aligned} \tag{31.6}$$

where the function G is still defined by formulas (30.13) or (30.17) depending on the nature of material hardening, whereas the parameter ϱ must be replaced with ϱ_p if using formula (30.17).

31.4   Pure Shift

In this case, $\sigma_1 = -\sigma_3 = T$, where T is the shear tangential shift. Then,

$$\sigma_m = T, \quad p = 0, \quad \varrho = 1, \quad \tau_o = T\sqrt{\frac{2}{3}}, \quad SG(\sigma_n) = 1.$$

Then the dependency gives

$$\varepsilon_1 = T\left[\frac{1}{2G} + \eta\left(\frac{1}{G} - \frac{1}{G_o}\right)\right], \quad \varepsilon_2 = 0,$$

$$\varepsilon_3 = -T\left[\frac{1}{2G} - \eta\left(\frac{1}{G} - \frac{1}{G_o}\right)\right],$$

(31.7)

where the function G is calculated using formulas (30.14)–(30.16) or (30.19) at $\varrho = 1$, whereas v still designates the change rate of octahedral tangential stress.

31.5 Determination of Model Parameters

To verify the model, it is required to find the parameters b (or a), η, c, and τ from experiments. They can be defined, for example, if we have experimental material test diagrams for elongation, shear, compression testing, as well as one of creep curves. After selecting one point from these diagrams and using the dependencies (31.2), (31.5), (31.6), and (31.7), we will have four equations relative to the sought parameters. By identical transformations, they can be brought to the system of two transcendent equations relative to η and τ:

$$\eta = F_1(\eta, \tau), \quad \tau = F_2(\eta, \tau),$$

(31.8)

where the functions F_1 and F_2, apart from their arguments, depend on the coordinates for four selected points on experimental curves.

A solution to the system (31.8) can be obtained by the method of iteration [1]. The process convergence depends on the zero approximation quality. The experience showed that for the sought materials, good zero approximation was $\eta = 0.1 \ldots 0.3$ and $v\tau = 10^5 \ldots 10^7$ Pa.

Determining the parameters using the described methodology when setting G using formula (30.13), the loading rate of $v = 500$ Pa/s, $G_o = 60$ MPa and using the experimental results [2, 3] gave the following values of model parameters: $b = 4.25 \cdot 10^{-5}$ Pa^{-1}; $\eta - 0.14$; $\tau = 200^{-1}c$; $c = 0.4$.

31.6 Comparison of Experimental and Calculation Results

Figure 31.1 shows experimental results taken from [2, 3] with dashed lines and computational dependencies with solid lines for the above values of model

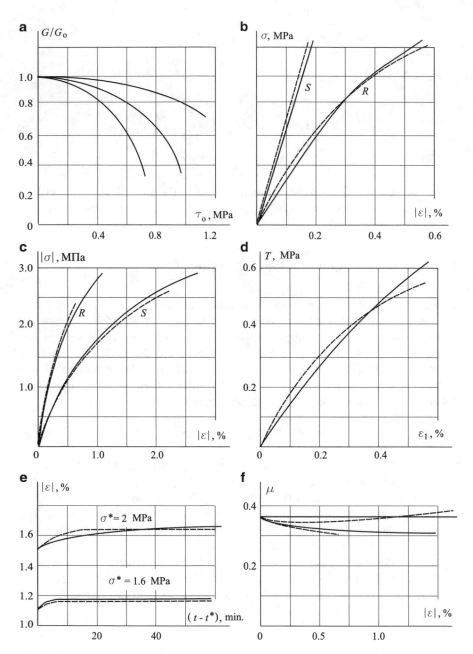

Fig. 31.1 Experimental and calculation dependencies for low-moistened clay loams

parameters. Position a shows the function G in compression (upper curve), shear (middle line), and elongation (lower curve). Positions b–d show stress diagrams in uniaxial elongation, uniaxial compression, and torsion. The absolute values of principal strains are laid on the X-axis and modules of principal stresses on the Y-axis. S designates curves belonging to compressing principal stresses and R designates elongating stresses. Position f shows curves of changes in the transverse strain coefficient (μ). It is characteristic that in compression, this coefficient remains almost unchanged, and in the case of elongation, in experiments, and upon computational dependencies, the coefficient falls down and strain increases. This phenomenon is also found in the paper by Panyaev and Rusinko [4].

To avoid confusion, Fig. 31.1d gives only the dependency $T \sim \varepsilon_1$ since the y-coordinates of the diagram $T \sim |\varepsilon_3|$ differ from the shown dependencies by a low value (increased). Position e gives creep results in uniaxial compression. As the figure shows, creep strains fade away quickly, and for the first half an hour when the creep rate is mostly notable, their value reaches 10% of the initial axial strain.

Graphical results shown in Fig. 31.1b–f, indicate the high coincidence of experimental and model dependencies with a rather representative set of studied phenomena found in soil strain.

References

1. I. Berezin, N. Zhidkov, *Metody vychislenii* (Calculation method). (Nauka Publ., Moscow, 1966)
2. V. Molotnikov, *Metody modelirovaniya v zemledel'cheskoi mekhanike* (Modeling methods in agricultural mechanics). PhD thesis, Sankt-Peterburg, ASFI, 1994
3. V. Molotnikov, A. Molotnikova, *Clozhnaya deformatsiya pochvogruntov* (A complex deformation of soils, Orenburg, pp. 321–327), in Prochnost' i razrushenie materialov i konstruktsii: materialy V mezhdunar. nauchn. konf. T. 2 (Strength and destruction of materials and structures: materials V Mezhdunar. Scientific Conf. vol. 2.)
4. V. Panyaev, K. Rusinko, *O deformatsiyakh i razrushenii polukhrupkikh tel* (On deformations and destruction of semi-brittle bodies). *Deformatsiya neuprugogo tela: sb. nauchn. statei* (Deformation of an inelastic body: sci. articles) (1970), pp. 98–100

Chapter 32
On Boundary Value Problems of Inelastic Body Mechanics

32.1 General Formulation of the Problem of Inelastic Solid Mechanics

Recall that, from a physical point of view, the complete deformation ε_{ij} is the result of a change in the distances between particles of a solid body and the order of location of these particles due to various structural disturbances in the body, changes in structural bonds, etc.

If the order in the arrangement of particles in a loaded body and the distances between them in an unloaded state are preserved, then the material experiences a purely elastic deformation ε_{ij}^{y}. This deformation is connected with the stress σ_{ij} by Hooke's law [8]

$$\sigma_{ij} = \lambda \varepsilon_{kk}^{y} \delta_{ij} + 2G_0 \varepsilon_{ij}^{y}, \quad (i, j = 1, 2, 3), \tag{32.1}$$

where the shear modulus G_0 and the Lame constant λ are expressed in terms of Young's modulus E and Poisson's coefficient ν according to the formulas [8]

$$G_0 = \frac{E}{2(1 + \nu)}, \quad \lambda = \frac{\nu}{(1 + \nu)(1 - 2\nu)}. \tag{32.2}$$

In formula (32.1) and later in this chapter, we use the notation introduced in Chap. 12 and adopted in tensor analysis, namely, over repeated indexes (except for the indexes i and j running through the values $1, 2, 3$), summation is performed; the character δ_{ij} is 1 or 0, depending on whether i and j are equal or not.

As before (p. 309), the difference between full and elastic deformations will be called inelastic deformation e_{ij}:

© The Author(s), under exclusive license to Springer Nature Switzerland AG 2021
V. Molotnikov, A. Molotnikova, *Theory of Elasticity and Plasticity*,
https://doi.org/10.1007/978-3-030-66622-4_32

$$e_{ij} = \frac{1}{2}(u_{i,j} + u_{j,i}) - \varepsilon_{ij}^y,$$ (32.3)

where u_i is the component of the offset in the direction of the coordinate axis x_i; the indexes separated by a comma denote differentiation by the spatial coordinate corresponding to the second index.

It is known [8] that the stress σ_{ij} must satisfy the equilibrium equation of any element of a solid body

$$\sigma_{ik,k} + f_i = 0,$$ (32.4)

where f_i is the specified volume forces. Substituting the stress (32.1) into Eq. (32.4) and taking into account the dependencies (32.2)–(32.3), we get

$$G_0 \left(u_{i,kk} + \frac{1}{1 - 2v} u_{k,ki} \right) + f_i + \overline{f}_i = 0,$$ (32.5)

where

$$\overline{f}_i = -\lambda e_{kk,i} - 2G_0 e_{ik,k}.$$ (32.6)

When formulating boundary conditions for the u_i functions, these conditions may also include summands that depend on inelastic deformation. In fact, let the surface forces p_{iv} be set on the surface Γ

$$p_{iv} = \sigma_{ik} v_k,$$ (32.7)

and v_k is the cosine of the angle between the normal v to the surface Γ and the k-th axis. Let us substitute here instead of tension its expression (32.1), in which the elastic deformation is replaced by the difference between full and inelastic deformations according to (32.3). Let us find

$$p_{iv} = \lambda \delta_{ik} v_k u_{m,m} + G_0 v_k (u_{i,k} + u_{k,i}) + \overline{p}_{iv} \qquad \text{on } \Gamma,$$ (32.8)

where

$$\overline{p}_{iv} = -\lambda \delta_{ik} v_k e_{mm} - 2G_0 v_k e_{ik}.$$ (32.9)

Thus, the complete deformation of an inelastic body can be formally [11] determined from the equations of elasticity theory by adding additional forces $\overline{f}_i$ and $\overline{p}_{iv}$. The specified system of volumetric $\overline{f}_i$ and surface $\overline{p}_{iv}$ forces is called fictitious. Fictitious forces are defined through inelastic deformation using formulas (32.6) and (32.9).

Let the plasticity condition be known for the material. Let us write this condition in the following symbolic form:

$$F(\sigma_{ij}) = 0. \tag{32.10}$$

Let us also assume that the relations that agree with the experience are found

$$e_{ij} = E_{ij}(\sigma_{mn}, I_1, I_2, \ldots), \tag{32.11}$$

making it possible by some operator E_{ij} from the stress at this point in time and from some parameters $(I_1, I_2, \ldots)$ that characterize the entire previous process of deformations to determine the inelastic deformation at the considered moment.

The problem can be formulated [4] as follows:

to find three functions u_i that satisfy in the area occupied by the body the equilibrium equations (32.5), on the border of Γ—the conditions (32.8), if in the elastic part of the region, the fictitious forces are zero, and in the region of inelastic deformations defined by the condition (32.10), the relations (32.11), (32.11), and (32.9) are true [7].

The formulated problem differs significantly from the classical problems of mathematical physics in that the relations (32.11) depends on the history of deformation. Apparently, for the first time, this fact was noted by Khill [5]. He is also the author of the idea of a method for solving inelastic problems by sequential step-by-step loading: "A process of plastic deformation has to be considered mathematically as a succession of small increments of strain, even where the overall distortion is so small that the change in external surfaces can be neglected" ([5], p. 90).

32.2   More About the Method of Elastic Solutions

In Chap. 16, we have already got acquainted with the essence and algorithm of the elastic solution method. Here we turn again to the described algorithm in relation to the formulation of the problem of plasticity theory, which is given in the previous paragraph. For the ease of reading, here is a repetition of some of the information set out earlier in Chap. 16.

Split the entire loading process in time into a series of sequential stages. The increment of the external loads at each stage of loading will be considered small enough that it is possible without significant errors to consider the trajectory of loading to be linear at any point of the inelastic region at each loading stage. Such splitting of the loading process into stages makes it easier to keep track of the history of deformation.

Suppose that at the end of the k-th stage of loading (in particular, at the moment of the occurrence of inelasticity at any point of the body), all the characteristics of the deformation process: $\sigma_{ij}^{(k)}$, $\varepsilon_{ij}^{(k)}$, $e_{ij}^{(k)}$, etc. are known. To find a solution to the problem, at the end of the $(k + 1)$-th stage, one can apply the iterative process proposed by A. A. Ilyushin [4], which he called the method of elastic solutions.

In Ilyushin's method, as a zero approximation, we accept deformations from an elastic solution, by which stresses are determined according to the deformation law of the type (32.11). Stresses found by this method will not satisfy the equilibrium equations. However, they can be considered as stresses that satisfy the equations of the theory of elasticity in the presence of additional (fictitious) mass forces. Solving these equations in displacements, deformations of the first approximation are found. Then, according to the found strains, according to the law of the type (32.11), the stresses of the first approximation are determined. Repeating this algorithm, the stresses and strains of subsequent approximations are found. The process can be completed as soon as the results of neighboring approximations coincide with the desired accuracy.

We have already said that the convergence of Ilyushin's elastic solution method has not yet been proven. One can also specify a class of problems for which the stress distribution in an inelastic body is close to the distribution of stresses in the elastic state of the same body, while deformations can differ from elastic ones by tens of times. Examples of such tasks are: a disk, compressible in diameter, a rectangular plate, squeezed diagonally, etc. For such problems, it is advisable to conduct the approximation process by stress. In Russian literature, this technique is known as the Birger method of additional deformations [2].

With this in mind, we find at the end of the $(k + 1)$-th stage of loading stress increment $\Delta\sigma_{ij0}^{(k+1)}$, considering the material to be perfectly elastic. Let us take this increment as a zero approximation (the approximation number is denoted by the last icon in the lower index) to increment the stress $\Delta\sigma_{ij}^{(k+1)}$ at the $(k + 1)$-th stage. Further, by the stress in the zero approximation

$$\sigma_{ij0}^{(k+1)} = \sigma_{ij}^{(k)} + \Delta\sigma_{ij0}^{(k+1)}, \tag{32.12}$$

let us define the boundary of the area where the inelasticity condition is met (32.10), and calculate by the ratios (32.11), (32.6), and (32.9) fictitious forces in the zero approximation $\overline{f}_{i0}$, $\overline{p}_{iv0}$. Then after solving the boundary value problem (32.5), (32.8), we calculate using formulas (32.3) and (32.1) the first approximation for the stress $\sigma_{ij1}^{(k+1)}$ at the end of the $(k + 1)$-th loading stage. Given the known $\sigma_{ij}^{(k)}$ and $\sigma_{ij}^{(k+1)}$, the loading path at the $(k + 1)$-th stage is specified and the procedure is repeated. The iteration process at each stage can be completed as soon as the difference between the stresses in the adjacent approximations is within the desired accuracy. The algorithm diagram is shown in Fig. 32.1.

The described method has a wide generality due to the lack of any restrictions on the type of relations (32.11). The only exception is a perfectly plastic material; this case is specifically discussed in the next paragraph.

Note that for the above class of tasks, when the distribution of the stresses in an inelastic body is close to the stress distribution at the elastic state of this body, already the first approximations of the stresses may be good enough. In addition, when solving problems, in which the stress field is close to uniform, the first approximation also gives an almost exact result.

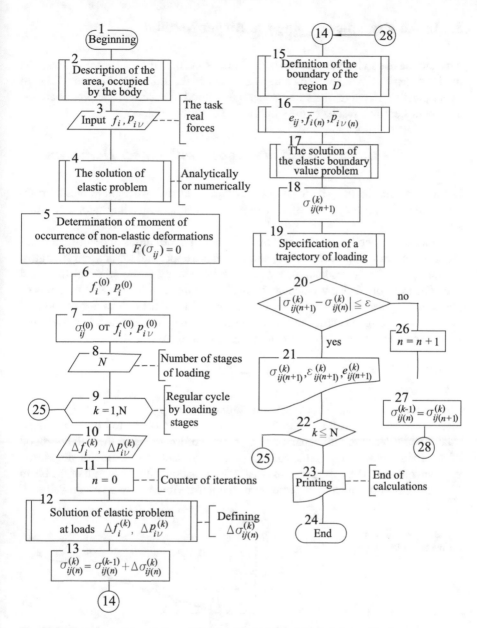

Fig. 32.1 Graphical representation of the algorithm of the elastic solution method

32.3   An Example of Using the Birger Method

An example of solving the problem using the method of additional deformations was implemented by Abdrakhmanov and Kozhobaev in the article [1]. For material with linear plastic hardening, the authors use the method of successive approximations. The main steps and results of the solution are described below.

32.3.1   The Initial Stage of the Process with Linear Hardening

Let us consider the uniaxial tension of a cylindrical rod of length l_0 and a cross-sectional area of F beyond the yield strength under static loading with a constant strain rate.

Suppose that the tensile diagram of the material under uniform deformation has the form shown in Fig. 32.2. Here σ_t and σ_s are the upper and lower yield strengths, ε_t and ε_s are the corresponding tensile strains, and σ_0 is the stress at the yield site. The symbols E and E_1 denote Young's modulus and the tangent modulus of the hardening section, respectively. In the case of the elastic behavior of the rod, deformation along the entire length of the rod is homogeneous, and when the upper yield strength is reached, the absolute elongation of the rod will be

$$\Delta l_t = \frac{\sigma_t}{E} l_0, \tag{32.13}$$

and the longitudinal force in the sections will be $P = \sigma_t F$.

As already mentioned in Chap. 24, further stretching of the rod by an infinitesimal amount leads to an appearance of plastic deformations in the neighborhood of a certain cross-section of the rod, and the length and the location of the plastic zone are random. The appearance of plastic deformation with the length of the beam $l_0 + \Delta l_t$ is accompanied by a decrease in the tensile force P_t to a certain value P.

Fig. 32.2 Tensile diagram for linear hardening

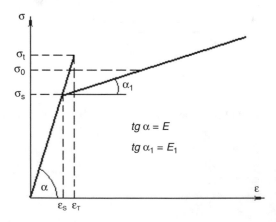

Fig. 32.3 The initial plastic zone

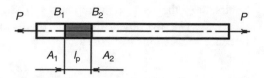

There is a redistribution of stresses. The drop in the tensile load can be significant [14, 15] and depends on the length of the plastic area of the bar.

We assume that the initial plastic zone covers some part of the rod of length l_p and is limited by planes perpendicular to its axis (Fig. 32.3). To study the stress and strain state of a rod after a drop in tensile load, consider the following.

32.3.2 Case of Semi-Infinite Plastic Zone

Let us find the stress field under tension of an infinitely long rod with the force P when one of the two parts of the rod went into a plastic state. We use the initial stress method. As the zeroth approximation, we take the stress $\sigma_0 = P/F$. In this case, the radial displacement u^y in the elastic part of the beam is determined by the formula

$$u^y = -\nu \cdot \frac{\sigma_0}{E} r, \tag{32.14}$$

where ν is Poisson's ratio, and r is the radius vector of the cross-sectional point of the rod.

Deformations in the plastic zone are represented as the sum of their elastic and plastic components

$$\varepsilon_j = \varepsilon_j^y + \varepsilon_j^{II}, \quad (j = r, \theta, z), \tag{32.15}$$

with

$$\varepsilon_r^{II} = \varepsilon_\theta^{II} = -\frac{1}{2}\varepsilon_z^{II}.$$

The plastic deformation component ε_z^{II} can be easily determined by the stretching diagram (Fig. 32.2). We have

$$\varepsilon_z^{II} = \frac{(\sigma_0 - \sigma_s)(E - E_1)}{E E_1}. \tag{32.16}$$

For the radial movement u^{II} of a point in the plastic part of the rod, taking into account formulas (32.15), (32.16), we get

$$u^{II} = \frac{\sigma_s(E - E_1) - \sigma_0(E - E_1 + 2\nu E E_1)}{2 E E_1}. \tag{32.17}$$

Subtracting the plastic components (32.17) from elastic radial displacements (32.14), we obtain the value of the incompatibility of the strains at the interface between the two zones

$$\Delta u(r) = u^{II} - u^{y} = \frac{(\sigma_s - \sigma_0)(E - E_1)}{2EE_1} r. \tag{32.18}$$

If R is the radius of the cross-section of an unloaded rod, then at the boundary of the elastic and inelastic zones there is a gap of radial displacements, defined by formula (32.18) for $r = R$ (Fig. 32.4).

The determination of the stress field in an elastic body from incompatible deformation (32.18) is performed as follows. Mentally, cut the rod in the section AB (Fig. 32.4) and apply uniform pressure p to the lateral surface of the right-hand side such that the radial displacement on the surface is $\Delta u(R)$. The solution to this problem will be

$$\sigma_r = \sigma_\theta = -p, \quad u = -\frac{1 - \nu}{E} pr. \tag{32.19}$$

Next, connecting both sides of the bar over the section AB, we neutralize the pressure p by applying the opposite sign of pressure to the surface of the right side (Fig. 32.5). Thus, the stress field caused by incompatible deformations is the sum of the stresses when the rod is loaded according to the diagram of Fig. 32.5 and the stresses defined by formula (32.19).

32.3.3 Auxiliary Task

Consider the problem of determining the stress field in a bar exposed to axisymmetric uniformly distributed forces applied to part of its surface (Fig. 32.5).

The desired solution can be obtained by a superposition of solutions for the same bar under loading according to the schemes depicted at the positions a and b in Fig. 32.6. Radial and circumferential normal stresses when loading a bar according to the scheme of Fig. 32.6a will be

Fig. 32.4 Rupture of the radial displacements in the cross-section of AB

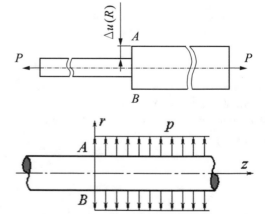

Fig. 32.5 Neutralizing pressure on the right side of the beam

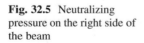

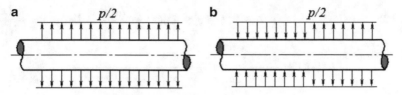

Fig. 32.6 Scheme of loading of a beam for the superposition of solutions

$$\sigma_r = \sigma_\theta = \frac{p}{2}.$$

The stress field during the loading of the beam according to the scheme of Fig. 32.6b was found in [15] by setting the bi-harmonic Lyav function φ in the form

$$\varphi = \int_0^\infty \left[\rho I_0(kr) - k^2 I_1(kr) \right] f(k) \cos kz dk, \tag{32.20}$$

where ρ is the constant to be determined, $I_0(kr)$, $I_1(kr)$ are the modified Bessel functions (first and second of the genus) of the zeroth and first order of the argument kr. The expression for the function $f(k)$ is chosen so that the stress function φ gives a solution to the problem. Omitting the intermediate calculations, we write down the formulas for the stress components in a round beam loaded according to the scheme of Fig. 32.5.

$$\sigma_r = \frac{p}{2} - \frac{p}{\pi} \int_0^\infty \left\{ \left[\frac{(\lambda\zeta)^2 + b + \lambda L_0(\lambda)}{\lambda\zeta} \right] L_1(\lambda\zeta) - \right.$$

$$\left. - [1 + \lambda L_0(\lambda)] L_0(\lambda\zeta) \right\} \Phi(\lambda) \sin \eta\lambda d\lambda;$$

$$\sigma_\theta = \frac{p}{2} + \frac{p}{\pi} \int_0^\infty \left[(2\nu - 1) L_0(\lambda\zeta) + \frac{b + \lambda L_0(\lambda)}{\lambda\zeta} L_1(\lambda\zeta) \right] \Phi(\lambda) \sin \eta\lambda d\lambda;$$

$$\sigma_z = \frac{p}{\pi} \int_0^\infty [2 - \lambda L_0(\lambda\zeta) + \lambda\zeta L_1(\lambda\zeta)] \Phi(\lambda) \sin \eta\lambda d\lambda; \tag{32.21}$$

$$\tau_{rz} = \frac{p}{\pi} \int_0^\infty [\lambda L_0(\lambda) L_1(\lambda\zeta) - \lambda\zeta L_0(\lambda\zeta)] \Phi(\lambda) \cos \eta\lambda d\lambda,$$

where the following notation is introduced:

$$\lambda = kR; \quad \eta = \frac{z}{R}; \quad \zeta = \frac{r}{R}; \quad b = 2(1-v); \quad L_0(\lambda) = \frac{I_0(\lambda)}{I_1(\lambda)}; \quad L_0(\lambda\zeta) = \frac{I_0(\lambda\zeta)}{I_1(\lambda)};$$

$$L_1(\lambda\zeta) = \frac{I_1(\lambda\zeta)}{I_1(\lambda)}; \quad \Phi(\lambda) = \frac{1}{\lambda^2 L_0^2(\lambda) - (b + \lambda^2)}.$$

Then the stress components from incompatible deformation are written in the form

$$\sigma_r = \pm\frac{p}{2} - \frac{p}{\pi} \int\limits_0^\infty \left\{ \left[\frac{(\lambda\zeta)^2 + b + \lambda L_0(\lambda)}{\lambda\zeta} \right] L_1(\lambda\zeta) - \right.$$

$$\left. - [1 + \lambda L_0(\lambda)] L_0(\lambda\zeta) \right\} \Phi(\lambda) \sin \eta\lambda d\lambda;$$

$$\sigma_\theta = \pm\frac{p}{2} + \frac{p}{\pi} \int\limits_0^\infty \left[(2v-1)L_0(\lambda\zeta) + \frac{b + \lambda L_0(\lambda)}{\lambda\zeta} L_1(\lambda\zeta) \right] \Phi(\lambda) \sin \eta\lambda d\lambda;$$

$$\sigma_z = \frac{p}{\pi} \int\limits_0^\infty \{[2 - \lambda L_0(\lambda)] L_0(\lambda\zeta) + \lambda\zeta L_1(\lambda\zeta)\} \Phi(\lambda) \sin \eta\lambda d\lambda;$$

$$\tau_{rz} = \frac{p}{\pi} \int\limits_0^\infty [\lambda L_0(\lambda)L_1(\lambda\zeta) - \lambda\zeta L_0(\lambda\zeta)] \Phi(\lambda) \cos \eta\lambda d\lambda. \qquad (32.22)$$

In formulas (32.3.3), the sign $(-)$ is taken for $z > 0$, and $(+)$ for $z < 0$.

The first approximation stresses are obtained by summing the field stresses from incompatible deformation defined by formula (32.3.3), with the zero approximation stress $\sigma_0 = P/F$.

32.3.4   Final Length of the Plastic Zone

Let, as before (p. 413), l_p be the length of the plastic section of the beam (Fig. 32.3). At the final length of the plastic area, we again take the tensile stress as the zero approximation $\sigma_0 = P/F$. The stress field from the incompatibility of deformations on the boundaries $A_1 B_1$ and $A_2 B_2$ of the plastic zone can be defined by the superposition of problem solutions according to the schemes of Fig. 32.7a and b.

Radial and circumferential normal stresses for the problem according to the scheme of Fig. 32.7a are known

$$\sigma_r = \sigma_\theta = -p. \qquad (32.23)$$

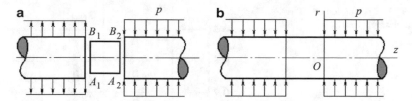

Fig. 32.7 To a superposition of solutions with a finite length of the plastic zone of the rod

The solution to the problem according to the scheme of Fig. 32.7a can be obtained using the solution to the auxiliary problem obtained in the previous section. Since the stress field in the left side of the beam from the load on the right side coincides with the stress field in the right side of the beam from the load on the left side, we can write for normal stresses with a finite length of the plastic zone

$$\sigma_j = \sigma_j(z) + \sigma_j(-z - l_p), \quad (j = r, \theta, z). \tag{32.24}$$

For tangent stresses, we obtain the same result

$$\tau_{rz} = \tau_{rz}(z) - \tau_{rz}(-z - l_p). \tag{32.25}$$

In formulas (32.24)–(32.25), normal and tangent stresses are determined by formulas (32.21).

Adding the corresponding stress components according to formulas (32.24)–(32.25) to zero approximation stresses, we get the first approximation for the stress field of the stretched rod at a finite length of plastic zone.

The results of computer calculations based on the specified formulas are presented in Figs. 32.8, 32.9, 32.10, and 32.11. These figures show the dependencies of stress components along the length of the bar at different values of the r/R relationship.

In calculations, the length of the plastic zone is assumed to be $0.8R$. It can be seen from the graphs that the inhomogeneity of the stress field is local in nature, and when the distance between the elastic and plastic zones is more than $\approx 1.5R$, the inhomogeneity can be neglected.

32.3.5 The Dependence of the Tensile Force and Pressure p on the Length of the Plastic Zone

Formulas that define the stress field in a stretched bar include the values of force $P = \sigma_0 F$ and pressure p, which depend on the length of the plastic zone l_p. Let us define these dependencies.

Fig. 32.8 Distribution of
radial normal stresses

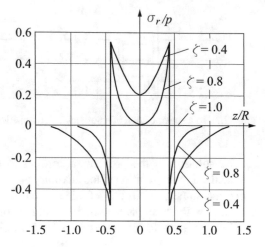

Fig. 32.9 Distribution of
tangential normal stresses

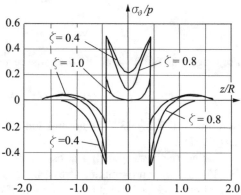

As already mentioned, when the tensile force reaches the value $P_t = \sigma_t F$, an
increase in the displacement of the end of the rod leads to a drop in the load to
P. Due to the infinitely small displacement of the ends of the beam, its absolute
elongations before and after the load drop should coincide

$$\frac{\sigma_t}{E} l_0 = \Delta l^e + \Delta l^p, \tag{32.26}$$

where Δl^e, Δl^p are the absolute elongation of the elastic and plastic parts of the
bar. Using the stretch chart (Fig. 32.2), it can be written as

$$\Delta l^e = \frac{\sigma_0}{E}(l_0 - l_p); \quad \Delta l^p = \left(\frac{\sigma_s}{E} + \frac{\sigma_0 - \sigma_s}{E_t}\right) l_p. \tag{32.27}$$

From formulas (32.26)–(32.27), the desired dependence follows:

Fig. 32.10 Distribution of
axial normal stress

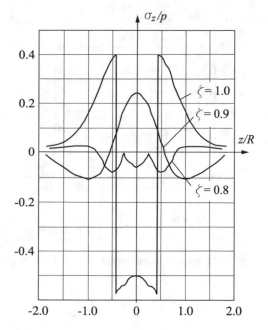

Fig. 32.11 Plot of shearing
stresses

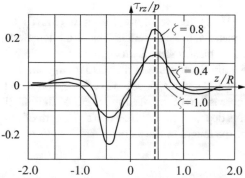

$$\frac{P}{F} = \frac{\sigma_t + \alpha k \sigma_s}{1 + \alpha k}, \quad \left(\alpha = \frac{l_p}{l_0}, \quad k = \frac{E}{E_1} - 1\right). \tag{32.28}$$

To eliminate the incompatibility of deformations, a uniform pressure p was
applied to the side surface of the beam. Equating the radial movement (32.19) to
the incompatible strain (32.18), we receive

$$p = \frac{k(\sigma_0 - \sigma\sigma_s)}{2(1 - v)}. \tag{32.29}$$

Given formula (32.28), we get the second of the required dependencies

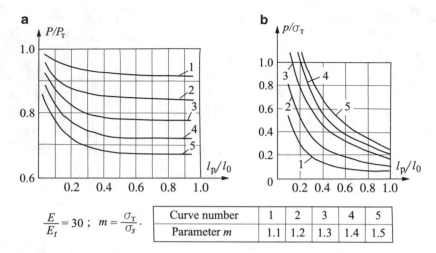

$$\frac{E}{E_t} = 30 \; ; \quad m = \frac{\sigma_{\mathrm{T}}}{\sigma_s} .$$

Curve number	1	2	3	4	5
Parameter m	1.1	1.2	1.3	1.4	1.5

Fig. 32.12 Dependence of power characteristics on the length of the plastic zones

$$p = \frac{k(\sigma_t - \sigma_s)}{2(1 - \nu)(1 + k\alpha)} . \tag{32.30}$$

Using the resulting formulas (32.28) and (32.30) in Fig. 32.12a and b, we constructed the dependencies of the tensile force and lateral pressure on the length of the plastic zone. Dimensionless coordinates are used here. Curves are constructed for different values of the yield tooth with the ratio of the modulus of elasticity to the tangent modulus equal to 30. As can be seen from the graphs, the depth of the tooth increases with increasing length of the plastic zone. The greatest manifestation of this dependence occurs when the length of the plastic zone does not exceed a quarter of the length of the sample.

32.4 Perfectly Plastic Body Case

The described iterative method for solving inelastic problems, taking into account the history of deformation, is most clearly demonstrated by the example of a perfectly plastic body. Here it gains independent fundamental importance.

Let τ_s denote the yield strength of a perfectly ductile material under shear. If at some point Q of the body, the maximum tangential stress τ_m reaches the yield point, then plastic deformation cannot occur due to the constraining action of the elastic material surrounding the point Q. Therefore, for the development of plastic deformation, it is necessary that the maximum tangential stress exceeds the yield strength in a certain neighborhood of the point Q. This requirement is satisfied if the yield condition is taken in the form of

$$T_m = \tau_m, \tag{32.31}$$

where T_m is the maximum tangent macro-stress [6].

Under the load $P_s + \Delta P_1$, the body will slip along the planes and directions of the idealized maximum tangential stresses [6]. In accordance with Axiom 20.1 (p. 303), plastic deformation is discontinuous in space and instantaneous in time. Consequently, as a result of the slip in the body, a region of plastic deformation is formed of a small thickness, extended in the direction of the minimum rate of change of the maximum tangential stress, since the sections of the planes of this region will be primarily involved in the slip process. Due to the small thickness, this region can be considered a surface (in the flat case, a line), which in what follows will be called a sliding surface (line).

The sliding system along the planes and directions of the maximum tangential stress can be replaced by a kinematically equivalent discontinuity of displacements distributed over the sliding surface. The density of this gap and the boundary of the sliding surface can be determined from the condition that the idealized maximum tangential stress on the indicated surface is equal to the yield strength of the material

$$\tau_m = \tau_s. \tag{32.32}$$

The latter condition is reduced to an integral equation with respect to the unknown gap density of displacements on the sliding surface. With a known solution of this equation, the determination of stresses and deformations does not present fundamental difficulties.

Next, the second stage of loading is considered. Such an increment of the load ΔP_2 is given, at which again the condition (32.31) is met at any point of the body. Then at the load $P_s + \Delta P_1 + \Delta P_2$, it is necessary to repeat the entire sequence of calculations described for the first stage, etc.

Thus, for an ideal plastic body, the method of successive loading is in a sense natural. Apparently, there is no other way to solve boundary value problems for a perfectly plastic body.

32.5   Using the Kröner Theory of Residual Stresses

The inelastic deformation e_{ij} is represented as a result of the continuous distribution of discontinuous deformations (slides) in some area of the elastic body. Tension in an inelastic body can be represented as a sum of two fields: the stress fields in an elastic body from a given external influence and the stress fields in an elastic body caused by slides. The second of these fields can be defined applying the Kröner residual stress theory [11].

Inelastic deformation will not satisfy, generally speaking, the conditions of compatibility [6]. To characterize the incompatibility of deformations, Kröner introduced the [11] so-called tensor incompatibilities

$$S_{ij} = -\varepsilon_{ikm}\varepsilon_{jln}\varepsilon_{kl,mn}. \tag{32.33}$$

Here, the symbol ε_{ijk} is 1 if the numbers 1, 2, 3 instead of the corresponding indexes form an even permutation, and -1 if this permutation is odd, and zero in all other cases.

The physical meaning of the components of the incompatibility tensor was shown by Eshelby [3]. If in the plane P normal to the i-th coordinate axis, a wedge with a vertex at some point Q is cut out from the body, and the edges of the cut are connected, then for the body in such a state of deformation, the component $S_{ii}(Q)$ is the angle of the specified wedge referred to the unit area P. The component S_{ij} with different indexes is the angle of twisting of the fiber parallel to the j-th axis, per unit of area. Another interpretation and a method for producing components of the incompatibility tensor are given by us in Sect. 32.7

According to Kröner, the stress field from incompatible deformations at a known incompatibility tensor is defined in the following order. Any solution $\tilde{\varepsilon}_{ij}$ to Eq. (32.33) is found, and the stress $\tilde{\sigma}_{ij}$, corresponding to the deformation $\tilde{\varepsilon}_{ij}$, is calculated by Hooke's law. Next, we find additional voluminous $\tilde{f}_i$ and surface $\tilde{p}_{ij}$ efforts, which are able to create in an elastic body the stress $\tilde{\sigma}_{ij}$

$$\tilde{f}_i = -\tilde{\sigma}_{im,m}, \quad \tilde{p}_{iv} = \tilde{\sigma}_{im}v_m. \tag{32.34}$$

Then we apply forces equal in magnitude and opposite in sign to the forces (32.34) and ε_{ij}^0 using the methods of linear elasticity theory and calculate the strain that satisfies the compatibility conditions. The sought-after stress from incompatible deformation of $\tilde{\varepsilon}_{ij}$ will be that which, according to Hooke's law, corresponds to the deformation $\tilde{\varepsilon}_{ij} + \varepsilon_{ij}^0$.

Since in real problems of mechanics, the incompatibility tensor, as a rule, is not set in advance, it is not possible to use the pure Kröner theory. However, it can be used in a slightly modified algorithm in combination with the above method of elastic solutions.

Split as before the loading process in time into a series of successive stages. To find the stress field from discontinuous deformations at the $(k+1)$-th stage, we apply the above-described procedure of successive approximations. As a zeroth approximation for the incompatibility tensor, we take its value calculated by formula (32.33), in which the deformation is determined from the relations (32.11) by stress (32.12). If the incompatibility tensor is known, then the stress is also known in a first approximation. Then the calculation procedure at this stage is repeated until the specified accuracy is reached.

Note that using the Kröner method requires that the strain components have continuous derivatives at least up to the first order. The described method is more visible and, in in a certain sense, elegant in the case of flat deformation.

32.6 Kröner Method for Plane Deformation

For flat deformation ($\varepsilon_{33} = 0$), the only component of the incompatibility tensor different from zero S_{33} will be a function of two variables. As already mentioned, this function depicts the density of wedge dislocations continuously distributed in the region of incompatible deformations.

Denote by $\Sigma_{ij}(x, y, x_0, y_0)$ the stress that occurs at an arbitrary point ($x_1 = x$, $x_2 = y$) of the elastic body from the wedge dislocation introduced at some point (x_0, y_0) of the considered region with an angular opening equal to unity. In what follows, we will call Σ_{ij} *Green's tensor function*. Then the additional stress $\tilde{\sigma}_{ij}$, which occurs in the body from discontinuous deformations, can be calculated by the formula:

$$\tilde{\sigma}_{ij} = - \iint\limits_{D} \Sigma_{ij}(x, y, x_0, y_0) S_{33}(x_0, y_0) dx_0 dy_0, \qquad (32.35)$$

where D is the area of incompatible deformations.

In accordance with the above, the stress σ_{ij} in an inelastic body will be equal to

$$\sigma_{ij} = \sigma_{ij}^0 + \tilde{\sigma}_{ij}, \qquad (32.36)$$

and σ_{ij}^0 is the stress in the body from the given external influences, calculated under the assumption of the ideal elasticity of the material.

To determine the stress σ_{ij}, we use the elastic solution method. At the ($k + 1$)-th loading stage, we determine the zero approximation for stress by formula (32.11) by solving the corresponding elastic problem. Using the calculated stress, from the condition (32.10) we refine the boundary of the region and use formulas (32.33) and (32.11) to calculate the zero approximation for the incompatibility tensor. Then, with the known function Σ_{ij}, it is easy to calculate the stress as a first approximation from the dependencies (32.35) and (32.36). Next, the iteration process is repeated until the required accuracy is obtained.

Thus, in the case of plane deformation, to solve the problem, it is necessary, in addition to finding a solution to the corresponding elastic problem, to construct Green's tensor function. If this function is found, the solution is reduced to formal calculations using the indicated algorithm.

The proposed method is especially effective in solving problems using the Kröner theory of residual stresses. Already the first approximation should give a fairly good result. If the region occupied by the body is rather simple (a circle, a plane with a hole, etc.), the construction of Green's function does not cause formal difficulties. Examples of its calculation are given in the articles [9, 10].

32.7   More About Incompatible Deformations

32.7.1   *Distributed Wedge Dislocations*

Imagine a rectangular contour $ABCD$ (Fig. 32.13) inside an elastic plane. Mentally free this contour (frame) from communication with the body and cut it (frame) (Fig. 32.14). If there were wedge-shaped dislocations inside the frame, then the vertical elements in the section nn (Fig. 32.13) will rotate relative to each other by some angle (ω). Assume that the frame in question is cut from a body subject to spatial deformation. Let us define the rotation angles of linear elements in the frame section in this general case. The frame element parallel to the Oy axis is rotated in the xOy plane by the angle whose tangent will be

at the point B

$$-\frac{\partial u}{\partial y}\left(x+\frac{\Delta x}{2}, y-\frac{\Delta y}{2}\right);$$

at the point C

$$-\frac{\partial u}{\partial y}\left(x+\frac{\Delta x}{2}, y+\frac{\Delta y}{2}\right);$$

Fig. 32.13 For the study of deformations of the contour

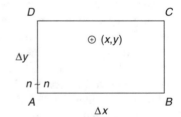

Fig. 32.14 Deformations of the cut frame

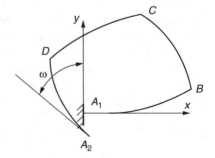

at the point D

$$-\frac{\partial u}{\partial y}\left(x - \frac{\Delta x}{2}, y + \frac{\Delta y}{2}\right);$$

at the point A_2

$$-\frac{\partial u}{\partial y}\left(x - \frac{\Delta x}{2}, y - \frac{\Delta y}{2}\right).$$

Here, as before, u, v, and w are the components of movement along the axes x, y, and z, respectively; indexes 1 and 2 mark the right and left sides of the $n - n$ section. The rotation angle of this element passing through the point B, relative to the parallel element passing through the point A_1, is written as follows:

$$\omega_{BA_1} = \frac{\partial^2 u}{\partial x \partial y}\left(x, y - \frac{\Delta y}{2}\right)\Delta x = -\frac{\partial \varepsilon_x}{\partial y}\left(x, y - \frac{\Delta y}{2}\right)\Delta x. \qquad (32.37)$$

In the same way, we define the rotation of the element passing through the point C, relative to the element passing through the point B

$$\omega_{CB} = -\frac{\partial^2 u}{\partial y^2}\left(x + \frac{\Delta x}{2}, y\right)\Delta y =$$

$$= \left[-2\frac{\partial \gamma_{xy}}{\partial y}\left(x + \frac{\Delta x}{2}, y\right) + \frac{\partial \varepsilon_y}{\partial x}\left(x + \frac{\Delta x}{2}, y\right)\right]\Delta y, \qquad (32.38)$$

where

$$2\gamma_{xy} = \frac{\partial u}{\partial y} + \frac{\partial v}{\partial x}.$$

By analogy, we can write

$$\omega_{CD} = \frac{\partial \varepsilon_x}{\partial y}\left(x, y + \frac{\Delta y}{2}\right)\Delta x,$$

$$\omega_{DA_2} = \left[2\frac{\partial \gamma_{xy}}{\partial y}\left(x - \frac{\Delta x}{2}, y\right) - \frac{\partial \varepsilon_y}{\partial x}\left(x - \frac{\Delta x}{2}, y\right)\right]\Delta y. \qquad (32.39)$$

Summing up the calculated relative rotations when traversing the frame contour A_1BCDA_2, we determine the relative rotation of the frame section A_2 relative to the fixed (see Fig. 32.14) section at the point A_1

$$\omega_{BA_1} + \omega_{CB} + \omega_{DC} + \omega_{A_2D} = \left(\frac{\partial^2 \varepsilon_x}{\partial y^2} + \frac{\partial^2 \varepsilon_y}{\partial x^2} - 2 \frac{\partial^2 \gamma_{xy}}{\partial x \partial y} \right) dx dy. \qquad (32.40)$$

The considered effect can be represented [7] as the result of the introduction of wedges continuously distributed with density

$$S(x, y) = \frac{\partial^2 \varepsilon_x}{\partial y^2} + \frac{\partial^2 \varepsilon_y}{\partial x^2} - 2 \frac{\partial^2 \gamma_{xy}}{\partial x \partial y}, \qquad (32.41)$$

and causing known residual stresses, which correspond to elastic deformations of ε_x, ε_y, and γ_{xy}.

32.7.2 Strain Incompatibility Tensor

Consider the case when the previously selected frame is cut out of a body subjected to spatial deformation with continuous components differentiated a sufficient number of times. We will consider this deformation as the result of the introduction or removal of material with possible shifts in its interlayers. We determine the rotation angles of the linear elements of the frame in this general case.

Partially repeating the calculations performed in the previous paragraph, we obtain the following results in this spatial case. The frame element parallel to the Oy axis rotates in the xOz plane by an angle whose tangent will be

at the point B

$$\frac{\partial w}{\partial y} \left(x + \frac{\Delta x}{2}, y - \frac{\Delta y}{2} \right);$$

at the point C

$$\frac{\partial w}{\partial y} \left(x + \frac{\Delta x}{2}, y + \frac{\Delta y}{2} \right);$$

at the point D

$$\frac{\partial w}{\partial y} \left(x - \frac{\Delta x}{2}, y + \frac{\Delta y}{2} \right);$$

at the point A_2

$$\frac{\partial w}{\partial y} \left(x - \frac{\Delta x}{2}, y - \frac{\Delta y}{2} \right).$$

The rotation angle of this element passing through the point B, relative to the parallel element passing through the point A_1, will be represented by the formula:

$$\omega_{BA_1} = \frac{\partial^2 w}{\partial x \partial y}\left(x, \, y - \frac{\Delta y}{2}\right)\Delta x. \tag{32.42}$$

Using the identity

$$\frac{\partial^2 w}{\partial x \partial y} = \frac{\partial}{\partial x}\gamma_{yz} + \frac{\partial}{\partial y}\gamma_{zx} + \frac{\partial}{\partial z}\gamma_{xy},$$

formula (32.42) can be written as

$$\omega_{BA_1} = \left[\frac{\partial}{\partial x}\gamma_{yz}\left(x, \, y - \frac{\Delta y}{2}\right) + \frac{\partial}{\partial x}\gamma_{zx}\left(x, \, y - \frac{\Delta y}{2}\right) + \right.$$
$$\left. + \frac{\partial}{\partial x}\gamma_{xy}\left(x, \, y - \frac{\Delta y}{2}\right)\right]\Delta x, \tag{32.43}$$

where x, y, z are components of the strain tensor.

The rotation of the element passing through the point C relative to the element passing through the point B will be

$$\omega_{BC} = \frac{\partial^2 w}{\partial y^2}\left(x + \frac{\Delta x}{2}, y\right)\Delta y,$$

or

$$\omega_{BC} = \left[2\frac{\partial}{\partial y}\gamma_{yz}\left(x + \frac{\Delta x}{2}, y\right) - \frac{\partial \varepsilon_y}{\partial z}\left(x + \frac{\Delta x}{2}, y\right)\right]. \tag{32.44}$$

In the same way, we find

$$\omega_{BD} = -\left[\frac{\partial}{\partial x}\gamma_{yz}\left(x, \, y + \frac{\Delta y}{2}\right)\right] + \frac{\partial}{\partial y}\gamma_{xz}\left(x, \, y + \frac{\Delta y}{2}\right) -$$
$$- \frac{\partial}{\partial z}\gamma_{xy}\left(x, \, y + \frac{\Delta y}{2}\right)\Delta x, \tag{32.45}$$

$$\omega_{DA_2} = \left[-2\frac{\partial}{\partial y}\gamma_{yz}\left(x - \frac{\Delta x}{2}, y\right) + \frac{\partial \varepsilon_z}{\partial z}\left(x - \frac{\Delta x}{2}, y\right)\right]\Delta y.$$

After traversing the frame contour $A_1 BCDA_2$, the vertical linear elements at the points A_1 and A_2 will rotate relative to each other in the plane yOz (twist) by an angle

$$\omega_{A_2A_1} = \left[\frac{\partial}{\partial y} \left(\frac{\partial \gamma_{xy}}{\partial z} + \frac{\partial \gamma_{yz}}{\partial x} + \frac{\partial \gamma_{xz}}{\partial y} \right) - \frac{\partial^2 \varepsilon_y}{\partial x \partial z} \right] dx dy.$$

The value in the square brackets defines the torsion component (p. 422) of the incompatibility tensor S_{ij}, $(i \neq j)$ and expresses, as already mentioned, the fiber twist angle per unit area parallel to the j-th axis, for a frame located in a plane with the normal i, i.e.

$$S_{zx} = \frac{\partial}{\partial y} \left(\frac{\partial \gamma_{xy}}{\partial z} + \frac{\partial \gamma_{yz}}{\partial x} + \frac{\partial \gamma_{xz}}{\partial y} \right) - \frac{\partial^2 \varepsilon_y}{\partial x \partial z}. \tag{32.46}$$

Similarly, we can obtain that the formula

$$S_{zz} = \frac{\partial^2 \varepsilon_x}{\partial y^2} + \frac{\partial^2 \varepsilon_y}{\partial x^2} - 2 \frac{\partial^2}{\partial x \partial y} \gamma_{xy} \tag{32.47}$$

determines the bending component of the incompatibility tensor S_{zz}, which is proportional to the rotation angle relative to the Oz axis of the linear elements of the frame located in the plane with the normal Oz.

By changing the axis indexes and bearing in mind notations like

$$\gamma_{xx} = \varepsilon_x, \quad 2\gamma_{xy} = \frac{\partial u}{\partial y} + \frac{\partial v}{\partial x},$$

we can obtain six components of the incompatibility tensor, and

$$S_{xy} = S_{yx}, \quad S_{xz} = S_{zx}, \quad S_{yz} = S_{zy}. \tag{32.48}$$

The components (32.46)–(32.48) can be written differently as

$$S_{mn} = e_{mki} e_{nsj} \frac{\partial^2 \gamma_{ks}}{\partial x_i \partial x_j}; \tag{32.49}$$

here, as mentioned above in this chapter, summation is carried out over repeated (dummy) indexes; e_{mki} and e_{nsj}—Levi–Civita symbols; they are equal to zero if there are duplicate indexes, equal to 1 in the following arrangement of indexes: 123, 321, 231, and equal to -1 for any other order of the indexes.

32.8 The Application of Kröner's Method to the Brazilian Test

In order to demonstrate the algorithm for solving inelastic problems described in the previous section, we consider the plane deformation of a circular disk, to which uniform normal pressure q is applied along two equal arcs symmetrically relative to its center (Fig. 32.15).

32.8.1 Zero Approximation

We will use the solution of the problem of elastic deformation of the indicated disk given in Sect. 9.6. The functions of Muskhelishvili [13] $\Phi(\zeta)$ and $\Psi(\zeta)$ for this task are represented by formulas (9.41). They can also be written as

$$
\begin{aligned}
\Phi(\zeta) &= -\frac{q}{2\pi i}\left[\ln\frac{(\sigma_2-\zeta)(\sigma_4-\zeta)}{(\sigma_1-\zeta)(\sigma_3-\zeta)} - 2i\theta_0\right], \\
\Psi(\zeta) &= -\frac{q}{2\pi i\zeta}\left[\frac{1}{\sigma_4-\zeta} + \frac{1}{\sigma_2-\zeta} - \frac{1}{\sigma_1-\zeta} - \frac{1}{\sigma_3-\zeta}\right]
\end{aligned}
\tag{32.50}
$$

with the appropriate choice of the branch of the logarithm in the first of the formulas (32.50).

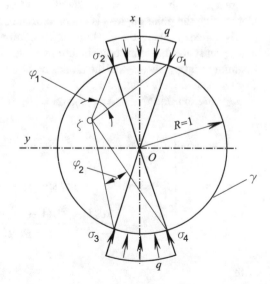

Fig. 32.15 For the inelastic problem of testing brittle materials using the Brazilian method

The functions (32.50) in the polar coordinates (ρ, φ) correspond to stresses

$$\sigma_r = -\frac{q}{\pi} \left\{ \varphi_1 + \varphi_2 - 2\theta_0 - (1 - \rho^2) \times \right.$$

$$\left. \times \left[\frac{\sin 2(\alpha_1 - \varphi)}{R_1^2} - \frac{\sin 2(\alpha_2 - \varphi)}{R_2^2} \right] \right\},$$

$$\sigma_\varphi = -\frac{q}{\pi} \left\{ \varphi_1 + \varphi_2 - 2\theta_0 + (1 - \rho^2) \times \right.$$

$$\left. \times \left[\frac{\sin 2(\alpha_1 - \varphi)}{R_1^2} - \frac{\sin 2(\alpha_2 - \varphi)}{R_2^2} \right] \right\},$$

$$\tau_{r\varphi} = \frac{q(1 - \rho^2)}{\pi} \left[\frac{\rho^2 - \cos 2(\alpha_1 - \varphi)}{R_1^2} - \frac{\rho^2 - \cos 2(\alpha_2 - \varphi)}{R_2^2} \right]. \qquad (32.51)$$

Here it is indicated

$$\varphi_1 = arg \ln \frac{\sigma_2 - \zeta}{\sigma_1 - \zeta}, \quad \varphi_2 = arg \ln \frac{\sigma_4 - \zeta}{\sigma_3 - \zeta},$$

$$R_1^2 = (1 + \rho^2)^2 - 4\rho^2 \cos^2(\alpha_1 - \varphi),$$

$$R_2^2 = (1 + \rho^2)^2 - 4\rho^2 \cos^2(\alpha_2 - \varphi),$$

where α_1, α_2 are the angles corresponding to the ends of the arc $\sigma_1 \sigma_2$, and the beam from which the polar angle φ is counted divides the arc $\sigma_1 \sigma_2$ in half.

The stresses inside the circle must be continuous everywhere, so the angles φ_1 and φ_2 must also be continuous functions of the point ζ. We satisfy this requirement, assuming that φ_1 and φ_2 are the angles at which the arcs $\sigma_1 \sigma_2$ and $\sigma_3 \sigma_4$ are visible from ζ, respectively (Fig. 32.15).

Keeping in mind that $\alpha_1 = -\theta_0$, $\alpha_2 = \theta_0$, we get

$$\sigma_r = -\frac{q}{\pi} \left[\theta - 2\theta_0 + 2(1 - \rho^2) \sin 2\theta_0 \frac{(1 + \rho^4) \cos 2\varphi - 2\rho^2 \cos 2\theta_0}{R_1^2 R_2^2} \right],$$

$$\sigma_\varphi = -\frac{q}{\pi} \left[\theta - 2\theta_0 - 2(1 - \rho^2) \sin 2\theta_0 \frac{(1 + \rho^4) \cos 2\varphi - 2\rho^2 \cos 2\theta_0}{R_1^2 R_2^2} \right],$$

$$\tau_{r\varphi} = \frac{2q(1 - \rho^4)(1 - \rho^2) \sin 2\theta_0 \sin 2\varphi}{\pi R_1^2 R_2^2},$$

$$(32.52)$$

with

$$\theta = \varphi_1 + \varphi_2 = \begin{vmatrix} \theta_1 + \theta_2 & \text{with } \rho|\cos\varphi| \leqslant \cos\theta_0, \\ 2\pi - \theta_3 - \theta_4 & \text{with } \rho|\cos\varphi| \geqslant \cos\theta_0, \end{vmatrix} \qquad (32.53)$$

and the angles $\theta_{1,...,4}$ are calculated by the formulas:

$$\theta_1 = \arccos \frac{\rho^2 + \cos 2\theta_0 - 2\rho \cos \theta_0 \cos \varphi}{\left[1 + \rho^2 - 2\rho \cos(\theta_0 - \varphi)\right]^{1/2} \left[1 + \rho^2 - 2\rho \cos(\theta_0 + \varphi)\right]^{1/2}},$$

$$\theta_2 = \arccos \frac{\rho^2 + \cos 2\theta_0 + 2\rho \cos \theta_0 \cos \varphi}{\left[1 + \rho^2 + 2\rho \cos(\theta_0 - \varphi)\right]^{1/2} \left[1 + \rho^2 + 2\rho \cos(\theta_0 + \varphi)\right]^{1/2}},$$

$$\theta_3 = \arccos \frac{\rho^2 - \cos 2\theta_0 - 2\rho \sin \theta_0 \sin \varphi}{\left[1 + \rho^2 - 2\rho \cos(\theta_0 - \varphi)\right]^{1/2} \left[1 + \rho^2 + 2\rho \cos(\theta_0 + \varphi)\right]^{1/2}},$$

$$\theta_4 = \arccos \frac{\rho^2 - \cos 2\theta_0 + 2\rho \sin \theta_0 \sin \varphi}{\left[1 + \rho^2 + 2\rho \cos(\theta_0 - \varphi)\right]^{1/2} \left[1 + \rho^2 - 2\rho \cos(\theta_0 + \varphi)\right]^{1/2}}.$$

$$(32.54)$$

In the future, we will also need the principal stresses (σ_1, σ_2, σ_3). We define them according to the known formulas:

$$\sigma_{1,2} = \frac{\sigma_r + \sigma_\varphi}{2} \pm \frac{1}{2}\sqrt{(\sigma_r - \sigma_\varphi)^2 + 4\tau_{r\varphi}^2}.$$

Substituting the expressions (32.52) here, we obtain after elementary transformations

$$\sigma_{1,2} = -\frac{q}{\pi}\left[\theta - 2\theta_0 \mp\right.$$

$$\left. \mp \frac{2(1 - \rho^2)\sin 2\theta_0}{\sqrt{\left[(1 + \rho^2)^2 - 4\rho^2 \cos^2(\theta_0 - \varphi)\right]\left[(1 + \rho^2)^2 - 4\rho^2 \cos^2(\theta_0 + \varphi)\right]}}\right],$$

$$\sigma_3 = -\frac{2q\nu}{\pi}(\theta - 2\theta_0). \qquad (32.55)$$

By direct verification, one can verify that the solution obtained for $\theta_0 \to 0$ coincides with the known [13] solution of the disk problem, compressed by diametrical concentrated forces.

32.8.2   Green's Tensor Function for a Circle

If a radial wedge with an angle equal to unity is removed from a circular ring $R_1 \leqslant r \leqslant R_2$, then we say that a single wedge dislocation is formed in the ring. For such a ring, the functions of Muskhelishvili will be [9, 10]

$$\Phi_1(z_1) = c\left(\frac{1}{2} - \frac{R_2^2 \ln R_2 - R_1^2 \ln R_1}{R_2^2 - R_1^2} + \ln z_1\right),$$

$$\Psi_1(z_1) = -2c \cdot \frac{R_1^2 R_2^2}{R_2^2 - R_1^2} \ln \frac{R_2}{R_1} \cdot \frac{1}{z_1^2}; \quad R_1 \leqslant |z_1| \leqslant R_2;$$

$$c = \frac{E}{8\pi(1 - v^2)},$$

where z_1 is the affix of an arbitrary point of the ring in the plane $x_1 O_1 y_1$, with the origin of O_1 aligned with the center of the ring.

Aiming $R_1 \to 0$ in these formulas, we find the Muskhelishvili functions for a continuous cylinder with an integrated unit wedge dislocation

$$\Phi_2(z_1) = \lim_{R_1 \to 0} \Phi_1(z_1) = c\left(\frac{1}{2} + \ln \frac{z_1}{R_2}\right),$$

$$\Psi_2(z_1) = \lim_{R_1 \to 0} \Psi_1(z_1) = 0.$$

In the latter formulas, we pass to the new coordinates xOy, which we choose parallel to the old axes, and place the beginning of the new system at the point O so that the affix of the point O_1 in the new axes is z_0, $(z = z_0 + z_1)$. Denoting the Muskhelishvili functions in the xOy axes by $\Phi^0(z)$, $\Psi^0(z)$, we find

$$\Phi^0(z) = c\left(\frac{1}{2} + \ln \frac{z - z_0}{R_2}\right), \quad \Psi^0(z) = -\frac{c\bar{z}_0}{z - z_0},$$

$$(z = x + iy = re^{i\varphi}, \quad z_0 = x_0 + iy_0 = r_0 e^{i\varphi_0}).$$

We introduce the transformation

$$z = R\zeta, \quad (R = const; \ |z| < R < R_2 - |z|). \tag{32.56}$$

Then the functions Φ_0 and Ψ_0 will take the form

$$\Phi_0(\zeta) = c\left[\frac{1}{2} + \ln \frac{R(\zeta - \zeta_0)}{R_2}\right], \quad \Psi_0(\zeta) = -\frac{c\bar{\zeta}_0}{\zeta - \zeta_0}, \tag{32.57}$$

where it is indicated

$$\zeta_0 = z_0/R. \tag{32.58}$$

Using the expressions (32.57) and also the known formulas [13] of the plane theory of elasticity, we find the following combination of normal (N) and tangent (T) forces on the contour of a circle of a unit radius:

$$N - iT = \Phi^0(\sigma) + \overline{\Phi^0(\sigma)} - \sigma\Phi^{0'}(\sigma) - \sigma^2\Psi^0(\sigma),$$

or

$$N - iT = c \left\{ 1 + 2\ln\frac{R}{R_2} + \ln\left[(\sigma - \zeta_0)\left(\frac{1}{\sigma} - \zeta_0\right) \right] - \frac{\sigma(1 - \sigma\overline{\zeta}_0)}{\sigma - \zeta_0} \right\},$$

(32.59)

where, as before, $\sigma = e^{i\varphi}$ is an arbitrary point on the contour of the unit circle. By transition in formula (32.59) to conjugate quantities, we obtain

$$N + iT = c \left\{ 1 + 2\ln\frac{R}{R_2} + \ln\left[(\sigma - \zeta_0)\left(\frac{1}{\sigma} - \overline{\zeta}_0\right) \right] - \frac{\sigma - \zeta_0)}{\sigma(1 - \sigma\overline{\zeta}_0)} \right\}.$$

(32.60)

We now solve the first main problem of the theory of elasticity for a circle with contour loading (32.59). The functions of Muskhelishvili $\Phi_3(\zeta)$, $\Psi_3(\zeta)$ for this task are already given in formulas (16.19) and (16.22). Substituting the expressions (32.59)–(32.60) into these formulas, we find after calculating the integrals

$$\Phi_3(\zeta) = c\left[1 - \frac{\rho_0^2}{2} + \ln\frac{R}{R_2} + \ln(1 - \overline{\zeta}_0\zeta) - \frac{1 - \rho_0^2}{1 - \zeta\overline{\zeta}_0} \right],$$

(32.61)

$$\Psi_3(\zeta) = \frac{c\overline{\zeta}_0^2(2 - \rho_0^2 - \zeta\overline{\zeta}_0)}{(1 - \zeta\overline{\zeta}_0)^2}.$$

We obtain the Muskhelishvili functions $\Phi(\zeta)$, $\Psi(\zeta)$ for the circle $|z| \leqslant R$, which has a single wedge dislocation at the point with polar coordinates (ρ_0, φ_0). To do this, subtract the corresponding functions (32.61) from the functions (32.57). As a result, we get

$$\Phi(\zeta) = c\left(\frac{\rho_0^2 - 1}{2} + \ln\frac{\zeta - \zeta_0}{1 - \zeta\overline{\zeta}_0} + \frac{1 - \rho^2}{1 - \zeta\overline{\zeta}_0} \right),$$

$$\Psi(\zeta) == c\overline{\zeta}_0\left[\frac{1}{\zeta - \zeta_0} + \frac{\overline{\zeta}_0}{1 - \zeta\overline{\zeta}_0} + \frac{(1 - \rho_0^2)\overline{\zeta}_0}{(1 - \zeta\overline{\zeta}_0)^2} \right].$$

(32.62)

The functions (32.62) correspond to the stresses

$$\frac{1}{c}\tau_{r\varphi} = \frac{2\rho_0(1 - \rho\rho_0)V_1[\rho(1 + \rho_0^2) - \rho_0 V_2]}{V_3 V_4^2},$$

$$\frac{1}{c}\sigma_r = \rho_0^2 - 1 + \ln\frac{V_3}{V_4} + \frac{2(1 - \rho_0^2)V_5}{V_4} - \frac{(1 - \rho_0^2)^2(V_6^2 - \rho_0^2 V_1^2)}{V_7 V_4^2},$$

(32.63)

$$\frac{1}{c}\sigma_\varphi = \rho_0^2 - 1 + \ln\frac{V_3}{V_4} + \frac{2(1-\rho_0^2)V_5}{V_4} + \frac{(1-\rho_0^2)(V_6^2 - \rho_0^2 V_1^2)}{V_3 V_4},$$

where $\rho = r/R$ and for the compactness of notation, it is indicated

$$V_1 = (\rho^2 - 1)\sin(\varphi - \varphi_0); \quad V_2 = (\rho^2 + 1)\cos(\varphi - \varphi_0);$$

$$V_3 = \rho^2 + \rho_0^2 - 2\rho\rho_0\cos(\varphi - \varphi_0); \quad V_4 = 1 + \rho^2\rho_0^2\cos(\varphi - \varphi_0);$$

$$V_5 = 1 - \rho\rho_0\cos(\varphi - \varphi_0); \quad V_6 = \rho(1+\rho_0^2) - \rho_0 V_2;$$

$$V_7 = \rho^2 - \rho_0^2 - 2\rho\rho_0\cos(\varphi - \varphi_0).$$

$$(32.64)$$

Formulas (32.63)–(32.64) determine in polar coordinates (ρ, φ) the stress components at an arbitrary point inside the circle of a radius R from the action of a single wedge dislocation at the point (ρ_0, φ_0). These stresses are the components of the sought tensor Green function for the circle.

32.8.3   Definition of Deformation in a First Approximation

For definiteness, we assume that for a material subjected to diametrical compression of a cylindrical sample, the dependence (32.11) in the principal axes has the form [12]

$$\varepsilon_j = \frac{p}{K} + \frac{\sigma_i - p}{2G} + \lambda|\sigma_j - p|\left(\frac{1}{G} - \frac{1}{G_0}\right), \quad (j = 1, 2, 3), \qquad (32.65)$$

where $K/3$ is the elastic modulus of volumetric expansion, G_0 is the shear modulus, G is a function of stress, λ is the parameter that we will consider constant here, and p is the medium stress

$$p = \frac{1}{3}(\sigma_1 + \sigma_2 + \sigma_3).$$

Since the dependencies (32.65) contain the components of the stress deviator in the third term, the deformations $(\varepsilon_r, \varepsilon_\varphi, \varepsilon_{r\varphi})$ will be expressed by different formulas depending on the sign of the specified component. From the solution (32.55), we determine the average stress and the components of the stress deviator

$$p = -\frac{2q(1+v)}{3\pi}(\theta - 2\theta_0),$$

$$\sigma_1 - p = \frac{q}{\pi}\left[-\frac{(1-2v)(\theta - 2\theta_0)}{3} + \frac{2(1-\rho^2)\sin 2\theta_0}{\sqrt{w_1 w_2}}\right],$$

$$\sigma_2 - p = \frac{q}{\pi}\left[-\frac{(1-2\nu)(\theta - \theta_0)}{3} - \frac{2(1-\rho^2)\sin 2\theta_0}{\sqrt{w_1 w_2}}\right],$$

where it is indicated

$$w_1 = (1 + \rho^2)^2 - 4\rho^2 \cos^2(\theta_0 + \varphi); \quad w_2 = (1 + \rho^2)^2 - 4\rho^2 \cos^2(\theta_0 - \varphi).$$

It is easy to see that the first term inside the square brackets of the latter two formulas is negative inside the circle. This implies that the second component of the stress deviator is negative everywhere, since $\theta_0 < \dfrac{\pi}{2}$. The boundary of the region, upon transition through which the first component of the deviator changes the sign, can be obtained from the condition that this component is equal to zero, i.e.

$$\theta - 2\theta_0 = \frac{6(1-\rho^2)\sin 2\theta_0}{(1-2\nu)\sqrt{w_1 w_2}}.$$

Calculations show that for small θ_0 (up to $\approx 7^o$), the first component of the deviator is positive everywhere in the circle for any $\nu \in [0; 0.5]$. In the future, we will assume that θ_0 does not exceed the specified value. With this in mind, one can get from (32.65) the following formulas for the strain components:

$$\varepsilon_r = \frac{\sigma_r - p}{2G} + \frac{p}{K} + \lambda\left(\frac{1}{G} - \frac{1}{G_0}\right)\frac{(\sigma_r - p)(\sigma_r - \sigma_\varphi) + 2\tau_{r\varphi}^2}{\sqrt{(\sigma_r - \sigma_\varphi)^2 + 4\tau_{r\varphi}^2}},$$

$$\varepsilon_\varphi = \frac{\sigma_\varphi - p}{2G} + \frac{p}{K} + \lambda\left(\frac{1}{G} - \frac{1}{G_0}\right)\frac{(\sigma_\varphi - p)(\sigma_r - \sigma_\varphi) - 2\tau_{r\varphi}^2}{\sqrt{(\sigma_r - \sigma_\varphi)^2 + 4\tau_{r\varphi}^2}}, \qquad (32.66)$$

$$\varepsilon_{r\varphi} = \tau_{r\varphi}\left\{\left(\frac{1}{G} - \frac{1}{G_0}\right)\left[1 + 2\lambda\frac{\sigma_r + \sigma_\varphi - 2p}{\sqrt{(\sigma_r - \sigma_\varphi)^2 + 4\tau_{r\varphi}^2}}\right] + \frac{1}{G_0}\right\}.$$

In a first approximation, the strain components $(\varepsilon_{r1}, \varepsilon_{\varphi1}, \varepsilon_{r\varphi1})$ are obtained by substituting the stress components $(\sigma_r, \sigma_\varphi, \tau_{r\varphi})$ in formulas (32.66), taken from the solution (32.52) of the elastic problem. After performing the above calculations, we find

$$\varepsilon_{r1} = -\frac{q}{\pi}\left\{\frac{1-2\nu}{3}(\theta - 2\theta_0)\left(\frac{1}{G_0} - \frac{1}{2G}\right) + \right.$$

$$+ \frac{(1-\rho^2)\sin 2\theta_0[(1+\rho^4)\cos\varphi - 2\rho^2\cos 2\theta_0]}{Gw_1 w_2} -$$

$$\left. - \lambda\left(\frac{1}{G} - \frac{1}{G_0}\right)\frac{T - 2(1-\rho^2)\sin 2\theta_0}{\sqrt{w_1 w_2}}\right\},$$

$$\varepsilon_{\varphi 1} = -\frac{q}{\pi} \left\{ \frac{1-2v}{3}(\theta - 2\theta_0)\left(-\frac{1}{2G} - \frac{1}{G_0}\right) - \right.$$

$$-\frac{(1-\rho^2)\sin 2\theta_0[(1+\rho^4)\cos\varphi - 2\rho^2\cos 2\theta_0]}{Gw_1w_2} +$$

$$\left. + \lambda\left(\frac{1}{G} - \frac{1}{G_0}\right)\frac{T + 2(1-\rho^2)\sin 2\theta_0}{\sqrt{w_1w_2}} \right\}, \qquad (32.67)$$

$$\varepsilon_{r\varphi 1} = \frac{2q(1-\rho^4)(1-\rho^2)\sin 2\theta_0 \sin 2\varphi}{\pi w_1 w_2}\left\{\left(\frac{1}{G} - \frac{1}{G_0}\right) \times \right.$$

$$\left. \times \left[1 - \frac{\lambda(1-2v)(\theta - 2\theta_0)}{3(1-\rho^2)\sin 2\theta_0}\sqrt{w_1w_2}\right] + \frac{1}{G_0}\right\},$$

where indicated

$$T = (1-2v)(\theta - \theta_0)[(1+\rho^4)\cos 2\varphi - 2\rho^2\cos 2\theta_0]/3.$$

With known deformations, (32.67), we can calculate, in a first approximation, the incompatibility tensor. In the case under consideration, we have

$$S_{33} = \frac{1}{r}\left[\frac{1}{r}\frac{\partial^2(r\varepsilon_{r\varphi})}{\partial r\partial\varphi} + \frac{\partial\varepsilon_r}{\partial r} - \frac{1}{r}\frac{\partial}{\partial r}\left(r^2\frac{\partial\varepsilon_\varphi}{\partial r}\right) - \frac{1}{r}\frac{\partial^2\varepsilon_r}{\partial\varphi^2}\right]. \qquad (32.68)$$

After calculating the components of (32.68), the stresses can be calculated in a first approximation using formulas (32.35)–(32.36), since the components of the Green function (32.63) are already known. These calculations are very cumbersome, although simple in nature. Therefore, apparently, it is more expedient to carry out calculations by numerical methods.

We also note that after calculating the strains in a first approximation, one can find fictitious loads in a first approximation by a single differentiation. Then, the determination of stresses in a first approximation can be performed based on the integration of the equilibrium equations (32.1), as described above.

References

1. S. Abdrakhmanov, D. Kozhobaev, *Plastichnost' i prochnost' materialov i konstruktsii: sb. nauch. tr.* (Plasticity and strength of materials and structures: collection of scientific works) (Izd-vo Frunzensk, Frunze, 1981). politekhn. in-ta Publ., pp. 3–16. Frunze, Izd-vo Frunzensk. politekhn. in-ta Publ., Nachal'nyi protsess plasticheskogo deformirovaniya brusa na predele tekuchesti (Initial process of plastic surgery deformation of the beam at the yield point), pp. 3–16

2. I. Birger, B. Shorr, G. Iosilevich, *Raschet na prochnost' detalei mashin: spravochnik. 4-e izd., dorab.* (Calculation of the strength of parts machines: reference), 4th ed. (Moscow, Mashinostroenie Publ., 1993)

3. D. Ehshelbi, *Kontinual'naya teoriya dislokatsii* (Continual theory of dislocations) (IL Publ., Moscow, 1963)

4. A. Il'yushin, *Mekhanicheskie svoistva i ispytanie metallov* (Mechanical properties and testing of metals) (OGIZ Publ., Leningrad, Moscow, 1948)

5. R. Khill, *Matematicheskaya teoriya plastichnosti* (Mathematical theory of plasticity) (IL Publ., Moscow, 1956)

6. M. Leonov, *Osnovy mekhaniki uprugogo tela* (Fundamentals of elastic body mechanics) (Izd-vo AS Kirg. SSR., Frunze, 1963)

7. M. Leonov, *Mekhanika deformatsii i razrusheniya* (Deformation and fracture mechanics) (Izd-vo Ilim Publ., Frunze, 1981)

8. A. Lyav, *Matematicheskaya teoriya uprugosti* (Mathematical theory of elasticity) (Leningrad, ONTI NKTP SSSR Publ., Moscow, 1935)

9. V. Molotnikov, *Dislokatsionnye napryazheniya v uprugoi ploskosti s krugovym otverstiem* (Dislocation stresses in an elastic plane with circular hole) (Frunze, Izd-vo Ilim Publ., 1978). Nachal'nyi protsess plasticheskogo deformirovaniya brusa na predele tekuchesti (Initial Process of Plastic Surgery Deformation of the Beam at the Yield Point), pp. 77–80

10. V. Molotnikov, *Materialy konferentsii molodykh uchenykh (Frunze, Izd-vo MSKh Kirg. SSR)* (Materials of the Conference of Young People Scientists) (Publishing house of the Ministry of Agriculture of the Kirg. SSR, Frunze, 1981). *Klinovye dislokatsii v ploskoi zadache mekhaniki neuprugogo tverdogo tela* (Wedge Dislocations in a Plane Problem of Mechanics an Inelastic Solid), pp. 18–23

11. V. Molotnikov, *Plastichnost' i prochnost' materialov i konstruktsii* (Plasticity and strength of materials and structures) (Izd-vo Frunz. politekhn. in-ta Publ., Frunze, 1981). O kraevykh zadachakh mekhaniki neuprugogo tv"erdogo tela (On Boundary Value Problems in Inelastic Solid Mechanics), pp. 71–79

12. V. Molotnikov, A. Molotnikova, Clozhnaya deformatsiya pochvogruntov (a complex deformation of soils, orenburg), in *'Prochnost' i razrushenie materialov i konstruktsii: materialy V mezhdunar. nauchn. konf. T. 2* (Strength and Destruction of Materials and Structures: Materials V Mezhdunar. Scientific Conf.), vol. 2 (2008), pp. 321–327

13. N. Muskhelishvili, *Nekotorye osnovnye zadachi matematicheskoi teorii uprugosti* (Some main problems of mathematical theory elasticity) (Nauka Publ., Moscow, 1966)

14. A. Nadai, *Plastichnost' i razrushenie tverdykh tel* (Plasticity and destruction of solids) (IL Publ., Moscow, 1954)

15. S. Timoshenko, J. Gud'er, *Teoriya uprugosti* (Theory of elastic strength) (Nauka Publ., Moscow, 1975)

Index

A
Aftereffect, 5
    elastic, 10
Aging, 331
    deformation, 316
Analogy
    Prandtl, 236
Autofrettage, 230
Axioms, 303
Azimuth, 115

B
Bend
    wedge's, 89
Bessel functions, 415
Biclination
    mathematical, 106
Biklination
    mathematical, 106
Body, 7
    barrel, 64
    ideal plastic, 420
    semi-fragile, 394
    uniform, 7
Boundary value problems, 146

C
Charge
    single, 41
Circle
    single, 93
Cloud
    blocking, 321
    Cottrell, 321, 327

Coefficients
    gives, 243
    Lame, 29
    Poisson's, 29
Compaction, 393, 394
Compression
    barrel-shaped bodies, 63
    cylinders, 59
        parallel, 67
    uniaxial, 401
Concept
    slides, 151
Condition for continuity, 384
Conditions
    balance
        in displacements, 30
        into stresses, 23
    boundary, 28, 91
    Huber-Mises, 185
    incompressibilities, 148, 214
    plasticities, 181
        full, 148
        Tresca, 184
    unloadings, 202
    yield, 328
Cone
    normals', 262
Conversion
    reflections, 250
    rotations, 250
Core
    dislocations, 114
Corner
    plasticities, 264
Cottrell cloud, 334

Creep, 146
   soil, 401

D
Deductions, 111, 112
Defect
   linear, 101
   structures, 101
Deformation, 5
   is almost simple, 331
   axisymmetric, 18
   cast iron
      gray, 396
   complete, 407
   creep, 362
   difficult, 145
   flat, 16, 27, 48, 270
   flat-plastic
      monotonous, 381
   full, 407
   homogeneous, 6, 8, 14
   hydrostatic, 16
   incompatible, 423
   inelastic, 407
   inhomogeneous, 8
   monotonous, 382
   purely elastic, 407
   purely plastic, 394
   relative, 8
   residual, 10
      stable, 10
      unstable, 10
   reversible, 5
   shift's, 10
   transverse, 8
   under compression, 360
   wedge's, 88
   when unloading, 360
Delay
   yield, 316
Dependencies
   Batdorf–Budyansky, 393
Deviator, 16, 268
Diffusion
   imperfections, 321
   processes, 327
Dislocation, 101
   boundary, 101
   disjunctive, 113
   edge, 103
   regional
      in a half plane, 117
   screw, 101

   lefthand, 102
   right, 102
   Somigliana, 106, 116
   uplift's, 113
   wedge, 105, 423

E
Eccentricity, 66
   ellipse's, 66
Effect
   Bauschinger, 178, 345
   Bausinger
      full, 345
   compaction, 395
   loosening, 395
Elasticity, 5
Ellipsoid, 62
Elongation
   negativity, 6
Energy
   changes
      shapes, 84
   springs, 79
Equations
   balances, 23, 210
   Campbell, 317
   Cauchy, 230
   Lagrange's, 208
   Laplace's, 31, 38
   Levy, 148, 149
   variation
      Lagrange's, 211
   Volterra, 355
Experience
   Hendrickson and Wood, 329
Experiments
   Coulomb's, 147
   Tresca, 147
Extension
   volumetric, 15

F
Flowability, 313
Forces
   fictitious, 408
Formula
   Chicala, 374
   Ecobory, 317
   Goursat, 90
   Ostrogradsky–Green, 209
Formulation
   Hill's, 409

Foundation
   pile, 118
Functions
   aging, 328
   bi-harmonic, 90, 415
   Erie, 88
   Galina, 111
   Green
      tensor, 423
      tensorial, 423
      3D, 39
   Greene's, 39
   harmonic, 31
      examples, 37
      properties, 43
   holomorphic, 91, 94
   Muskhelishvili, 92, 108, 116, 429, 431
   Neumann's, 41
   Prandtl, 236, 237
   softening, 341
   strengthenings, 197
   stresses, 87
   universal, 397

G
Gap
   displacements, 421
Grille
   crystal, 101

H
Half-plane
   elastic, 118
Half-space, 42
   elastic
      deformation, 48
Hardening
   isotropic, 270
   linear, 276
   monotonous, 398
   translational, 275
Helicoid, 102
History
   loadings, 262, 321
Hypothesis
   incompressibilities, 148
   Mises, 149
   structural softening, 327

I
Identities
   Saint-Venant, 26

Imperfections
   structural, 5, 101
Increment
   deformations, 261
Integrals
   Cauchy, 94
   of Cauchy-type integrals, 111
   elliptic, 61
   Kolosov, 91
   probabilities, 398
   Trefftz, 33
Intensity
   deformations', 174
   stress
      tangent, 73
Invariants, 70, 171
Isomorphism, 253

K
Kernel
   degenerate, 357
   dislocations, 105

L
Lame
   constant, 216
Law
   creep, 402
   delays, 254, 255
   distribution
      normal, 397
   Hooka
      generalized, 14
      when shifting, 12
   Hooke's, 198, 226, 407
   Hooke's tensile , 9
   Laning, 273, 275
   pairings, 13
   strengthenings, 195
   unloading
      first, 202
      second, 202
   volume elasticity, 195
Lengthening
   relative, 6
Limit
   elasticities, 10, 321
      conditional, 10
      natural, 10
   proportionalities, 9, 10
Line
   level's, 238

Linear
  movement, 25
Loading
  proportional, 346
Loam, 396
  slightly humid, 405
Loosening, 393, 394

M
Macrocracks, 393, 394
Material
  incompressible, 262
  polycrystalline, 7
Mechanism
  compaction-loosening, 395
Membrane, 240
Mesomechanics, 155
Method
  Birger, 410
  Brazilian, 429
  Cracker, 421
  elastic solutions, 240, 409
  iteration, 403
  Ritz's, 212
Model
  Klyushnikova, 287
  Prager, 277
  uniform, 17
Module
  Jung's, 29
  secant, 373
  shift's, 12
  tangent, 261
Modulus
  shear, 407

P
Parameter
  Lode–Nadai, 226
Path
  loadings, 273
Period
  loading
    conical, 262
Phenomenon
  plasticity, 393
Pile, 118
Place
  fluidity, 9
Plane
  atomic, 102
Platforms

contact
  circular, 58
  elongated, 66
  main, 70
Point
  loadings, 256
Poisson's ratio, 9
Postulate
  anti-isotropy, 323, 343
  Drucker, 257, 277
  isotropies, 157
    Ilushin, 151, 252
Potentials, 203
  individual masses, 41
  speeds', 41
Power
  deployment
    wedge, 105
  tangent, 10
Principle
  delays, 151, 156
  gradients, 259
  Haar–Karman, 148
  reciprocity
    movings', 81
    works', 80
  Saint-Venant, 22
Problem
  in displacements, 30
  first main, 23, 28, 29
Process
  diffusion, 318
  loadings, 195
  quasistatic, 319
  unloadings, 195
Properties
  tangential stress, 74

R
Rapper
  Frenet, 157
Recreation
  when unloading, 331
Region
  ring road
    flat, 432
Rock burst, 156

S
Shear resistance, 303, 304
  operator, 341
Shift

clean, 379, 402
octahedral, 366
Shifting, 25
Slides, 147
Slip, 303
freeze, 346
intensity, 306
Softening
deformation, 321
elastic, 322
Soils
clay, 394
loamy, 394
Solid angle, 305
Space
deviator's, 264
five-dimensional, 249
nine-dimensional, 249
representing, 262
Speed
loading, 317
Strain
normal, 8
Strength
additional, 408
fictitious, 409
surface, 6, 408
Stress
concentration, 21
full, 71
extreme, 71
intensity, 174
main, 69
octahedral
tangent, 73
tangent
maximum, 73
Stresses
radial, 18
ring, 18
Stretch
diagram, 8, 9
Stretching
uniaxial, 402
Surface
loadings, 262, 273
bulge, 260
smooth, 273
soap film, 238
Synergetics, 156
Synergy
principles, 295
System
slides, 284

T
Task
Dirichlet, 42
fictitious, 216
first
for a circle, 93
flat, 87
Hertz, 55
axisymmetric, 57
Lame, 18, 226
Neumann's, 42, 48
into stresses, 28
Tense state, 7
flat, 16
homogeneous, 13
Tension, 8
average, 394
contact, 59
hydrostatic, 15, 175
octahedral, 268, 366, 396
relaxation, 363
residual, 216
tangent, 10
tectonic, 113
Tensor
ball, 170
deformations', 166
deviator, 170
guiding, 151
incompatibilities, 421, 423
strain rate, 168
tension, 166
Test
Brazilian, 95, 429
Theorem
about simple loading, 213
about the minimum, 207
about unloading, 215
Betty, 81
Bezout, 70
Castigliano's, 82
pure shift, 377
Theory
Batdorf–Budyansky, 393
Cauchy–Navier, 393
dislocations', 101
endochronic, 291
flow, 267
Goldenblatt and Prager, 245
Ishlinsky, 276
Kadashevich–Novozhilov, 278
Kadashevich–Pomytkin, 294

Theory (*cont.*)
    Laning, 273
    plasticity
        deformational, 195
        endochronic, 157
        first, 148
        Mises, 149
        Prandtl, 149
    Prandtl, 149
    Prandtl–Reuss, 269
    processes', 151, 156
    Saint-Venant–Mises, 269
    slides
        Budiansky, 283
    Vakulenko–Backhaus, 294
    Valanis, 293
Tooth flow, 333
Tooth yield, 313

U
Uplift, 113

V
Values
    mated, 94, 97
Variations
    movements, 210
Vector
    burgers, 103
    stress, 7
Verification, 403

W
Warping, 236
Work
    inner strength, 78
    virtual, 77

Y
Yield strength, 9
    secondary, 345
Young's modulus, 9

Printed in the United States
by Baker & Taylor Publisher Services